D0875722

FUNCTION AND EVOLUTION
IN BEHAVIOUR

PROFESSOR NIKO TINBERGEN, F.R.S.

Photograph by Dr. L. Shaffer

Function and Evolution in Behaviour

ESSAYS IN HONOUR OF
PROFESSOR NIKO TINBERGEN, F.R.S.

Edited by

GERARD BAERENDS

COLIN BEER

and

AUBREY MANNING

CLARENDON PRESS · OXFORD
1975

Oxford University Press, Ely House, London W. I

GLASGOW NEW YORK TORONTO MELBOURNE WELLINGTON
CAPE TOWN IBADAN NAIROBI DAR ES SALAAM LUSAKA ADDIS ABABA
DELHI BOMBAY CALCUTTA MADRAS KARACHI LAHORE DACCA
KUALA LUMPUR SINGAPORE HONG KONG TOKYO

ISBN 0 19 857382 0

© OXFORD UNIVERSITY PRESS 1975

All rights reserved. No part of this publication may be reproduced, stored in a retrieval system, or transmitted, in any form or by any means, electronic, mechanical, photocopying, recording or otherwise, without the prior permission of Oxford University Press

PRINTED IN GREAT BRITAIN BY
REDWOOD BURN LIMITED, TROWBRIDGE AND ESHER

591.51
F979
175884

This book is dedicated to
Professor Niko Tinbergen, F.R.S.,
to whom it was presented
on the occasion of his retirement
from the Chair of Animal Behaviour,
University of Oxford

Contents

List of Plates

Introduction

IN THIS volume ethologists pay tribute to a man who has been described as 'a Grand Master of Ethology'. Ethology is not so well known as chess, but Professor Niko Tinbergen is now so famous, and ethology now so much a household word, that an introduction to either might well be regarded as superfluous. However there are some points about Tinbergen's kind of ethology that may not be common knowledge and which, in any case, need to be made to explain the composition of the book. These points, the editors decided, could best be drawn from a biographical sketch. Hence the form of this introduction.

Ethology is 'the biological study of behaviour' and therefore involves four classes of question, which concern respectively the causation, ontogeny, survival value, and evolution of behaviour. This was the way in which Tinbergen characterized ethology in a paper he published in 1963 entitled 'On the aims and methods of ethology'. His fourfold division of ethology's field of enquiry had already appeared in the organization of his now classic book *The study of instinct*, published in 1951. In that book the questions of causation or motivation received much more attention than the others. The answers he proposed to these causal questions had a profound and important influence on the development of ethological research and ideas, and in the course of this development they have been subjected to much revision and refinement. Tinbergen's own research effort since that time has tended to concentrate on the questions of survival value and evolution; his 1951 treatment remains as a clear foundation for all later work. In this work he has remained as close as any other ethologist to ethology's roots in the past: the comparative morphological tradition of pre-molecular biology; the Darwinian and neo-Darwinian synthesis and theory; and amateur natural history, with its cultivation of a sharp eye, disciplined curiosity about natural phenomena, and love of the out-of-doors. But these influences only confirmed and refined tastes and activities that the young Tinbergen had developed well before he became a professional scientist.

Nikolaas Tinbergen was born on 15 April 1907, at The Hague in Holland. He was the third of five children; he had an older brother and sister and two younger brothers. They were all bright and went into intellectual careers in which they achieved a remarkable degree of success. Jan, the oldest, became a famous and influential economist. He anticipated

Niko by winning a Nobel Prize, the first to be awared in economics. Lukas, the youngest became a Professor of Zoology at Groningen and acquired an international reputation for his studies in animal ecology. The father of these clever children was a Doctor of Dutch Letters and a teacher at a high school in The Hague. He, and consequently the rest of the family, made an interest in and discussion of current events and the world around a part of everyday family life. Scientific questions and points of view tended to predominate in the discussions, but aesthetic feeling and expression were also encouraged. Tinbergen Senior impressed upon his children the advantages of speaking and writing simply and clearly, a lesson which Niko carried over from his native language to his expression in German and English.

Prominent among the subjects discussed in the Tinbergen household, and among the activities for spare-time occupation, was natural history. This interest in natural history received impetus and direction from a series of popular books on nature study written by two Dutch schoolmasters: E. Heimans and J. P. Thijse. These books appeared between 1900 and 1925. In addition to opening the young naturalist's eyes to the wonders of the plant and animal worlds about him, the books pointed his curiosity towards questions that he could investigate, either by careful observation or by doing experiments. This way of getting education appealed strongly to Niko, for whom solving a problem by means of a single, neat experiment has remained one of the main delights of scientific endeavour. Heimans and Thijse also exploited the addiction for collecting, to which people in general but naturalists in particular are prone, and in so doing helped to sell the products of a well-known biscuit firm. They prepared albums into which coloured picture cards that came with the packets of biscuits could be pasted. The albums included a series on the seasons, another on different regions of the country, and another on different biotopes. The taste for biscuits and the taste for nature reinforced one another to widespread effect. Gerard Baerends recalls that his interest in Digger Wasps began with a picture in the album 'Summer'.

Of at least equal importance for the young Tinbergen was the founding of a Natural History Society for Dutch youth. This society was organized by the young people themselves. They arranged lecture meetings, excursions, and camps, and published a journal. Niko was an enthusiastic member and active contributor to the activities of the society. A lecture he gave to promote the society led Baerends to join it in 1928.

This lecture was about birds observed near the Hook of Holland. At high school, and as an undergraduate at Leiden University, Niko belonged to a small club of bird watchers (Club van Haagse trekwaarnemers) devoted to the study of migration. Led by G. J. Tijmstra, a local high-school teacher, the members of this club made numerous field trips, particularly to the dunes and beaches near The Hague. Part of the time was given to bird photography, an art in which Niko already excelled. Together with G. van Beusekom, F. Kooymans, and M. G. Rutten he collaborated in a book of observations and photographs entitled *Het Vogeleiland*. It was an illustrated account of the birds of what was then one of the ornithologically richest islands in the region.

Niko's activities as a naturalist also received encouragement at high-school, particularly from one of his teachers: Dr. A. Schierbeek. However, school studies were mixed in their appeal to Niko; in those subjects he liked he worked hard and did well, but in those that failed to catch his interest he tended to backslide. His ambivalence about academic work made him hesitate about the prospect of going on to university. To help settle this question Schierbeek and Tijmstra advised Niko's father to let him spend some time at the Vogelwarte Rossiten, an ornithological field station near Königsberg in Germany, the director of which was Professor Heinz Thienemann. This was arranged and the experience decided Niko to make biology his career. Accordingly he enrolled at Leiden University.

Even as an undergraduate, however, Niko continued his pattern of loafing in the subjects he found uninteresting. One of his professors complained that he was too 'playful'. Games certainly rivalled study for Niko's attention. Since his schooldays he had excelled at sports; at field hockey he reached international class.

Despite his apparent lack of seriousness Niko had a genuine and, indeed, profound interest in biology, and this was evident to those of his teachers who were closest to him. Of these Jan Verwey was of particular significance. Verwey was about seven years Niko's senior, and an assistant in zoology courses that Niko took. Because of his spell at Rossitten, Niko arrived late to take up his studies at Leiden, and Verwey was assigned to help the newcomer to catch up. Verwey was immediately impressed by Niko's enthusiasm for natural history. He recalls that Niko gave a talk about his Rossitten experience which enthralled the audience in a way that was a portent of things to come. Mutual admiration between the two men formed the basis of a friendship they still enjoy. Verwey was a pioneer of ethology in Holland. He was a methodical and astute observer of animal behaviour, and in his teaching at Leiden he steered his students into ethological paths. He knew and discussed the work of such proto-ethologists as Heinroth, Selous, and J. S. Huxley. He did important original research on the behaviour of Lapwings, Herons, and later, during a period he spent in Indonesia, on estuarine crabs. Niko accompanied Verwey on many field trips, and they talked together often and at length.

Niko's first formal research project was for the degree that was preliminary to the doctorate at Leiden. He did an ecological study of the food of owls and an ethological study of the breeding of terns. After successfully completing the requirements for this degree (the equivalent of an M.Sc. at a British university) he began work on the orientation behaviour of the Digger Wasp *Philanthus* for his doctoral thesis. An invitation to spend a year in Greenland as a member of a Dutch polar expedition brought this work to an early close. His mentor, Professor H. Boschma, persuaded the examiners for the doctorate to accept a report of the preliminary research, which amounted to an extraordinarily short thesis of about twenty-five pages. He pleaded that in Tinbergen's case the presence of outstanding scientific ability was so obvious that it would be pedantic and obstructive, in the circumstances, to insist on the usual amount of work. And so Dr. Niko Tinbergen packed his bags for the Arctic. Before leaving, however, he married Elizabeth Rutten, a chemistry student from Utrecht and a member of the Youth Natural History Society, whose brother, Martien Rutten, had long been one of Niko's close friends. For Niko and Lies the trip to Greenland was their honeymoon. They worked together at field studies of bird behaviour and spent a winter with the eskimos. The field work led to the writing of two monographs that are now classics of ethological literature, one on the Snow Bunting and one on the Red-necked Phalarope. In a less formal way Niko wrote also about the eskimos and the social behaviour of their dogs (for example, in *Curious naturalists*).

On his return to Holland (1933) Tinbergen took over the niche that Verwey had formerly occupied: he was appointed assistant in the zoological laboratory at Leiden, first under Professor Boschma and later under Professor van der Klaauw. This was a junior and poorly paid post, but it involved a considerable amount of work, including lecture courses to third-year biology students and the supervision of graduate research at both the M.Sc. and Ph.D. levels. Among the students who worked under him for the Ph.D. at this point were van Eck and Meynecht (colour vision in birds), van Beusekom (orientation of *Philanthus*), and Baerends (reproductive behaviour of *Ammophila*). Sticklebacks provided an easily obtained and convenient subject for laboratory investigation. For field study within easy reach of the university there was a colony of Herring Gulls located between Leiden and The Hague. And for work in an area of relatively undisturbed natural habitat there was Hulshorst, a place of sand dunes, heath, and poor pine woods on the edge of the polders where Tinbergen had done his *Philanthus* study. Summer camps in this area became a regular part of the programme for the Leiden ethologists. Much of the work done there was on insects, particularly the Digger Wasps and the *Satyrus* butterflies, but there was a wealth of other kinds of animals as well, the variety and interrelationships of which conveyed the scope and complexity of ethology and ecology.

The number of people who worked under or in collaboration with Tinbergen during this period is too large for mention of every one.

Outstanding among them, in addition to those mentioned already, were J. J. ter Pelkwijk, who helped in those now famous stickleback studies; D. J. Kuenen, who contributed to a study of the stimuli eliciting gaping in nestling songbirds; W. Kruyt, who helped in *Philanthus* orientation experiments; Tinbergen's brother Luuk, who worked on the predatory behaviour of Hobbies; and Miss J. van Roon who worked with Baerends on *Ammophila*, and later became his wife. Tinbergen's attraction for such people was in part due to his effectiveness as a teacher. As his father had taught him, he expressed himself simply and clearly, even about complex matters, but also with an enthusiasm that was contagious. To stir up thought and the interchange of ideas he began holding evening discussion sessions, and these became regular weekly meetings during term-time for the Leiden ethologists, and later for the group he assembled at Oxford after moving to that university. He also brought innovation to the undergraduate teaching by introducing a six-week practical course in ethology in which students worked in pairs on different and often untried research projects. In this, as elsewhere, he put into practice his belief that discovery of something by and for one's self is a most effective way of getting education, as he himself had learned in his childhood from the books of Heimans and Thijse.

Among the influences on Tinbergen's ideas during the thirties the writings of the British ornithologists, such as Eliot Howard, Edmund Selous, and Julian Huxley, occupied a prominent place. Also important were two Dutch animal psychologists: A. F. J. Portielje and J. A. Bierens de Haan. Both of these men adhered to McDougall's teachings about the role of purpose in behaviour, and they were critical of behaviourism and any form of behavioral analysis that left the animal's subjectivity out of account. Consequently, although they were interested in and admired much about Tinbergen's work, they often raised objections to both his methods and interpretations. In so doing they probably helped to refine his tough-minded approach to his subject. Of more direct positive benefit was the influence of the work of von Frisch, particularly the honey-bee studies and the use of conditioning techniques for investigating the sensory capacities of fish. From von Frisch Tinbergen learned valuable lessons about, among other things, how one could conduct experiments on animal behaviour in the field. He went on himself to extend the experimental approach in original ways and in new directions.

Tinbergen's work in animal behaviour, and Dutch ethology in general, were well into stride before the ideas of Konrad Lorenz became well known or began having an influence on them. But with the publication of Lorenz's Kumpan paper in 1935, and a visit that Lorenz made to Leiden in 1936, Tinbergen turned to interpreting his observations in Lorenzian terms. It became apparent that, although the two men had worked independently of one another, they were essentially in harmony as far as their conceptions of many aspects of animal behaviour were concerned. In 1937 Tinbergen spent three months at Lorenz's home in Germany, during which he and Lorenz worked together on the famous study of the egg-rolling of the Grey-lag Goose. In a recent reprinting of the account of this study it appears that Lorenz was the only one involved. This is misleading, for the study was a true collaboration, to which Tinbergen brought the experimental expertise. Lorenz has acquired the title of 'the father of ethology', and Tinbergen has recently been reported as saying that it was Lorenz who gave ethology its theoretical framework (*Psychology Today* **7**, No. 10). In a sense Lorenz did give ethology its start in life, and continues to occupy something like a parental role with regard to it; but Tinbergen took at least an equal share in guiding the science towards its coming of age, and if his has been the quieter voice it is not the less respected.

By 1938 Tinbergen's work was already becoming known abroad. In that year, at the behest of biologists such as Ernst Mayr and Kingsley Noble, Yale University invited him to the United States for three months. During this visit he made contact and began friendships with a number of Americans who were later to join with the Europeans in making ethology into a transatlantic science. Among these was Daniel S. Lehrman, who was then a youth working as a part-time assistant at the American Museum of Natural History. Tinbergen's interest in and encouragement of the young Lehrman helped to shape a career which, in its turn, helped to shape the development of ethology.

This development was more or less arrested during the war years. The Germans closed Leiden University after the faculty protested against the dismissal of the Jewish professors. Further refusal to conform to the dictates of the occupying command led to the arrests of several of the Leiden professors, and their confinement in a camp for hostages. Tinbergen was among them, along with Verwey and van der Klaauw. From September 1942 to the liberation of Holland in 1945, Tinbergen lived behind barbed wire, for a time with the dread that he would be the next to face a firing squad as a reprisal victim. Among other things he did to fill this time in limbo he drew pencil portraits of his fellow prisoners, prepared the first version of what became his *Social behaviour in animals*, and wrote storybooks for his children about the animals he had studied. Two of these books, *Kleew* and *The tale of John Stickle*, were later published.

At the end of the war Tinbergen returned to his position in the Zoology Department at Leiden, which was again headed by Professor van der Klaauw. He resumed the direction of the ethological programme, but with a new group of associates and students, including Jan van Iersel as his assistant, and P. Sevenster, A. Perdeck, and L. de Ruiter. The work consisted of the mixture as before—sticklebacks, Herring Gulls, summer camps, and so forth—with the addition of new topics that fitted in, such as experimental testing of the survival value of camouflage in certain insects. The group expanded rapidly as interest in the work spread; so much so that part of the gull research had to be moved to the island of Terschelling to accommodate the increasing demands.

In 1947 Tinbergen again visited the United States, this time at the invitation of the American Museum of Natural History and Columbia University. The series of lectures he wrote for the assignment he undertook for this trip formed the basis of what became *The study of instinct.* Although it is a quite short and relatively untechnical book, *The study of instinct* is more prominent in the landscape of ethology's history than any other piece of ethological writing. It forms a divide between the synthetic phase of theory-building that has been called 'classical ethology', and the succeeding analytical or revisionist phase of theory testing that might be called 'reformation ethology'. Its theory of instincts as hierarchically organized systems, though offered modestly as 'an attempt at a synthesis', was the culmination of the trends in ethological thought during the preceding fifteen years or so; it was, so to speak, the consummatory integration of ideas that had been appetitive towards a general theory of behaviour.

Before *The study of instinct* was published, however, Tinbergen moved from The Netherlands to England. Although he had been promoted to full professor in 1947, a new Chair of Experimental Zoology having been created for him at Leiden, he accepted the offer of a Lectureship in Professor (later Sir Alistair) Hardy's department at Oxford. Professor van der Klaauw and Leiden University were very disappointed at Tinbergen's decision to leave, which they tended to regard as ingratitude for what they had done for him. Tinbergen could have pleaded that he had received no more than his due for the many years he had worked at the rank of assistant while doing the job of a full professor. But Oxford had many positive attractions. There was the exciting challenge of life in a new country, together with a sense of mission to carry the word of ethology to the unconverted of the English-speaking world. More specifically, Oxford was strong in the kinds of evolutionary and ecological interests that resonated with Tinbergen's kind of ethology. He had much in common, intellectually, with such people as E. B. Ford and Arthur Cain, David Lack, and Charles Elton. They encouraged and reinforced his concern with questions about the evolution and adaptive significance of behaviour. Oxford ethologists came to be involved with such questions more actively and more exclusively than perhaps any other comparable group in the world.

Many of the studies that Tinbergen instigated at Oxford were continuations of those with which he had been involved at Leiden. For example experimental study of protective coloration in insects continued in David

Blest's work on the 'eye-spot' displays of butterflies and moths; and stickleback studies were carried on by Desmond Morris, Fae Hall, Beatrice Tugendhat, and others, right up to the work of David Wilz in the late sixties. Gull studies expanded in an ambitious programme covering numerous species and kinds of problem, even reaching out to take in terns and skuas. Among the people involved in this work were Martin Moynihan, Rita and Uli Weidmann, Michael and Esther Cullen, Gilbert Manley, Colin Beer, Hans Kruuk, Ian Patterson, Monica Impekoven, Heather McLannahan, Michael and Barbara MacRoberts, and Larry Schaffer. A new departure was the comparative and genetic study of behaviour in *Drosophila* species, to which Margaret Bastock, Aubrey Manning, Stella Crossley, Richard Brown, and others contributed. Robert Hinde, although not officially a member of Tinbergen's group, saw enough of it—and of Tinbergen himself—to become an ethologist in his Oxford doctoral research on the behaviour of Great Tits. The same species was later studied by Nick Blurton-Jones, who went into human ethology after completing his doctorate. Contact with Tinbergen's group influenced W. M. S. Russell's work on *Xenopus*. Brian Nelson brought the gannets of the Bass Rock, and then the gannets and boobies of the world, within the scope of the comparative ethological approach in work analagous to that on gulls. In the sixties Africa became a place of ethological pilgrimage, and as a conse·quence of Tinbergen's involvement with the Serengeti National Park several Oxford people went there to do their studies, including Hans Kruuk, who worked on hyenas, and Ian Douglas-Hamilton, who worked on elephants. Tinbergen's old interest in motivational models was apparent in a number of the Oxford studies, such as some of the early work of Hinde, and Moynihan's attempt to relate gull displays to underlying ratios of the strengths of attack and escape drives. Von Holst's reafference principle and the cybernetics it exemplified, computer analogs, set theory, probability theory, and other mathematical approaches progressively moved the model-builders in more sophisticted and abstract directions, as exemplified in the work of David McFarland and Richard Dawkins. Juan Delius started on the same road, with his stochastic analysis of skylark behaviour sequences, but a desire for more physiological answers to motivational questions turned him toward neurophysiology and electrical stimulation of the brains of gulls.

Some of this work extended outside the range of Tinbergen's informed interest. His usual way with a graduate student was to point him or her toward a problem, guide the work for as long as needed, and then stand back to let the student take command and make the discoveries. In this way he fostered self-reliance and independence in his students, sometimes to such a degree that they ended up with a thesis the soundness and value of which he felt himself incompetent to judge. Certain topics fully engaged his interest, however. One of these was the 'specific search image', a concept that his brother Luuk had formulated on the basis of observations of the foraging of titmice. Harvey Croze, Marion Dawkins, and others did experimental studies involving the search-image concept. The concern with feeding strategies extended to other questions, such as the relationships between feeding patterns and the distribution of food, and in developing this line of research he collaborated extensively with Michael Cullen, who had remained in Oxford. Their students studied several diverse feeding strategies, for example the work of James Smith on European Blackbirds and Great Tits, and Michael Norton-Griffiths' study of the ontogeny of two alternative modes of feeding in oystercatchers. Graham Phillips, using a combination of observation, comparison, and experiment, found feeding to be the key to the functional question of why most sea birds are white.

In part this interest in feeding behaviour developed out of work in which the focus was on the eaten rather than the eater. The Black-headed Gull colony at Ravenglass, in which Tinbergen and his students worked for nearly ten years, was frequently raided by foxes and other predators. Being the student of behavioural evolution that he was, Tinbergen had raised the question of how this predator pressure might be reflected in the species-typical behaviour of the gulls. The influence of such predator pressure in the evolution of gull behaviour had already been indicated by Esther Cullen's comparison of the relatively predation-free cliff-nesting kittiwakes with the more typical predation-prone ground-nesting gulls such as Black-headed Gulls. Tinbergen, together with a number of associates, launched a direct attack on the question with Black-headed Gulls, using comparative observation combined with experimental approaches. They began with an investigation of the causal basis and functional significance of egg-shell carrying; then continued with such studies as those of Kruuk and Patterson on other apparently anti-predator features in the reproductive behaviour of the gulls, including the spacing of nests and the timing of the cycle. Broadly speaking, functional questions such as these, and their bearings upon questions about the evolution of behaviour, have been at the heart of Tinbergen's ethology, at least during his time at Oxford. For this reason the editors chose 'the functions and evolution of animal behaviour' as the major theme for the contributions to this book in Professor Tinbergen's honour.

Our initial plan was to divide the book between the two kinds of evolutionary question—the phylogenetic kind of question and the survival-value kind of question—and subdivide each of these two divisions into a section comprising chapters on general conceptual, theoretical, and methodological matters and a section comprising chapters on particular pieces of research. Accordingly we invited people to contribute to the book, choosing mainly from those who had been students or close associates of Tinbergen. Of those contributions that could be labelled in accordance with the quadrants of our scheme, the phylogenetic group was such a minority that to have persisted with the scheme would have been to preserve symmetry in design at the expense of balance in the division of the substance of the book. We should perhaps have expected such a problem since the purely historical treatment of behavioural evidence has never had a large following, and may even be in decline at present. Moreover, even though he wrote some of the classical papers on the subject, such as the one on 'derived activities', Tinbergen's own active interest in it has waned in recent years as he has become progressively more concerned with functional matters. In any case an exclusive dichotomy between what might be described as the narrative and the executive aspects of behavioural evolution is more a taxonomic convenience than a working reality. Few investigations of the course of the evolution of a behaviour pattern leave aside entirely the question of what selection pressures might have determined this course; and few studies of the survival value of a behaviour pattern ignore entirely the question of how the selection pressures involved changed the behaviour from generation to generation. However, the means by which one arrives at narrative or phylogenetic interpretations of behavioural evolution—the comparative method—was well represented in the contributions that we received. This suggested that we arrange the chapters according to whether or not they rely mainly on the comparative method to make their points.

About half of the contributions dealt with function, either from some general point of view or focusing on a particular case, but with approaches other than the comparative, at least for the main business. We have grouped these essays together in the first part of the book and ordered them more or less in the sequence from general to specific. Hinde begins with a discussion of conceptual and methodological problems associated with the use of the term 'function' in animal behaviour studies. Is it, for example, necessary or even advisable to assume a function for every behavioural characteristic that an animal can be described as possessing? Beer writes about similar problems, but particularly as they arise in cases where a pattern of behaviour appears to serve more than one function, and specifically as he sees them in the communication behaviour of gulls. Both of these papers make the point that the meaning of the term function varies with context, ranging from the 'how it works' of physiological analysis to the 'what it is for?' of investigations of survival value. Roeder's contribution represents the 'how it works' end of the spectrum. Manning also looks at mechanisms, but at the genetic level, in a discussion of the contribution of behaviour genetics to the study of the how of the evolution of behaviour. Liley and Seghers illustrate selective pressure influences in the evolution of form and behaviour by applying correlation analysis and experimental testing to the closely related species of fish inhabiting an ecologically diverse river system. This study used comparison to obtain the correlations but we included it in this section because it represents the kind of experimental treatment that Tinbergen himself has pursued for testing hunches about survival value. Kruuk's contribution deals with how the members of a pack, in certain species of African carnivores, work together to bring down prey. It too trespasses into the comparative domain, but its emphasis is on the 'how' and 'how effective' of socially integrated hunting. Van Iersel presents a 'how does it do it?' study of insect orientation which recalls Tinbergen's early work on *Philanthus*. Patterson dons the recently fashioned cap of an etho-ecologist in a paper on the functions of fighting in rooks.

The second part of the book groups together the papers that share a strong reliance on some form of the comparative approach to make a case about either function or phylogeny. Again we have ordered the chapters more or less from general to specific. Baerends leads off with a discussion of a topic that has an important place in Tinbergen's thinking about both the causation and evolutionary origins of displays: the conflict hypothesis, which suggests that many displays are the ritualized derivatives of behaviour caused by simultaneous arousal of antagonistic motivational systems. Then, from an evolutionary perspective, Lindauer takes a comparative look at orientational learning in various animals; and Immelmann deals similarly with the effects of early experience on certain aspects of behavioural development. Marler continues the developmental theme with a report of comparative and experimental studies of factors affecting the acquisition of song in songbirds. Moynihan illustrates the more purely morphological approach in behavioural comparison with a study of behaviour and colour patterns in cephalopods. Robinson uses a similar approach to get at the evolutionary history of prey-catching in araneid spiders. Nelson combines morphological comparison with functional interpretation in his paper on the behaviour of gannets and boobies. McKinney uses comparison and adaptive correlation to elucidate the interrelated influences that have steered the courses taken by duck displays in their evolution. Finally Tschanz and Hirsbrunner-Scharf compare two closely related colonial and cliff-breeding species of birds and show how differences between them, in features of the behaviour and behavioural capacities of the chicks, are adaptive with respect to differences in details of family life and location on the cliff.

We believe this collection of essays to be representative of the range and variety in the field of evolutionary and functional studies of animal behaviour, at least insofar as Tinbergen has been concerned with it. Nevertheless there are some unfortunate omissions. For example, we have not included what has come to be called 'human ethology', although

Tinbergen, together with his wife, has recently become actively involved in this type of study, and some of his former students, such as Blurton-Jones, have pioneered the subject.

The number of Tinbergen's former students and associates is so large that we have been unable even to mention every one of them by name in this introduction. This number in itself is a mark of the extent of Tinbergen's contribution to ethology. But this contribution extends far beyond the conduct and direction of research, and the teaching and guiding of students. In his popular books, his film-making, and his talks on radio and television, he has carried ethology and the fascination of animal behaviour study to the general public. To this activity he has brought the same enthusiasm and grace as to his academic teaching and research, and with the same measure of effectiveness and success. For example his television film *Signals for survival* was the winner of an international award for documentary films.

Awards and honours have come to Tinbergen in increasing number with the recognition of his scientific achievements. Most notable among these have been his elevation to a personal Chair of Animal Behaviour at Oxford; his election to a Fellowship of the Royal Society; the Swammerdam Medal, which the Dutch award only once every ten years for outstanding achievement in biology; and, crowning them all, the 1973 Nobel Prize for Medicine, which he shared, most appropriately, with Lorenz and von Frisch.

We hope that Niko Tinbergen will be pleased by this volume from his former students and associates in ethology, which we offer with our best wishes for a long and active retirement. The editors and contributors constitute a small and, so to speak, self-appointed delegation from a large constituency, but we are sure that we speak for the whole when we say that we wish to express pride and gratitude for having been a part of the continuing endeavour of Professor Tinbergen's distinguished career.

Selected Bibliography of the works of Niko Tinbergen

1931 Die Paarungsbiologie der Flußseeschwalbe, *Ardea* **20**, 1–20.
1932 Ueber die Orientierung des Bienenwolfes, *Philanthus triangulum, Z. vergl. Physiol.* **16**, 305–35.
1932 Ethologische Beobachtungen am Baumfalken, *Falco subbuteo, J. Orn. Lpz.* **80**, 40–52.
1932 Ueber die Ernährung einer Waldohreulenbrut, *Beitr. Fortpfl. biol. Vogel* **8**, 54–5.
1933 Die ernährungsökologischen Beziehungen zwischen *Asio otus otus* L. und ihren Beutetieren, insbesondere den Microtus-Arten, *Ecol. Monogr.* **3**, 443–92.
1935 Field observations of East Greenland birds I. The behaviour of the Red-necked Phalarope in spring, *Ardea* **24**, 1–42.
1935 Ueber die Orientierung des Bienenwolfes. II. Die Bienenjagd, *Z. vergl. Physiol.* **21**, 699–716.
1936 Zur Soziologie der Silbermöwe, *Beitr. Fortpfl. Biol. Vogel* **12**, 89–96.
1936 The function of sexual fighting in birds, and the problem of the origin of 'territory', *Bird-Banding* **7**, 1–8.
1937 Ueber das Verhalten kämpfender Kohlmeisen, *Ardea* **26**, 222–3.
1938 Ergänzende Beobachtungen über die Paarbildung der Flußseeschwalbe, *Ardea* **27**, 247–9.
1938 Why do birds behave as they do? Part I, *Bird-Lore* **40**, 389–95.
1939 Why do birds behave as they do? Part II, *Bird-Lore* **41**, 23–30.
1939 Field observation of East Greenland birds. II. The behaviour of the Snow Bunting in spring, *Trans. Linn. Soc. N.Y.* **5**, 1–94.
1939 On the analysis of social organisation among vertebrates, with special reference to birds, *Am. Midl. Nat.* **21**, 210–34.
1940 Die Uebersprungbewegung, *Z. Tierpsychol.* **4**, 1–40.
1941 Ethologische Beobachtungen am Samtfalter, *J. Orn. Lpz.* **89**, 132–44.
1942 The objectivistic study of the innate behaviour of animals, *Biblthca biotheor.* **1**, 39–98.
1946 Die Struktur der Wirbeltiergemeinschaften, *Rev. suisse Biol.* **53**, 427–31.
1948 Social releasers and the experimental method required for their study, *Wilson Bull.* **60**, 6–52.
1948 Physiologische Instinktforschung, *Experientia* **4**, 121–33.
1950 The hierarchical organisation of nervous mechanisms underlying instinctive behaviour, *Symp. Soc. exp. Biol.* **4**, 305–12.
1950 Einige Beobachtungen über das Brutverhalten der Silbermöwe. In *Ornithologie als biologische Wissenschaft (Stresemann-Festschrift)*, p. 162–7.
1951 Recent advances in the study of bird behaviour, *Proc. int. orn. Congr.* **10**.
1952 'Derived' activities: their causation, biological significance, origin, and emancipation during evolution *Q. Rev. Biol.* **27**, 1–32.
1952 On the significance of territory in the Herring Gull, *Ibis* **94**, 158–9.

1952 A note on the origin and evolution of threat display, *Ibis* **94**, 160–2.
1952 The curious behavior of the Stickleback, *Scient. Am.* (December), pp. 22–6.
1953 Fighting and threat in animals, *New Biol.* **14**, 9–23.
1953 Ein ethologischer Beitrag zur Tierpsychologie, *Arch. neerl. Zool.* **10**, Suppl., 121–6.
1954 The origin and evolution of courtship and threat display, In *The evolutionary process* (ed. J. S. Huxley).
1955 Some aspects of ethology, the biological study of animal behaviour, *The advancement of science*, p. 17–19.
1955 Some neurophysiological problems raised by ethology, *Br. J. Anim. Behav.* **2**, 115.
1955 Psychology and ethology as supplementary parts of a science of behavior, In: *Group processes*, 1st *Conference sponsored by the Josiah Macy, Jr. Foundation*, New York, p. 75–167.
1956 The activation, extinction and interaction of instinctive urges. *Royal Institution, Friday Evening Discourses.*
1956 On the function of territory in gulls, *Ibis* **98**, 401–11.
1957 The functions of territory, *Bird Study* **4**, 14–27.
1957 The study of behaviour. In *The ornithologists' guide*, p. 60–5. B.O.U., London.
1957 On anti-predator responses in certain birds—a reply, *J. comp. Physiol. Psychol.* **50**, 412–14.
1958 Bauplanethologische Beobachtungen an Möwen, *Arch. neerl. Zool.* **13**, 369–82.
1959 Recent British contributions to scientific ornithology, *Ibis* **101**, 126–31.
1959 Behaviour, systematics, and natural selection, *Ibis* **101**, 318–30. (Reprinted in *Evolution after Darwin*, Vol. 1, pp. 595–616. Chicago University Press, 1960.)
1959 Film on the reproductive behaviour of the Black-headed Gull, *Ibis* **101**, 503–4.
1959 Einige Gedanken über Beschwichtigungsgebärden, *Z. Tierpsychol.* **16**, 651–65.
1959 The Ruff, *Br. Birds* **52**, 302–6.
1959 Comparative studies of the behaviour of gulls (Laridae): a progress report, *Behaviour* **15**, 1–70.
1960 Kampf und Balz der Lachmöwe, *J. Orn. Lpz.* **101**, 238–41.
1960 The evolution of behavior in gulls, *Scient. Am.* (December), 118–130.
1961 *Larus ridibundus* (Laridae)—fighting between males, Encyclop. Cinematographica E 334.
1961 *Larus ridibundus* (Laridae)—pair formation, Encyclop. Cinematographica E 335.
1961 *Larus ridibundus* (Laridae)—agonistic displays, Encyclop. Cinematographica E 336.
1962 Foot-paddling in gulls, *Br. Birds* **55**, 117–20.
1962 Sub-Family *Larinae*. An introduction to the behaviour and displays of British gulls. In *The birds of the British Isles*, Vol. XI. (ed. David A. Bannerman), pp. 191–200.
1962 The evolution of animal communication—a critical examination of methods, *Symp. zool. Soc., Lond.* **8**, 1–8.
1963 On aims and methods of ethology, *Z. Tierpsychol.* **20**, 410–33.

1964 On adaptive radiation in Gulls (Tribe Larini), *Zool. Meded. Leiden* **34,**
      209–23.
1964 The search for animal roots of human behaviour. Lecture 'Social studies
      and biology' given at Oxford University.
1964 Aggression and fear in the normal sexual behaviour of some animals, In *The*
      pathology and treatment of sexual deviation (ed. I. Rosen), pp. 1–23.
1964 Review of *The natural history of aggression*—Proceedings of Symposium—
      held at British Museum 1963, Technical Book Review.
1965 Behaviour and natural selection, In *Ideas in modern biology* (ed. J. A.
      Moore), pp. 521–40 National History Press, New York.
1965 Some recent studies of the evolution of sexual behavior. In *Sex and behavior*
      (ed. F. A. Beach), pp. 1–34. Wiley, New York.
1965 Von den Vorratskammern des Rotfuchses (*Vulpes vulpes* L.), *Z.*
      Tierpsychol. **22,** 119–49.
1966 Ritualization of courtship postures of *Larus ridibundus* L., *Phil. Trans. R.*
      *Soc. B***251,** 457.
1967 Adaptive features of the Black-headed Gull *Larus ridibundus* L., *Proc. int.*
      orn. Congr., **14,** 43–59.
1968 Masses on the move, *Animals* **11,** 418–21.
1968 On war and peace in animals and Man, *Science, N.Y.* **160,** 1411–18.
1969 Ethology, In *Scientific thought 1900–1960* (ed. R. Harré), pp. 238–68.
      Clarendon Press, Oxford.
1969 *Signals for survival.* Script for B.B.C. Television film.
1969 Cracking the footprint code. In *Look* (ed. J. Boswell), pp. 28–37. B.B.C.
      Publications, London.
1969 Von Krieg und Frieden bei Tier und Mensch. In *Kreatur Mensch* (ed. G.
      Altner), pp. 163–78. Heinz Moos, Munich.
1970 Umweltbezogene Verhaltensanalyse—Tier und Mensch, *Experientia* **26,**
      447–56.
1971 Clever gulls and dumb ethologists—or the trackers tracked, *Die Vogelwarte*
      26, 232–8.
1972 The Croonian Lecture, 1972: Functional ethology and the human sciences,
      *Proc. R. Soc. B***182,** 385–410.
1974 Ethology and stress diseases, In *Les Prix Nobel en 1973*, pp. 197–218. The
      Nobel Foundation, Stockholm.

JOINT PAPERS

1936 with G. Schuyl and L. Tinbergen
      Ethologische Beobachtungen am Baumfalken, *J. Orn. Lpz.* **84,** 387–433.
1937 with J. J. ter Pelkwijk
      Eine reizbiologische Analyse einiger Verhaltensweisen von *Gasterosteus*
      aculeatus L., *Z. Tierpsychol.* **1,** 193–200.
1938 with K. Lorenz
      Taxis und Instinkthandlung in der Eirollreaktion der Graugans L., *Z.*
      Tierpsychol. **2,** 1–29.
1938 with W. Kruyt
      Ueber die Orientierung des Bienenwolfes. III. Die Bevorzugung bestimm-
      ter Wegmarken. *Z. vergl. Physiol.* **25,** 292–334.
1938 with R. J. van der Linde
      Ueber die Orientierung des Bienenwolfes. IV. Heimflug aus unbekanntem
      Gebiet, *Biol. Zentralbl.* **58,** 425–35.

1939   with D. J. KUENEN
Ueber die auslösenden und die richtunggebenden Reizsituationen der
Sperrbewegung von jungen Drosseln, *Z. Tierpsychol.* **3**, 37–60.

1942   with B. J. D. MEEUSE, L. K. BOEREMA, and W. W. VAROSSIEAU
Die Balz des Samtfalters, *Z. Tierpsychol.* **5**, 182–226.

1947   with J. J. A. van IERSEL
'Displacement reactions' in the three-spined Stickleback, *Behaviour* **1**,
56–63.

1950   with A. C. PERDECK
On the stimulus situation releasing the begging response in the newly
hatched Herring Gull chick (*Larus argentatus argentatus* Pont), *Behaviour*
3, 1–39.

1952   with M. MOYNIHAN
Head flagging in the Black-headed Gull: its function and origin, *Br. Birds*
45, 19–22.

1954   with G. J. BROEKHUYSEN
On the threat and courtship behaviour of Hartlaub's Gull, *Ostrich* **25**,
50–62.

1957   with R. HOOGLAND and D. MORRIS
The spines of Sticklebacks (*Gasterosteus* and *Pygosteus*) as means of defence
against predators (*Perca* and *Esox*), *Behaviour* **10**, 205–36.

1958   with R. A. HINDE
The comparative study of species-specific behavior. In *Behavior and
evolution* (eds A. Roe and G. G. Simpson), pp. 251–68. Yale University
Press.

1962   with G. J. BROEKHUYSEN, F. FEEKES, J. C. W. HOUGHTON, H. KRUUK, and E.
SZULC
Egg shell removal by the Black-headed Gull *Larus ridibundus* L.; a behaviour
component of camouflage, *Behaviour* **19**, 74–117.

1962   with H. KRUUK and M. PAILLETTE
Egg shell removal by the Black-headed Gull (*Larus ridibundus* L.) II. The
effects of experience on the response to colour, *Bird Study* **9**, 123–31.

1962   with H. KRUUK, M. PAILLETTE, and R. STAMM
How do Black-headed Gulls distinguish between eggs and egg-shells? *Br.
Birds* **55**, 120–9.

1964   with M. NORTON-GRIFFITHS
Oystercatchers and mussels, *Br. Birds* **57**, 64–70.

1967   with M. IMPEKOVEN and D. FRANCK
An experiment on spacing-out as a defence against predation, *Behaviour*
28, 307–21.

1972   with E. A. TINBERGEN
Early childhood autism—an ethological approach, *Z. Tierpsychol. Suppl.* **10**.

BOOKS

1930   (with G. VAN BEUSEKOM, F. P. J. KOOIMANS, and M. G. RUTTEN) *Het
Vogeleiland*
Schoonderbeek, Laren.

1935 *Eskimoland*
     Van Sijn, Rotterdam.
1946 *Inleiding tot de Diersociologie*
     Noorduyn, Gorkum.
1951 *The study of instinct*
     Clarendon Press, Oxford.
1953 *The Herring Gull's world*
     Collins, London.
1953 *Social behaviour of animals*
     Methuen, London.
1954 *Bird life*
     Clarendon Press, Oxford.
1958 *Curious naturalists*
     Country Life, London.
1965 *Animal behavior*
     Life Nature Library, New York.
1967 (with E. A. R. ENNION) *Tracks*
     Clarendon Press, Oxford.
1970 (with H. FALKUS and E. A. R. ENNION) *Signals for survival.*
     Clarendon Press, Oxford.
1972 *The Animal in its world. Explorations of an ethologist 1932–1972.* Vol. I, *Field Studies*
     Allen and Unwin, London.
1973 *The Animal in its world. Explorations of an ethologist 1932–1972,* Vol. II, *Laboratory experiments and general papers*
     Allen and Unwin, London.

(Various translations in German, French, Italian, Spanish, Swedish, Danish, Japanese, Russian, Dutch)

CHILDREN'S BOOKS

1948 *Kleew*
     Oxford University Press, New York.
1954 *The tale of John Stickle*
     Methuen, London.

PRINCIPAL EDUCATIONAL FILMS

Insect studies
Evolution in progress
The kittiwake
Fighting, threat, and courtship of the Black-headed Gull
Gannet city
Eggshell removal in the Black-headed Gull
Crafty predators and cryptic prey
(with H. FALKUS)   *The gull watchers*
                The beachcombers
                The sign readers
                Signals for survival

List of Contributors

G. P. BAERENDS   Zoologisch Laboratorium der Rÿksuniversiteit
Groningen,
Kerklaan 30,
Postbus 14,
Haren (Gr.),
The Netherlands

C. G. BEER   Institute of Animal Behavior,
Rutgers University,
101 Warren Street,
Newark,
New Jersey, 07102,
U.S.A.

R. A. HINDE   Sub-Department of Animal Behaviour,
M.R.C. Unit on the Development and Integration of
Behaviour,
Madingley,
Cambridge,
England.

M. HIRSBRUNNER-SCHARF   Zoologisches Institut,
Universität Bern,
3000 Bern,
Sahlistraße 8,
Switzerland.

J. J. A. VAN IERSEL   Zoologisches Laboratorium der Rijksuniversiteit Leiden,
Kaiserstraat 63,
Leiden,
The Netherlands

K. IMMELMANN   Arbeitstelle Biologie,
Universität Bielefeld,
48 Bielefeld,
Postfach 8640,
Germany.

H. KRUUK   Institute of Terrestrial Ecology,
Banchory Research Station,
Hill of Brathens
Glassel, Banchory,
Kincardineshire, AB3 4BY,
Scotland.

R. LILEY — Department of Zoology,
University of British Columbia,
Vancouver 8,
British Columbia,
Canada.

M. LINDAUER — Zoologisches Institut,
Universität Würzburg,
87 Würzburg,
Röntgenring 10,
Germany.

F. McKINNEY — Department of Ecology and Behavioral Biology,
University of Minnesota,
Minneapolis,
Minnesota, 55455
U.S.A.

A. MANNING — Department of Zoology,
University of Edinburgh,
Kings Buildings,
West Mains Road,
Edinburgh, EH9 3JT,
Scotland.

P. MARLER — Field Research Center for Ecology and Ethology,
The Rockefeller University,
Tyrrel Road,
Millbrook,
New York, 12445,
U.S.A.

M. MOYNIHAN — Smithsonian Tropical Research Institute,
P.O. Box 2072,
Balboa,
Canal Zone,
Panama.

J. B. NELSON — Department of Zoology,
University of Aberdeen,
Aberdeen, AB9 1AS,
Scotland.

I. J. PATTERSON — Culterty Field Station,
University of Aberdeen,
Newburgh,
Ellon,
Aberdeenshire, AB4 OAA,
Scotland.

M. ROBINSON   Smithsonian Tropical Research Institute,
P.O. Box 2072,
Balboa,
Canal Zone,
Panama.

K. D. ROEDER   Department of Biology,
Tufts University,
Medford,
Massachusetts, 02155,
U.S.A.

B. H. SEGHERS   Department of Zoology,
University of Manitoba,
Winnipeg,
Canada, R3T 2N2.

B. TSCHANZ   Zoologisches Institut
Universität Bern,
3000 Bern,
Sahlistraße 8,
Switzerland.

Part I
Function

1. The concept of function

R. A. HINDE

1.1. Introduction

NIKO TINBERGEN, more than any other ethologist, has stressed the importance of the problem of function. He has shown not only that its understanding contributes to the distinct but yet related questions of causation and evolution, but also, and more fundamentally, that in its own right it is an important, interesting, and too frequently neglected question. In particular, the paper Tinbergen presented at the 1966 Ornithological Congress (Tinbergen 1967) contained a vivid demonstration of the increased understanding of the biology of a species which studies of function provide, an evaluation of the types of evidence relevant to it, and a demonstration of the effectiveness of the experimental analysis of function which he had pioneered. But in spite of his hard-headed example, the problem of function is still too often treated as a matter for speculative asides (Tinbergen 1965). Therefore, although it means some retreading of old ground, it seems worthwhile to give further consideration to the nature of the concept, and to the sorts of evidence used in its discussion.

1.2. The behavioural nexus

Each item of behaviour forms a link in a nexus of events which precede and follow it. If we focus on the behaviour, we refer to some of the former as causes and some of the latter as consequences. The consequences can in principle be classified according to their effects on the organism's subsequent reproductive success: some are detrimental, some neutral, and some beneficial.[†] For example, when a gull incubates its eggs it (1) perhaps exposes itself to predation and loses time that it could spend feeding (harmful consequences); (2) slightly expands the egg shell (presumably a neutral consequence);[‡] and (3) keeps the embryos warm and thus enhances its chance of reproductive success (beneficial consequence) (Tinbergen

---

[†] Strictly, success should be assessed in terms of gene survival rather than progeny survival (Hamilton 1964), but that does not affect the present argument.

[‡] Characters are said never to be neutral (Cain 1964), though that depends in part on what is meant by a character. But the present discussion is concerned not with characters (for example, incubation behaviour) but with their consequences. That some of these can be neutral is self-evident.

1942). In a broad way, beneficial consequences are referred to as 'functions'. A character of structure or behaviour is said to be adaptive if its beneficial consequences outweigh the deleterious ones.

Now according to this formulation, the function of all adaptive characters is ultimately the same, namely, contributing to eventual reproductive success. In practice, of course, the interesting question is how they so contribute. In the case of behaviour each act has ramifying consequences: if the behaviour is adaptive, one or more of the interlinked chains of consequences contributes to reproductive success. The term 'function' is often applied to items on such chains following fairly soon after the behaviour itself, and preceding the subsequent enhancement of reproductive success. For instance, if territorial behaviour spaces out a breeding population, and spacing out reduces predation, and thus improves reproductive success (Tinbergen 1956), either spacing-out or the reduction in predation may be called the function of the territorial behaviour.†

1.3. Weak and strong meanings of the term 'function'

Some characters of structure or behaviour are present in some individuals of a population but not others.‡ More usually, however, a character is present in all, but to varying degrees. In such a case, its mere presence could (in theory at least) be sufficient to ensure that a given beneficial consequence occurs with full efficiency in each individual. That consequence could not then provide material on which natural selection could act to maintain the character in the population. Natural selection can operate only through consequences that are achieved to a different extent in different individuals. (It is, of course also necessary that the differences have a genetic basis.) For example, a bird's body feathers provide insulation, and also carry patterns and colours that are biologically advantageous. But if the natural variation in the number of feathers is within the range which would affect insulation but not coloration, natural selection cannot act through the latter to maintain the number of feathers at its present level. Thus the effectiveness of the colour pattern may not be

---

† This usage of function by evolutionarily minded biologists to refer forward in time must be distinguished from two others. One closely related usage implies the question 'How does it work?'. For instance, the bastard wing of a bird has the function of smoothing the airflow over the wing when stalling conditions are approached. This usage is closely related to that under discussion in that it also refers to a beneficial consequence of the character, but differs in that that consequence is more or less immediate. A quite different usage is the more mathematical one implying merely the existence of a relationship between two events. 'Behaviour Y is a function of X' or 'Y is a function of behaviour X' can mean merely that a mathematical relation exists between X and Y, without implication that the causal relations, if any, are direct or indirect.

‡ The range of meanings of the term 'character' will not be pursued here: in the present context it is used both for the properties of a behaviour pattern (for example, the precise form or frequency of a signal movement) and for the pattern itself.

the consequence through which natural selection is acting to maintain the present density of plumage. The coloration of the feathers could also affect insulation, but if the variation in coloration is such that insulation is virtually unaffected, natural selection cannot affect coloration through its heat-radiating consequences. Again, the preservation of biological requirements and the prevention of epidemic diseases have both been suggested as beneficial consequences of avian territorial behaviour. If the variability in territory size is sufficient to affect the availability of necessities, but not the spread of diseases, then it could be said that the former is a function and the latter is not (Hinde 1956).

Thus not all beneficial consequences provide material for the action of natural selection. In practice, of course, the dichotomy just implied between beneficial consequences of a character through which natural selection does and does not act is an oversimplification. The extent to which the various beneficial consequences provide material for natural selection will vary along a continuum, some consequences being achieved so nearly fully in all individuals that the effectiveness of natural selection operating through them is overridden by its effects through others. But the argument does indicate that a concept of function equated with 'any beneficial consequence' does little more than answer the question, 'What is the character good for?', and need tell us nothing about the dynamics of the evolutionary process. How then can the concept be hardened up?

First, if function is to have empirical relevance it must refer ultimately to the consequences of a *difference*. Thus if we speak of the function of flocking, we refer to the beneficial consequences of associating with others not enjoyed by individuals living alone. If we speak of the function of 'the cliff-nesting behaviour of kittiwakes', we imply 'as compared with the ground-nesting behaviour of gulls'. Sometimes the comparison is with a hypothetical organism: when we speak of 'the function of bird song', we refer to events consequential upon the birds' singing which would not occur if birds did not sing, and thus imply a hypothetical population which does not sing.

Second, we must recognize that the concept of function carries a spectrum of meanings, A weak meaning answers the question 'what is it good for?'. In so far as a comparison is implied, it is with a hypothetical organism similar in every respect except that it lacks the character in question. Furthermore, a weak use of function may contain no indication of precisely how the character in question contributes to reproductive success. The statement that the function of birds' wings is flight is a weak one in so far as it implies at most comparison with a hypothetical bird without wings, or an avian ancestor not yet possessing them, and because it contains no indication of how flight contributes to reproductive success. (This is not so obvious as it seems: on oceanic islands birds tend to become flightless, presumably because they are liable to be blown out to sea (Lack 1947).) Discussion of function at this level has little meaning in terms of the dynamic processes of evolution: only when the differences between this wing and that are discussed does the problem begin to relate to the operation of natural selection.

By contrast, 'function' in a strong sense attempts to answer the question 'through what consequences does natural selection act to maintain this character?'. Now, to answer this question, it would strictly be necessary to assess reproductive success with and without the character. Cross-species and even cross-population studies are seldom strictly adequate for this, simply because the different characters of each form a co-adapted complex. For example, the young of hole-nesting passerine birds tend to spend longer in the nest than those of open-nesting species, presumably because they are less vulnerable to predation. This means that they need less food (weight for weight) each day, and the parent can thus rear more individuals. Furthermore, there is less premium on female crypticity, and thus the sexes are more likely to be coloured similarly. Thus nest-site, fledgling period, clutch size, and sexual dimorphism are adaptively related characters. This argument could be continued almost indefinitely: the point at issue is that, from the point of view of natural selection, characters cannot be considered in isolation, but only in relation to each other.† Thus it is usually impossible to find two populations differing in only one character relevant to the consequence under examination. Sometimes, however, it is possible to use existing variations within a population, assessing the consequences of that variation. Such a procedure can lead to hard evidence on function in this strong sense—provided, of course, that it can be assumed that the variations in the character in question are not correlated with variations in other characters which affect its consequences.

All this seems rather pedantic. Indeed it leads at first sight to the improbable view that it is meaningless to ask about the function (in the strong sense) of stable characters equally present in all members of the population. But it is in fact the case that as we move from gross characters (for example, bird's wings) to finer ones (for example, long narrow wings in this species as opposed to short broad ones in that) to yet finer ones (for example, longer wings in this population than that) the questions that we can ask about beneficial consequences move from gross ones (what is it good for?) to fine ones about how natural selection actually acts. We shall return in a moment to the nature of the data available for answering such questions: the issues here are that when we ask 'what is its function?', 'it' implies a difference and 'function' can have a spectrum of meanings.

---

† Well worked-out examples are provided by the studies of Tinbergen and his pupils on gulls (for example, Cullen 1957) or Crook (1964) on reproduction in weaver-birds.

Characters   Chains of consequences

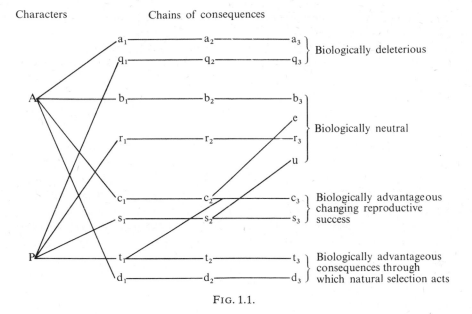

FIG. 1.1.

1.4. The concept of function: summary

The argument so far is summarized in Fig. 1.1. A and P represent two items of behaviour. A leads to biologically deleterious, neutral and advantageous chains of consequence a, b, c, and d, only the last leading to consequences through which natural selection acts. P leads to comparable chains q, r, s, and t. Consequences c and s are functional in a weak sense, d and t in a strong sense. The biologically advantageous chain c also leads to a neutral consequence e, and s to u. The adaptive nature of the further consequences of c_2 (itself a consequence of A) are affected by t_1 (itself a consequence of P). Consequences c, d would represent functions of possessing the character A in the weak sense but only d in the strong sense.

1.5. Is all behaviour functional?

Only characters which confer a positive biological advantage can survive for long against the forces of mutation and selection. Biologists are thus prone to consider that characters of behaviour are biologically advantageous unless proved otherwise: this assumption is one reason why the problem of function has not been studied more assiduously in the past. In any case, it is a natural point of view when to prove function in a strict sense is a task of such complexity. The climatic, vegetational, and zoogeographic circumstances in which a species lives are not constant. To the extent that the variations are orderly, the patterning will exert a selective influence. To

assess the function of a character subject to shifting selective forces could
be an impossible task.

Nevertheless, certain categories of exceptions are recognized as account-
ing for the presence of characters that appear not to be biologically
advantageous.

1. The character in question may be a by-product of a character
   adaptive in another context. Thus Tinbergen (1969) regards the
   aggressiveness of a male Black-headed Gull to its mate as a by-
   product of aggressiveness adaptive in the context of territorial
   disputes. Similarly, Kruuk (1972) argues that the phenomenon of
   surplus killing by carnivores is a relatively trivial disadvantageous
   consequence of behavioural characteristics otherwise adaptive in
   obtaining food.

2. Just as a structure may be biologically advantageous but not used all
   the time, so every expression of a character of behaviour does not
   have to be functional. Thus when Blue Tits (*Parus caeruleus*) enter
   houses and tear paper, this may be a non-functional expression of a
   motor pattern biologically adaptive for feeding, and normally used in
   that way (Hinde 1953).

3. A structure or item of behaviour may be a relict of a formerly
   adaptive character which has not yet been lost. For example, many
   birds extend their ipsilateral wing downwards when scratching their
   head; this is often held to be a phylogenetic remnant from tetrapod
   ancestors (but see discussion in Wickler (1961)). The persistence of
   neutral characters is perhaps particularly likely in view of the
   co-adapted nature of the character complex.

In addition a character may be adapted but not perfect because of
conflicting requirements. Thus the extent to which a bird will display at or
attack a predator near its nest is the result of the conflicting requirements
of protecting its young and protecting itself (Tinbergen 1965, 1967).

1.6. Observational evidence for 'functions'

We may now consider some of the types of evidence by which 'functions'
are identified, bearing in mind the distinction between weak and strong
meanings of 'function'.

The first indication of function often comes from contextual information.
For example, it was primarily (though not solely) because Hooker and
Hooker (1969) found that, in the shrikes they studied, 'duetting occurs
throughout the year between members of an established pair occupying
and defending a territory', that they concluded that duetting functions 'in
maintaining the bond between family groups in dense vegetation and in
joint aggressive vocal display during territorial disputes'.

In addition to contextual information, statements about function are usually based on observations that the behaviour in question has a certain consequence, and that that consequence appears to be a 'good thing'. Thus the observation that an intruder onto a robin's territory flies away when it hears the owner sing (Lack 1939) is evidence that song functions in maintaining territorial ownership, additional to the contextual evidence that a robin sings only when on its territory. The hypothesis that gulls roost on open beaches up to the time when incubation starts because they are safer from predators there was confirmed by data showing that the number of gulls killed by foxes is much greater in the breeding areas than on the beaches (Kruuk 1964; Tinbergen 1965).

A further type of evidence, perhaps to be regarded as a form of contextual evidence, concerns correlations between characters of structure or behaviour and environmental factors. For example, if body size in a number of distantly related groups tends to become larger the nearer the poles that they breed, this suggests that body size is related to some environmental factors that change with latitude. If conspicuous coloration and behaviour are common in male birds living in areas where the ranges of closely related species overlap, whereas where only one species is present both sexes tend to be cryptic, this suggests that the conspicuous features have a function related to the presence of closely related species. A wealth of further examples of this sort has been given by Lack (1966, 1968). When evidence of this sort is added to evidence about consequences that appear to be advantageous—for instance, that large animals withstand cold better, or conspicuous colours are used in pair formation—the evidence for function even in the strong sense becomes powerful.

It would probably be true to say that most statements about functions of behaviour are based on evidence of these types. Often further evidence is impossible to obtain. However, it is as well to recognize not only that they can provide at most indirect evidence about the action of natural selection, but also that a number of other difficulties may arise.

One of these is that even the characters chosen for discussion may have little biological relevance. Thus there has recently been considerable speculation about the functions of the different types of social structure to be found among non-human primates, and attempts have been made to relate the size or composition of groups to ecological variables (Crook and Gartlan 1966). As more information has become available, the inadequacy of these attempts has become increasingly apparent (for example, Struhsaker 1969; Crook 1970). This is not only because of the complexity of the problems, but because social structure depends on the behaviour of individuals, and it is the behaviour of the individuals for which functions must be identified. Characters of group size or structure may be by-products of behaviour adaptive through other consequences. Of course if such characters were adaptively disadvantageous, they might be selected against; the point here is that they may be near-neutral.

Another type of difficulty can arise if the observer focuses his attention on the wrong consequence, which may be a mere by-product of a nexus of consequences leading also to a quite different event that is adaptively important. For example, it is often said that the function of threat displays is to reduce actual physical combat, and it is in fact the case that physical contact often does not occur while animals are displaying. However, it seems reasonable to suppose that an animal that is definitely going to attack or to flee would do best to do so immediately, without giving warning by displaying first. On this view, threat displays would be useful only in moments of indecision. If what one individual will do depends in part on the probable behaviour of the other, threatening by one or both individuals may convey the information necessary for a decision. Such a view is indeed prompted by Stokes's (1962) finding that the threat displays of Blue Tits do not give an absolute prediction of whether the bird will subsequently attack, stay, or flee; and by Simpson's (1968) description of the fighting of the Siamese Fighting Fish (*Betta splendens*), which involves reciprocal displays of mutually similar and gradually increasing intensity, often until one individual suddenly capitulates before either has given a bite. On this view, then, the function of a display lies in communication with the opponent, which maximizes the chances of either a successful attack or escape; a reduction in the actual amount of combat is a by-product. This view, of course, requires substantiation; the issue here is that it is a reasonable alternative to that often advanced. The observation that display has a consequence which appears to be advantageous is not proof of function in a strong sense.

A related example concerns the formation of dominance hierarchies. This is accompanied by a reduction in aggressive behaviour, and this is often said to be its function. But examination of the behaviour of individuals suggests that the subordinate individuals behave as they do because under natural conditions (and perhaps in the long run) they stand a better chance of obtaining access to the resources they need if they do not attempt to obtain them in the presence of the dominants (for example, Rowell 1966). A reduction in aggressive behaviour is one consequence of the interactions between individuals, a pattern of relationships which can be described as a dominance hierarchy is another, but to say that the former is a function of the latter in a biological sense is both imprecise and misleading. It would be nearer the truth to conjecture that the reduction in fighting has consequences which are beneficial in different ways for dominants (for example, access to food without risk of injury) and subordinates (for example, reduction in risk of injury by dominant and possibility of finding food elsewhere); and that the hierarchy is a consequence of the reduction in fighting, rather than vice versa.

Furthermore, there are often difficulties in proving that the supposed function really is biologically advantageous, and really is a consequence of the behaviour in question. An example is provided by a point in Marler's (1968) important review of the mechanisms of group spacing in primates. He suggests that the signals used can be classified as distance-increasing, distance-maintaining, distance-reducing, and proximity-maintaining signals, and that their properties conform with these functions. Marler himself points out that the distinction between distance-increasing and distance-maintaining functions is not always clear, but there are also further difficulties with the latter category. First, one may ask whether there is indeed evidence that it is advantageous for a primate group to have another group just so far away but no further; one can speculate that it may be, but that is all, and unless distance-maintenance has biologically advantageous ramifications, it cannot be designated as a function. Second, when a particular signal does result in the maintenance of distance, this may be because it either attracts or repels other groups, the effect being balanced by a tendency to withdraw or approach with a quite different basis. The consequence of maintaining distance may thus be a joint consequence of the signal and some other factor and cannot be designated as a function of the signal *per se*. Third, Marler himself points out that distance-maintaining calls may sometimes either increase or decrease the distance between troops, depending on their spatial relations. For example, Ellefson (1968) reports that if a gibbon group locates the source of a call outside the area it occupies, it may reply in kind, but if the call comes from a point within or near that area, it may approach and then give distance-increasing calls. This suggests that the important consequence of such calls is that of facilitating localization and that, depending on this, other mechanisms are called into play which promote over-dispersal.

Another type of difficulty which sometimes arises involves the confusion of consequences at different points in the chains ramifying from the behaviour in question. For example, Barnett (1969), generalizing about the functions of agonistic behaviour says that '(1) it influences population density; (2) it determines the structure of the groups of colonial species, and (3) it regulates interactions between groups, or, in solitary species, between individuals.' But (1) is a consequence of (2) and (3); and (2), in so far as agonistic behaviour also regulates interactions between individuals of colonial species, of (3).

1.7. Experimental evidence

So far we have considered evidence derived from field observations. Can experimentation help? It can, but again its limitations must be recognized. Van den Assem (1967) demonstrated in the laboratory that the territorial behaviour of Three-spined Sticklebacks could limit density, that the possession of an adequate territory increased reproductive success in a number of ways, and that owners of large territories were more successful in rearing young than owners of small ones. From these experiments he concluded that the reduction of interference by conspecifics is the main function of territorial behaviour. But, as Tinbergen (1968) points out, the interesting results obtained in this study cannot be extrapolated beyond the experimental situation; other variables, such as food availability and predator pressure, in relation to which territorial behaviour might have important consequences in nature, were not manipulated. Although the validity of deductions from laboratory experiments about selection in nature varies from case to case, it must be acknowledged that, in the absence of evidence of other sorts, laboratory experiments of this sort can provide evidence about beneficial consequences only in the circumstances in which they were conducted, and thus lead only to suggestions about function in a fairly weak sense.

To a lesser degree, of course, the same is true of field studies; the selective forces that impinge on a species are not uniform throughout its range. However, much stronger evidence for function in the strong sense can be obtained in the natural situation. Tinbergen and his pupils have pioneered this work, and we may consider three of their experiments.

1. Patterson (1965) was concerned to show whether the aggressive displays of territory-owning Black-headed Gulls (*Larus ridibundus*) do or do not have a deterrent effect on others. By comparing the behaviour of intruders onto territories in which the owners were present but immobilized by a stupefying drug with that of intruders on territories in which the owners were behaving normally, it was shown that the displays do have a deterring effect. This experiment by itself (and Patterson, of course, used also other types of evidence) showed that the difference between territory-owning gulls that display normally and individuals (anaesthetized or hypothetical) that occupy territories but do not display has consequences for territorial ownership. Accepting that the latter is adaptively important (see below) this demonstrates a function in the weak sense for territorial display, but does not demonstrate that natural selection acts through this consequence of the display to maintain it at its present level.

2. Black-headed Gulls remove fragments of egg shells from the nest soon after the young have hatched. Tinbergen and his colleagues (Tinbergen, Broekhuysen, Feekes, Houghton, Kruuk, and Szulc 1962; Tinbergen, Kruuk, and Paillette 1962) suspected that this functions in reducing predation on unhatched eggs or chicks. This was tested by laying out eggs in the colony with or without a broken egg shell nearby. The former were taken by predators more rapidly than the latter. This demonstrates that the difference between gulls that do and do not remove egg shells, or that remove them with different

latencies, is adaptive. Furthermore, since individual gulls remove egg shells with different latencies, so that the difference between eggs with a broken shell nearby and eggs without is a difference that actually occurs in the colony, and since it can be assumed that the difference has some genetic basis, the experiment provides compelling evidence for function in the strong sense at least in this colony.

3. As indicated above, Patterson (1965) showed that territorial behaviour promotes the spacing out of nests. By taking advantage of the natural variation in spacing in the colony he showed that spacing out reduced predation on pipped eggs and newly hatched chicks by other gulls and by other species (see also Kruuk 1964; Tinbergen 1967). Again, since Patterson demonstrated differential success in relation to natural variation within the species, and since it can be assumed that this variation has some genetic basis, this comes as near as possible to showing that natural selection is operating to maintain spacing out at its present level.

In his discussion of the adaptive features of the Black-headed Gull, Tinbergen (1967) is careful not to overrate the value of his experimental approach. In some cases it is not necessary, sometimes it is not possible, and often it is not sufficient. Furthermore, description and interpretation must always precede experimentation. However, the descriptive–comparative method and the experimental method can together carry the study of function far beyond the reach of armchair speculation.

1.8. Conclusion

I have argued that the concept of function can properly be applied only to the consequences of differences between characters, and that it is advisable to recognize that it covers a spectrum of meanings. At the weak end it is equivalent to beneficial consequences, while at its strongest it has reference to the action of selective forces.

In discussing the sources of evidence adduced for function, I have emphasized the difficulties of proof. Just because hard evidence is so difficult to obtain, it has become respectable to speculate about the function of behaviour in a manner that would never be permissible in studies of causation. This is counterproductive. But I hope that this emphasis on its dangers will not tend to inhibit those types of speculation that may suggest new avenues to explore or new experiments to conduct— or indeed speculation that feeds the sense of wonder that is the mainspring of biological research.

Acknowledgements

I am grateful to Professor N. Tinbergen for the help and advice which he has given me on many occasions, and in particular when I was first starting in research.

The present essay has been improved as the result of critical comments from P. P. B. Bateson, T. Clutton-Brock, C. Goodhart, and N. W. Humphrey.

References

VAN DEN ASSEM, J. (1967). Territory in the three-spined stickleback, *Gasterosteus aculeatus*. *Behaviour, Suppl.* **16**, 1–164.

BARNETT, S. A. (1969). Grouping and dispersive behaviour among wild rats. In *Aggressive behaviour* (eds. S. Garattini and E. B. Sigg). Excerpta Medica, Amsterdam.

CAIN, A. J. (1964). The perfection of animals. In *Viewpoints in biology* (eds. J. Carthy and R. Duddington), Vol. 3, pp. 36–62. Butterworth, London.

CROOK, J. H. (1964). The evolution of social organisation and visual communication in the weaver birds (Ploceinae). *Behaviour, Suppl.* **10**, 1–178.

—— (1970). (ed.). The socio-ecology of primates. In *Social behaviour in birds and mammals*. Academic Press, London.

—— and GARTLAN, J. S. (1966). Evolution of primate societies. *Nature, Lond.* **210**, 1200–3.

CULLEN, E. (1957). Adaptations in the kittiwake to cliff-nesting. *Ibis* **99**, 275–302.

ELLEFSON, J. O. (1968). Territorial behaviour in the common white-handed gibbon, *Hylobates lar*. In *Primates* (ed. P. C. Jay). Holt, Rinehart, and Winston, New York.

HAMILTON, W. D. (1964). The genetical evolution of social behaviour. *J. theor. Biol.* **7**, 1–52.

HINDE, R. A. (1953). A possible explanation of paper-tearing behaviour in birds. *Br. Birds* **46**, 21–3.

—— (1956). The biological significance of the territories of birds. *Ibis* **98**, 340–69.

HOOKER, T. and HOOKER, B. I. (1969). Duetting. In *Bird vocalizations* (ed. R. A. Hinde). Cambridge University Press.

KRUUK, H. (1964). Predators and ant-predator behaviour of the Black-headed Gull (*Larus ridibundus* L.). *Behaviour, Suppl.* **11**, 1–130.

—— (1972). Surplus killing by carnivores. *J. Zool. Lond.* **166**, 233–44.

LACK. D. (1939). The behaviour of the robin: I and II. *Proc. zool. Soc. Lond.* A**109**, 169–78.

—— (1947). *Darwin's finches*. Cambridge University Press.

—— (1966). *Population studies of birds*. Clarendon Press, Oxford.

—— (1968). *Ecological adaptations for breeding in birds*. Methuen, London.

MARLER, P. (1968). Aggregation and dispersal: two functions in primate communication. In *Primates* (ed. P. C. Jay). Holt, Rinehart, and Winston, New York.

PATTERSON, I. J. (1965). Timing and spacing of broods in the black-headed gull, *Larus ridibundus*. *Ibis* **107**, 433–59.

ROWELL, T. E. (1966). Hierarchy in the organization of a captive baboon group. *Anim. Behav.* **14**, 430–43.

SIMPSON, M. J. A. (1968). The display of the Siamese fighting fish, *Betta splendens*. *Anim. Behav. Monogr.* **1**, No. 1.

STOKES, A. W. (1962). Agonistic behaviour among blue tits at a winter feeding station. *Behaviour*, **19**, 208–18.

STRUHSAKER, T. T. (1969). Correlates of ecology and social organization among African cercopithecines. *Folia. Primat.* **11**, 80–118.

TINBERGEN, N. (1942). An objective study of the innate behaviour of animals. *Biblthca Biotheor.* **1**, 39–98.

TINBERGEN, N. (1956). On the functions of territory in gulls. *Ibis* **98**, 401–11.
—— (1959). Comparative studies of the behaviour of gulls (Laridae): a progress report. *Behaviour* **15**, 1–70.
—— (1965). Behaviour and Natural Selection. In *Ideas in modern biology*, (ed. J. A. Moore). *Proc. int. Congr. Zool.* **6**, 521–42. New York.
—— (1967). Adaptive features of the black-headed gull, *Larus ridibundus* L. *Proc. Int. orn. Congr.* **14**, 43–59.
—— (1968). Book review. *Anim. Behav.* **16**, 398–9.
—— BROEKHUYSEN, G. J., FEEKES, F., HOUGHTON, J. C. W., KRUUK, H., and SZULC, E. (1962). Egg shell removed by the black-headed gull, *Larus ridibundus*, L.; a behaviour component of camouflage. *Behaviour* **19**, 74–117.
—— KRUUK, H., and PAILLETTE, M. (1962). Egg shell removed by the black-headed gull *Larus r. ridibundus*: II. *Bird Study* **9**, 123–31.
WICKLER, W. (1961). Über die Stammesgeschichte und den taxonomischen Wert einiger Verhaltensweisen der Vögel. *Z. Tierpsychol.* **18**, 320–42.

2. Multiple functions and gull displays

C. G. BEER

2.1. Introduction

'IN BIOLOGY that consequence which a system appears as though designed
to achieve is usually termed the system's "function"' (Bowlby 1969, p.
125). There may be more than one such consequence, however, especially
when one is dealing with behaviour. Consider bird song, for example. In a
hypothetical case there might be evidence for each of the following
consequences of singing by a male: (1) deterrence of approach by rival
males and consequent spacing out of males so that the density of the
breeding population is commensurate with food supply, the number of
nest-sites, or some other environmental factor affecting breeding success;
(2) attraction of an unmated female, and consequent formation of the pair
bond necessary for breeding; (3) stimulation of the female's reproductive
system and its consequent synchronization with that of the male to the
degree necessary for effective intergration of the breeding activities of the
two birds; (4) stimulation of other birds of the same species in the vicinity,
and consequent effects on the timing and synchrony of breeding in the
population, which answer to temporal patterns of food supply, predator
pressure, or some other environmental factor affecting breeding success;
(5) prevention of pair formation between members of different species and
hence the maintenance of the genetical integrity of the species and the
avoidance of waste of breeding potential in the production of hybrids with
a relatively low chance of survival. Thus, the male's song may be
functionally related to territoriality, pair formation, breeding synchronization and timing, and reproductive isolation. It may appear as though
designed for each and all of these functions.

In a case such as this the assumption that each biological characteristic is
answerable to but one natural selection pressure would be difficult to
sustain. It is not necessary to be hypothetical to make the point. Investigations of genetic polymorphism have revealed interaction effects of different
and divergent selective tensions on one and the same system. The work of
Niko Tinbergen and his associates on the breeding adaptations of gulls has
led to the view that a number of the characteristics of the reproductive

behaviour of these birds are 'compromises' to opposing selection pressures (see, for example, Tinbergen 1964). But selection pressures need not be opposed to exert joint effects on the evolution of structure or behaviour. The design and appearance of a fish's fin or a bird's wing may answer to the joint requirements of locomotion, camouflage, and display.

In a case like this last, however, it could be argued that one of these functions is primary and the others secondary—that the fin or the wing was initially designed for locomotion and that this remains its basic function, the characteristics serving the other functions being later and super-imposed additions. Considerations of comfort, safety, and economy influence the design of a car, but only after discovery of the means of effecting the basic function of the machine, namely transportation. Similarly, the reproductive isolation function of a bird's song must be regarded as secondary to the pair-formation function of the song, since the song presents a means for effecting reproductive isolation only because of its role in pair formation.

However, these examples also suggest another way in which we might deal with cases of apparent multiple function. Instead of considering the functional significance of a structure or behavioural system as a whole, we might distinguish different aspects and seek the function of each in abstraction from the others. For example, the coloration of the fish's fin or the bird's wing might be functionally related to camouflage or display, but not to locomotion, A reproductive isolation function might attach to the *differences* between the songs of closely related sympatric species of birds, not to other aspects of the songs.

Yet another way to attempt analysis of cases of apparent multiple function is to look for variation in the senses in which the term 'function' has been used. The term has several meanings in biological contexts, not all of which are sharply distinguished from one another. One of the distinc-tions that should be observed is that between immediate consequences and ultimate consequences (cf. Baker 1938). When a male songbird sings, the immediate consequence is often the repulsion of the approach of a rival, and the singing 'appears as though designed to achieve' this end. But then one can ask why this end needs to be achieved. It results in the establishment and maintenance of territorial boundaries, and hence in the spacing-out of breeding pairs. But this observation only shifts the question. Why should the birds have to space themselves out? What is the point—the adaptive significance—of territoriality? The answer sought will be in terms of survival value—the ultimate consequences—the ways in which the behaviour optimizes the chances of successful breeding in the population, compared to possible alternatives. An ultimate consequence of the singing might thus be its contribution to regulation of the population size in accordance with what food supply dictates as the optimum level. To

determine the ultimate consequences of behaviour thus entails identification of the selective pressures to which the behaviour answers. This is not
so, or at least not necessarily so, in determining the immediate consequences of the behaviour.

However, there is an intermediate class of consequences that is, so to
speak, short of being ultimate yet to which selection pressure can be
directly referred. The courtship and mating behaviour of many kinds of
animals consists of sequences in which, at each point, the act of one of the
partners is response to the preceding act of the other, and stimulus for the
succeeding act of the other, different kinds of acts thus being strung
together to reach the culminating act of copulation. If completion of such a
sequence requires each step to be performed in a specific manner and in a
specific place in the order, failure of an individual to follow the script will
deny it parenthood. Even if the requirements are not so rigorous there will
tend to be discrimination against deviants from the standard patterns of
reproductive and social behaviour. In general, the existence of ritualized
features in the social behaviour of a species will carry with it a conservative
bias—an *intrinsic selection pressure* tending to exclude radicals from
contributing to the next generation. Of course, within the 'acceptable'
limits, there may be some variants that succeed better than others in
reproducing their kind because their behaviour provides the more effective
stimulation. Given such a situation the standard patterns will shift towards
those of such variants, as generation follows generation. Thus do we
envisage the processes of ritualization and sexual selection. But I
include these processes within what can be distinguished as intrinsic
selection pressures: selection resulting from biases within the behavioural
characteristics of a species that affect the chances of successful reproduction.

Thus, in answer to a question about the function of a display, for
example, one can point to the way in which the effects of performance of
the display co-ordinate the activities of the members of a pair, or of a
group, and hence how absence of these effects would be detrimental to the
chances of reproductive perpetuation. However, this self-preserving, or
self-propelling, selection pressure, intrinsic to the social communication
system of a species, will almost certainly not account for all of the features
of the system in adaptive terms. It may not explain why the system is as
complicated as it is, why one sensory modality rather than another is the
main channel of communication, or why the social organization is of one
type rather than another. To answer questions such as these we should
have to look beyond the system itself to environmental or ecological
factors, such as those discussed by McKinney (1973, and in this volume).
These ultimate factors constitute what will be designated *extrinsic selection
pressures.*

Corresponding to some of the distinctions in categories of function that have been mentioned are differences in methods of investigation and hence in the kinds of evidence upon which judgements about function depend. Investigation of the immediate consequences of behaviour and, where applicable, their intrinsic selection pressures, involves observation and description of sequences in which the behaviour occurs, and, when possible, experimental tests of whatever order is observed. A typical experimental approach is to observe the course of events when the feature in question is deliberately added to or removed from the relevant situations. Thus, Lehrman and his associates have been analysing the functional roles of the various components of pre-laying behaviour in the induction of incubation in Ring Doves (for example, Erickson and Lehrman 1964; Barfield 1971). In such investigation the distinction between the study of function and the study of causation or motivation tends to disappear, since behaviour discerned to be a consequence of some other is, at the same time, revealed in relationship to an antecedent. If the distinction can be made it will be in the focus of attention rather than in a difference of method. Thus investigation of the function of the broken-wing display of a duck or plover will focus on the effects of the displays on the movements of potential predators, but will be concerned with the causation of these movements only in so far as it is relevant to understanding the effects.

Discovery that performance of the broken-wing display has the immediate consequence of leading predators away from the clutch or brood suggests at once what the ultimate function of the behaviour must be: defence of the eggs or young from predation. There are many such cases in which inferences about survival value are drawn from observation of immediate function. But from the fact that distraction displays distract, it does not necessarily follow that they are effective anti-predator adaptations. To demonstrate their supposed survival value, it would be necessary to obtain evidence of correlation between use of such behaviour and survival against predator pressure. In general, the investigation of survival value requires approaches in addition to and different from investigation of immediate function. Such approaches fall into two classes: the experimental and the comparative.

The typical experiment on survival value involves the setting up of two groups, the characteristic in question being present in one and absent in the other, with all other conditions the same for both. The experimenter subjects the two groups to the supposed situation of selection and records and compares survival. Thus, for example, de Ruiter (1956) investigated the survival value of counter-shading in caterpillars of the Eyed Hawk Moth; Blest (1957) the survival value of the eye-spot displays of various saturnid moths; and Tinbergen and his associates (Tinbergen,

Broekhuysen, Feekes, Houghton, Kruuk, and Szulc 1962) the survival value of egg-shell carrying in Blackheaded Gulls. Though preferred, the experimental approach is not always feasible, however. In its absence one can resort to the less conclusive evidence of adaptive correlation (cf. Beer 1973a) that can come from the comparative approach. By comparing cases in which the characteristic in question is present and cases in which it is not, one tries to find correlated circumstances that could account for the distribution of the characteristic in terms of survival value. Such comparison is typically made across closely related species, as in the now classic study of Cullen (1957) in which many of the differences between the breeding behaviour of kittiwakes and the breeding behaviour of most other gulls were shown to be correlated with differences between the hazards of cliff-nesting and the hazards of ground-nesting. But such comparison can sometimes also be made within the behaviour of a species, as in Phillips's work (cited in Tinbergen (1964)) on the functional significance of whiteness in sea-birds, in which he argued for adaptive correlation between the differences in plumage colour that distinguish juvenile and adult Herring Gulls and the differences in the methods of obtaining food in these two age groups.

Many of the examples I have used to illustrate the distinctions and aspects that can be recognized in the subject of the functional significance of behaviour are from studies in which Niko Tinbergen was directly involved, or in which his was the guiding spirit. My selection reflects more than the fact that this book is dedicated to Professor Tinbergen. Throughout his career, questions of function and survival value have been in the forefront of Tinbergen's work and thought, and his contribution to the development of this area of ethological enquiry is second to none. Hence reference to his work is virtually inevitable whenever functional issues arise in discussions of behaviour. Such is especially the case as far as functional aspects of the behaviour of gulls are concerned.

In 1950, shortly after he moved to Oxford, Tinbergen initiated a programme of comparative studies of gull behaviour which was so broad in its scope and productive in its pursuits that it became the point of departure and return for gull work throughout the world. Most of the contributors to gull studies during the last twenty years were trained by Tinbergen or by one of his former students, or were influenced to some degree by his programme in their choices of problems, methods, and concepts. The aim of the programme was to obtain detailed descriptions of the behaviour of the gulls of the world so as to accumulate sufficient comparative information for reconstruction of the evolution of social displays in the group. To this end it was considered necessary to include work on causation, function, and ecological aspects of the behaviour '...because it had become increasingly clear that an understanding of the origin and evolution of displays can be considered enhanced by a wide knowledge of entire behaviour patterns and by some insight in the functions and the causation of displays' (Tinbergen 1959, pp. 1–2).

The paper from which I have just quoted was subtitled *A progress report*. In it Tinbergen summarized the gull work of the preceding 10 years and set out the generalizations that appeared to be emerging about the causation, functions, and evolutionary origins and histories of the displays. He also discussed concepts and methods with a view to clarifying and distinguishing the questions at issue, '...since the problems of causation, function and evolutionary history are so closely interrelated that the danger of circular argument is great' (ibid., p. 3). This danger is still with us. Hence, in part, my reason for beginning this essay with a tour of the options presented by questions about the function of behaviour.

More than 10 years have passed since Tinbergen published his progress report and in this time detail, refinement, and some new directions have been added to the gull story. Much remains to be done however, even at the descriptive level. Indeed the new knowledge has tended to bring along new questions rather than final answers. The matter of multiple function has emerged as one source of such questions, I think, at least as far as the immediate consequences of the displays are concerned. It is with this matter, as it has emerged in my studies of social communication in Laughing Gulls, that I shall be concerned in what follows. I shall contend that even when one bears in mind the distinctions I mentioned earlier, and restricts one's attention to immediate functions, the displays of Laughing Gulls show multiple function to a degree that is problematical, at least for most currently held conceptions of animal communication.

2.2. Laughing Gull displays

In a display case in one of the bird rooms of the American Museum of Natural History there is a diarama of Laughing Gulls posturing to one another in a marsh on the New Jersey coast. The birds have been fixed in these attitudes of suspended action ever since Kingsley Noble and his assistant M. Wurm contrived the scene about 30 years ago. The observations upon which the scene was based were published in a paper which became one of the source documents for students of gull behaviour (Noble and Wurm 1943). It provided a categorization of gull displays that has been followed to some extent in all the subsequent descriptions. For example, 'head-flagging' and 'choking' were christened in this paper.

Laughing Gulls continue to breed on the New Jersey coast in marshes that look much as they did to Noble and Wurm; but only in the museum scene has time stood still. In all probability the gulls have not changed their patterns of behaviour substantially since 1940, but observers have changed their accounts of this behaviour as well as the questions and interpretations that have guided their studies of it. New distinctions and categories have been added to the catalogue of displays. Ideas about the motivation and function of displays have undergone revision in the wake of ethologists' doubts about instinct.

Initially the prevalent assumption was that each type of display or variant of a display expresses but one motivational state, which is exclusive to it, and serves but one communication function, which is also exclusive to it. Employing form analysis, situation analysis, and sequence analysis (cf. Tinbergen 1959), Moynihan (for example, 1955) attempted to interpret the gull displays motivationally and functionally in terms of varying mixtures of attack and escape tendencies. Manley (1960), grappling with the fact that some of the displays of Black-headed Gulls occur in more than one type of interaction (for example, in both territorial conflict and in courtship sequences), invoked sexual attraction or receptivity as a third source of motivational influence—which either could be combined with attack and escape tendencies, or be expressed on its own—in the production of displays, with corresponding differences in functional significance. Manley's evidence led him to propose that the effects produced by performance of a display are often influenced by the context in which it occurs, the same type of display signifying hostility in one context, and readiness to engage in pair formation or mating behaviour in another. It appeared that the relevant context includes the social relationships between the birds concerned—for example, whether they are rivals of the same sex, potential mates, or birds between which a pair-bond has been established—and also the other behaviour that precedes, follows, or accompanies performance of a display. For example, the types of displays that occur in the 'greeting ceremony' of courting gulls are performed in a sequence, and with details of orientation and other features, that distinguish these displays from their occurrences in hostile interactions. Manley argued that for a full understanding of the causation and immediate functions of displays in gulls it is necessary to take into account that the displays may be both 'context determined' and 'context interpreted' as far as the gulls are concerned.

Unfortunately, Manley's fine work has not been published. Reference to it will be found in Tinbergen (1959). However, the importance of context in animal communication was independently discovered by John Smith in his studies of American flycatchers, and he has published his account of it (Smith 1963, 1965, 1966, 1968).

Restriction of a display to one type of social situation is no more the rule in Laughing Gulls than it is in the Black-headed Gulls studied by Manley. In Table 2.1 I have listed display postures and calls of the Laughing Gull and indicated for each the situations in which it occurs, according to my observations. Neither the list of displays nor the list of situations is

TABLE 2.1

Patterns of social behaviour of Laughing Gulls and contexts in which they occur

	Rivalry	Nest-defense	Pair-formation	Courtship feeding	Pre-copulation	Nest-site selection and nest-building	Nest-relief	Parent–chick (filial)	Adult–chick (hostile)	Non-social disturbance	Flock flight	Communal feeding
Upright	×		×	×	×		×		×	×		
Oblique	×		×	×			×	×	×			
Horizontal	×		×	×								
Facing-away	×		×	×			×					
Choking	×					×	×					
Stooping	×		×	×	×	×	×	×				
Head-toss	×		×	×	×		×					
Long-call	×		×	×			×	×	×			
Crooning	×		×	×	×	×	×	×				
Ke-hah	×		×	×			×	×	×			
Kow	×	×								×	×	×
Kek-kek										×	×	×
Gackering		×										
Quavering wail		×									×	
Regurgitation				×	×			×				

exhaustive, but the intersections of the two lists show enough for my point that exclusive one-to-one correspondence between display and situation does not obtain.

Of course it could be argued that the categorization of both the displays and the situations is too coarse to reveal one-to-one correspondence between signals and function. Indeed I shall try to show later how the drawing of finer distinctions can lead to more refined conjectures about when and where and why certain of the displays occur. But taken in isolation from their associations with other features of behaviour that accompany them or occur in sequence with them, several of the displays defy the attempt to pin them down to a single position in the social commerce of the species, or to identify a common ingredient of any high degree of specificity in the variety of their occurrences. I have selected three types of display to exemplify this point: crooning, facing-away, and the long-call.

Crooning

The call to which I give this name is similar and probably homologous to what has been referred to as the 'mew call' in descriptions of the behaviour of other species of gulls (for example, Tinbergen 1953, 1959; Evans 1970a, 1970b). I prefer my label for the Laughing Gull call because it is the more descriptive of the quality of the call, and also because I do not wish to appear to be begging the question about homology. In any case there is a precedent for 'crooning' that is older than that for 'mew call'; Kirkman (1937) used 'crooning' in his description of the behaviour of Black-headed Gulls.

As the name suggests, crooning in the Laughing Gull is a relatively soft, smooth call; it sometimes includes a yodel-like step up in pitch but lacks any harsh or really abrupt features. It occurs between bouts of choking in hostile interactions, and between bouts of choking during nest-site selection and nest-building. It is performed by the male prior to and during courtship feeding of his female. During the laying, incubation, and hatching periods it can occur at nest-relief, when it might be performed by either or both birds, sometimes in association with choking. Parents croon prior to, and during, their feeding of their chicks, when they lead their chicks on excursions away from the nest, and when they intice their chicks to approach them from a distance.

If we compare these different contexts in which crooning can occur we can see that some of them have features in common but that these features differ between different selections: the courtship and parental contexts share an association with regurgitation feeding; in both nest-site selection and luring of the chicks performance of crooning has the effect of inducing approach; in nest-site selection and nest-relief, crooning has to do with occupation of the nest; in hostile interactions, nest-site selection, and nest-relief crooning occurs in sequence with choking. There are thus what Wittgenstein, in a discussion of how the meanings of a word can differ from context to context, referred to as 'family resemblances': '…a complicated network of similarities overlapping and criss-crossing: sometimes overall similarities, sometimes similarities of detail' (Wittgenstein 1953, p. 32).

An over-all similarity in the contexts in which crooning occurs is that in each case the activity of the crooning gull is related to spatial configurations in some way, either the distance between the individuals concerned, or a place such as the nest or a defended area. This may tell us no more than that we are dealing with a piece of social behaviour however. McBride (1966, 1971) has construed social behaviour as the dynamics of spacing between the individuals of a group. At any rate most of the displays of Laughing Gulls are related in some way to spatial configurations. Hence to invoke spatial configurations as a means of setting crooning apart, motivationally and functionally, from the other displays, we have to specify in what ways the relationships of crooning to space are exclusive to crooning. We can narrow the field by noting that in all the situations in which crooning occurs the birds concerned are either in close proximity to one another or their coming into such proximity is imminent. This feature is not peculiar to crooning, but at least some of the other displays that share it have additional special features which differentiate them; for example, the copulation call of the male occurs only during attempts at copulation. Moreover, the relatively soft quality of crooning is consistent with communication at close quarters. However, when we look more closely at this feature for a clue to the motivation and function of crooning, we find ourselves still with the variety with which we started out. In the case of crooning in hostile interactions the behaviour apparently contributes to deterring the other bird from approaching; but in the cases of crooning during courtship-feeding, parental interactions, and nest-site selection the behaviour serves to induce the other bird to approach or stay close. The crooning associated with the nest prior to laying is an enticement to the mate to join its partner in the nest or prospective nest-site; but during the incubation period such crooning signifies contention over occupation of the nest, the sitting bird expressing reluctance to leave and the mate expressing eagerness to take over. Crooning during courtship feeding and feeding of chicks appears to contribute to the stimulation of the bill-pecking response, but has no such effect in the other contexts.

This crooning associated with regurgitation feeding, and the related use of crooning in the luring of chicks, are also exceptional in that they lack the company of choking. Here I am drawn to another of Manley's ideas. Manley (1960, cited in Tinbergen (1959)) contended that choking is a ritualized derivative of regurgitation feeding movements. If we identify choking with regurgitation feeding movements in this evolutionary sense we can argue that the association between crooning and choking preserves an ancestral association between crooning and unritualized regurgitation, which is present in unaltered form in the courtship-feeding and chick-feeding contexts, and thus presents a feature common to all occurrences of crooning. Evidence of a common motivational basis or a common immediate function remains as elusive as before, however, for the variety in apparent causes and consequences attending occurrences of choking and regurgitation is the same as that attending crooning.

Investigation of crooning in Laughing Gulls has, as yet, advanced little beyond the kind of qualitative observations I have presented here. Hence, it would be premature to maintain that search for a common motivational basis, and even a common communication function, for all occurrences of crooning should be abandoned. However, I think the evidence we have at present raises a doubt about the assumptions that lie behind such a search. Why should it be taken for granted that whenever a particular type of behaviour occurs it must always be for the same reasons? Is it not conceivable that the behaviour might be employed for different reasons on different kinds of occasions? I think mechanistic conceptions of animal communication have tended to prevent these questions from being taken seriously. Even W. J. Smith, who made respectable the view that the meaning of a display can vary with the context, argued, in effect, that all occurrences of such a display probably express the same thing about the state of the displaying animal (Smith 1968). Admittedly, Smith presented this as an hypothesis rather than as an established principle, and he was able to support it with data from his flycatcher studies. Also I appreciate that investigation guided by this hypothesis has been fruitful and may continue to be so. But there is an alternative, and I think it is worthy of consideration, especially when a pattern of behaviour occurs in a range of contexts as various as that covered by the crooning of Laughing Gulls. I suggest that, in cases like these, it is at least a possibility that the same display may be used to *express* different things in different contexts, as well as signify different things because of context differences. This may be so even when one can find a 'common denominator' in the contexts in which the behaviour pattern occurs. Crooning appears to express something about the immediate vicinity of the calling gull, in all the contexts in which it occurs, but it also appears that a gull can express a number of different messages about its immediate vicinity by varying what other behaviour it performs in concert or in sequence with crooning and because contexts qualify signals. Thus, for example, crooning unaccompanied by choking can be interpreted as part of an invitation to the mate or to the chicks to come close and get fed; crooning accompanied by a version of choking that includes a ruffling of the back feathers and a predominantly horizontal stance, when it occurs in a hostile context, can be interpreted as part of an expression of readiness to oppose further approach with fight.

That the use of displays can be varied in this sort of way seems to me to be no less conceivable than that a non-communicatory act can be used as a means to more than one kind of end. It is not hard to find instances of the latter. Running can occur in both attacking and fleeing. Great Tits (*Parus major*) use the same pecking movement for opening a nut and for delivering a blow in a fight (Blurton Jones 1968).

However, cases like these involving motor patterns may be construed differently depending upon the criteria used to designate acts. If instead of form of motor pattern alone, context and means–end configuration were also included in the specifications of acts, then running, for example, would be regarded as a constituent of different kinds of acts, and hence as no more motivationally autonomous than a joint flexion that is a constituent of a variety of different types of movement. An hierarchical principle obtrudes here, although one that differs from Tinbergen's (1950, 1951) in that lower units are not uniquely related to upper units—a particular type of movement can be included in a variety of types of action, and a particular type of action can be included in a variety of 'plans' (cf. Miller, Galanter, and Pribram 1960).

A similar type of hierarchy is to be found in linguistics in the relationships between phonemes, morphemes, and the categories of intelligible utterance (phrases, clauses, sentences). The flexibility and limitless semantic universe of language place language in a class apart from all animal communication systems, of course (cf. Chomsky 1966). Nevertheless, the levels of linguistic analysis may provide a useful model for hierarchical classification of the components of at least some animal communication systems. At least in the short run, this sort of model could well prove to be more heuristic than have the causal and stochastic models presupposed in most investigations of animal communication so far. Such an approach would involve reconsideration of the categories in terms of which animal communication should be described and quantified, and of the criteria for categorial assignment. As in linguistics, the marking of distinctions within and between levels (cf. Twaddell (1963), on the definition of a phoneme) pose hard questions. But I am inclined to the view, at least in regard to the kind of communication that goes on among Laughing Gulls, that these questions need to be posed, and that, until they are tackled, attempts to apply sophisticated quantitative analysis to such communication will be premature and hence conducive to the production of stillborn issues. My reasons for this view have their most impelling basis in investigations of facing-away and the long-call.

Facing-away

Noble and Wurm (1943) described two types of display in which a Laughing Gull turns its head away from the other bird. In *erect posturing*, '...the body is kept horizontal, the neck is stretched vertically to a maximum and the head is kept horizontal, facing away from the bird for which the activity is intended. The stance is maintained for a number of seconds during which the wings are drooped, mantle feathers are smooth, and the animal appears rigid and strained' (ibid., pp. 189–90). In *head-flagging*: 'While the gull maintains an erect posture, the head is slowly and deliberately turned from side to side. The performing bird directs the back of its head toward the bird for which the display is intended. When two birds simultaneously perform, their bodies form an angle facing away from each other' (ibid., p. 190). According to Noble and Wurm erect posturing and head flagging serve a pair-formation function and are exchanged between the members of a pair 'throughout the breeding season'; but they did not say how these two types of display differ motivationally or functionally.

In my experience movement of the head from side to side while facing away in what Noble and Wurm called the erect posture does not occur with any marked distinctness or regularity in the social interactions of Laughing Gulls. Such movement is quite prominent in Black-billed Gulls, *Larus bulleri* (unpublished personal observation). It is virtually lacking in Black-headed Gulls and Herring Gulls, but Tinbergen, Moynihan, and others (see, for example, Tinbergen and Moynihan 1952) have used 'head-flagging' for the rigid form of facing away shown by individuals of these species. The more recent practice, however, has been to refer to any turning of the back of the head to the other gull as 'facing-away', and I shall adopt this terminology here. I shall also follow current convention by referring to the posture in which the head is held high, neck vertical and carpel joints lifted from the sides of the body, as an 'upright'. When facing-away is superimposed on an upright I refer to the posture as 'upright facing-away' or 'facing-away in the upright'.

As the singling out of this combination implies, Laughing Gulls can superimpose facing-away on more than just the upright posture. In the greeting or meeting ceremonies of courting Laughing Gulls one or other of the birds—either the male or the female but not both together, although one may follow the other—runs round or in front of its partner in a posture in which it holds its head down more or less in line with the long axis of the body, its neck partially extended, its bill pointing horizontally or slightly downwards (cf. the 'pairing charge' of Noble and Wurm (1943)) and bends its neck and head to the side away from the other bird (Fig. 2.1). Less distinctive and often less complete facing-away can also be superimposed on the oblique postures in which long-calls are performed and the bent-down, stooping postures and squatting postures in which choking and crooning are performed. The situations in which these various forms of

FIG. 2.1. 'Meeting ceremony' of courting Laughing Gulls. The bird on the right runs with its head lowered and bent away from its partner.

facing-away occur include agonistic encounters between adults and between adults and chicks, pair-formation sequences, pre- and post-
copulation sequences, and nest-reliefs. Facing-away can be performed by
stationary birds—standing, sitting, or floating—or by moving birds—
walking, running, or swimming—but I have not seen it clearly displayed by
birds in flight. I have seen facing-away by chicks as young as 7 days
post-hatching; but all the instances on record of facing-away by chicks
occurred in agonistic interactions, either with other chicks or with adults.

We thus have variety as great as that for crooning in the situations in
which facing-away occurs in Laughing Gulls, and also variety in the forms
in which the display is presented However, I shall not attempt here to deal
with the full range. It will be sufficient for my present purpose to
concentrate on facing-away in the upright posture.

In my first season of field work on Laughing Gulls, I took a large number
of photographs of the various displays. In those I took of facing-away in the
upright I discovered two extreme forms: in one the black hood and face are
completely concealed from the view of the other bird; in the other the rear
portion of the hood and a back view of the prominent white feathers just
above and below the eyes are visible to the other bird. In subsequent field
observation and viewing of cine-films the distinction between these two
forms of upright facing-away seemed so obvious that I marvelled that I and
others had missed it before. There are intermediates and transitions
between the two forms, but I have found that I can assign most occurrences
of upright facing-away to either a 'white' category (hood hidden) or a
'black' category (hood visible).

In white upright facing-away the gull holds its bill horizontal or pointing
downwards to some degree, sometimes almost vertically downwards.
Downward pointing of the bill pulls the rear margin of the hood to a
vertical edge, probably with the assistance of contraction of muscles that
draw the skin of the head forward over the skull. The bird also at least
partially erects the white neck feathers behind the margin of the hood,
which thus form a low ruff concealing the face from a rear view and fringing
the face with a white halo from a front view (Fig. 2.2). In black upright
facing-away the bird usually extends its neck more, and therefore holds its
head higher than in the other form. It may point its bill slightly upwards,
horizontally, or downwards to some degree, but never as acutely as the
extreme in the white form. The margin of the hood is more or less
horizontal, and the feathers of hood and neck are usually depressed. From
a rear view one sees a black hemisphere or stocking of hood, with the white
eye-margin tufts projecting on either side (Fig. 2.3). In photographs of
non-displaying gulls with their heads elevated to the same extent and held
at the same angle as in a typical black upright facing-away, the eye-margin
tufts are not visible from behind.

FIG. 2.2. White upright facing-away of Laughing Gulls.

Three functions have been suggested for facing-away in gulls: appeasement, allaying fear, and 'cut-off'. For appeasement the display is supposed
to decrease the likelihood of attack by the other bird by removing from its
view stimuli that provoke its hostility (cf. Tinbergen and Moynihan 1952).
For the allaying of fear the display is supposed to decrease the likelihood of
fleeing by the other bird by removing from its view stimuli that cause fear
(cf. Tinbergen 1964). Both of these possibilities place the direction of
immediate effect from the displayer to the other bird; and they recall
Darwin's (1872) principle of antithesis in that contrasting messages—
threat and denial of threat—are associated with 'movements of a directly

FIG. 2.3. Black upright facing-away of Laughing Gulls.

opposite nature'; facing towards and facing away. The third possibility, taken in isolation, implies that facing-away is not really a display at all, for the immediate effect of consequence is supposed to be reflexive—on the facing-away bird itself. By turning its head away from the other bird, the performer 'cuts-off' from its own vision stimuli likely to provoke *it* to attack or flee, and thus uses the manoeuvre to control its own behaviour tendencies (Chance 1962).

These three possibilities do not exclude each other, of course. It is conceivable that reflexive effects of facing-away combine with effects on the other bird for the containment of aggression, or fear, or both, depending upon the context. At least in the case of Laughing Gulls it would appear that 'cut-off' cannot be the whole story, for this suggestion gives no answer to the question why there are two forms of upright facing-away in this species.

My initial thinking about this question was that the black form of upright facing-away could be used to effect cut-off while, at the same time, maintaining the threat signal vested in the black hood displayed in the upright to induce the other bird to leave or stay its approach; and that the white form could be used to cancel the threat signal in situations where it might interfere with an intent to attract and keep the other bird close.

Upright facing-away occurs during hostile interactions and during court-ship interactions, two contexts for which the behavioural vectors are opposite in just the way that is concordant with the suggestion of opposite functions for the two forms of upright facing-away. If my thinking were correct, then the black form of the display would be exclusive to, or at least prevalent in, hostile interactions; and the white form would similarly predominate in courtship interactions. I therefore made field observations and studied cine-films of these two classes of interaction to test the implications of my ideas about the functions of upright facing-away in Laughing Gulls.

Before giving the results of this work, however, I have to make some comment about how they were obtained. It is a commonplace that the field-worker on animal behaviour often faces methodological problems for which the precepts and routines of his laboratory colleague provide, at best, only partial solutions. Instead of contriving situations to test his ideas, the field-worker often has to take the natural situation as he finds it, which can mean without cleanly separated categories, balanced numbers, or controls for all relevant variables. So it has been with my studies of social interactions of gulls, and I have to confess that my criteria for sorting observations for purposes of comparison are not as rigorous as ideally I should prefer them to be.

Nature does not present a clean separation between hostile interactions and courtship interactions. Many sequences of social interactions between Laughing Gulls include elements of both hostility and courtship. However, if one makes the reasonable assumption that mixed and courtship interac-tions are between individuals of opposite sex, one can distinguish as purely hostile interactions those in which agonistic behaviour occurred between individuals of the same sex. Laughing Gulls, however, are not sexually dimorphic to any marked degree. There are slight differences between males and females in bill and head dimensions and in over-all body size, but I have found these to be usually insufficient for a confident judgement of a bird's sex unless it was standing next to another of opposite sex, and even then I have often been unsure which was which. But some gulls carry a mark by which they can be individually identified—such as a smudge of dirt on the plumage, a peculiar pattern of creases in the head plumage, or a leg-band—and if they can be watched for long enough their sexes can be judged from their behaviour. For example, males do the mounting in copulation, except in very rare cases.

Only a proportion of the gulls in a group under observation can be so identified and sexed reliably, however. By limiting comparison to interac-tions involving only such birds one can make one's selection of groups as rigorous as the conditions allow, but at the cost of reduced group sizes and the rejection of observations at places and at times where and when numbers and turnover of birds were so great as to preclude the following of individuals; for example, as on the communal display areas during the very early days of the breeding season. Nevertheless this is the strategy I chose. As a consequence the comparisons I shall make involve only small numbers of birds and do not include observations from the earliest stages of the breeding season. What these comparisons show is consistent with what appear to be the general trends in the many observations for which I was not sure of the sexes of the birds concerned.

For my category of hostile interactions, then, I chose encounters between birds of the same sex, usually males, consisting of attacking, fleeing, and associated displays. As a comparable category of courtship interactions I chose encounters between birds of opposite sex during which one of the birds performed the display of running around or past the other in the low, horizontal posture, with the head bent away, to which I referred earlier. In my experience this display occurs only as a constituent of sequences that correspond to the greeting or meeting ceremonies that have been described for other species of gulls (cf. Tinbergen 1959). These sequences are sometimes interrupted by attack of one of the birds on the other—usually the male attacking the female—but such attacks are rare and transitory compared to those of hostile encounters; they are not retaliated and so do not lead into fights such as often erupt in purely hostile encounters. In addition to hostile and courtship interactions involving only two birds I observed and filmed interactions in which two courting birds became embroiled with a third, and I shall have something to say about some of these as well. I shall also comment upon upright facing-away at nest-relief. My discussion will be based mainly on filmed sequences, for these provide the most detailed and precise information, but I shall supplement this information with observations noted in the field.

In my sample of 30 or so performances of upright facing-away in what I judged to be purely hostile interactions, all but two were black or tending to black. I leave the number vague because in some instances the turning away of the face was so fleeting, so slight, or so blended with turning away of the whole body, that I could not make up my mind whether they should count or not. In general, facing-away in these hostile encounters was brief compared to that in courtship. According to frame counts of 17 filmed occurrences (24 frames per second) of black or tending-to-black upright facing-away in hostile encounters, the longest was held for only one second and the mean duration was 0.65 seconds. The bill was inclined downwards throughout most of these performances. In 4 cases a slightly bill-up attitude was adopted just before fleeing. The sequences were too variable and the numbers of observations too small for statistical analysis to show anything definite about the immediate consequences that performance of black upright facing-away is most likely to have in hostile interactions. It was most often followed by other displays, such as mutual choking, 'kow' calling, or long-calling; but attacking or fleeing, by either the displayer or its opponent, were not uncommon. The bird facing-away was usually oriented side-on to the other bird or angled towards it. Only during sequences with choking was parallel orientation maintained for any length of time, and even then there was a tendency for the head ends of the two birds to become angled apart. If a bird followed facing-away by fully turning its back on the other it did not remain stationary in that position for more than a moment; it usually fled almost immediately. Supplanting attacks ended with the attacker landing side-on to its line of approach and then facing-away, on a number of occasions; and I observed a few instances of very brief and repeated facing-away by a gull as it was making a running charge at another.

Variations and details aside, the general fact was that the form of upright facing-away predominating in the hostile interactions I observed was the black one, and hence accorded with my prediction. However, the exceptions to this rule added an interesting and unexpected observation. The two cases of white upright facing-away in my filmed sample of hostile interactions were given by gulls that were attacked but did not themselves show any signs of hostility. In each case the bird facing-away was walking past the other, *en route* to its nest or display area, and took flight when attacked. The posture was held for 1·92 seconds in one case and 1·88 seconds in the other (estimated from frame counts). Thus each was nearly twice as long in duration as the longest black facing-away measured in the hostile context. Other occurrences of white upright facing-away during hostile encounters, involving gulls whose sex I was not sure of, also conformed to this pattern: the gull facing-away adopted and held the posture while walking or running towards what appeared to be its site of attachment; it was attacked by a bird it passed by but did not itself show hostile behaviour. Such attacks were only occasional however. I have many more records of gulls facing-away in the manner I have just described but which were allowed to go their way unmolested. In one case the gull had to pass three residents in succession, two on its right and one on its left. It switched the direction toward which it faced-away in accordance with the positions of the birds as it passed them, and they took no apparent notice of it. Performances of white upright facing-away in this context give the impression of negating or denying any challenge that the displaying bird's mere presence might imply to another regarding it as a trespasser. This possibility is consistent with the fact that this form of facing-away completely hides the features that are so prominently presented in threat displays: the black hood, red bill, and eye tufts. At any rate, white upright facing-away in this context was not accompanied by any aggression by the displaying bird, and seldom met with aggression from other birds, whereas black upright facing-away was associated with aggression both given and received.

In contrast to the hostile interactions, the courtship interactions failed to accord with my expectations at all. Instead of showing only or predominantly the white form of upright facing-away, the courting gulls faced-away even more often in the black form than in the white form. For example, 12 of the courtship sequences I filmed yielded 37 instances of the black form and 25 of the white form.

At present these sequences are all I have that qualify for detailed analysis. It is out of the question that such analysis can draw generally conclusive results from so small a sample, but I shall report some results nevertheless because they show what I think are interesting trends that encourage further investigation. I wish them to be entertained for purposes of argument, rather than as definitive data. Seven different males and 10 different females participated in the twelve sequences; hence some individuals contributed to more than one sequence. However, I shall treat the sequences as though they were independent of one another.

The males contributed slightly more of the occurrences of upright facing-away than did the females (36:26). The males also tended to hold the posture longer than did the females, but the difference is not statistically significant (by Mann–Witney U test, $p > 0 \cdot 05$). The ratio of white to black facings-away did not differ between males and females. The males tended to hold the white posture longer than the black, but again not significantly so ($p > 0 \cdot 05$). The figures are summarized in Table 2.2.

White facing-away predominated early in an encounter, black facing-away later. This observation is illustrated by Table 2.3 in which the 12 encounters are sorted according to the form of the first facing-away performed by each of the two birds, and then according to the forms of the last facing-away performed by each of the two birds. Encounters in which the first facings-away of both birds were white were, on the average, longer in duration (median duration: 39·3 seconds) than the encounters in which

TABLE 2.2

Upright facing-away in courtship interactions: their distributions between types and sex, and their durations, in twelve interaction sequences.

	White facing-away			Black facing-away			Totals		
	N†	Median duration‡	IQR§	N†	Median duration‡	IQR§	N†	Median duration‡	IQR§
Males	15	1·46	1·29	21	1·17	1·71	36	1·37	1·25
Females	10	0·92	0·42	16	0·92	0·71	26	0·92	0·84
Totals	25	1·17	1·14	37	1·08	1·06	62	1·10	1·04

† Number of occurrences of upright facing-away.
‡ In seconds—estimated from frame counts of movie films taken at 24 frames per second.
§ IQR = interquartile ranges.

TABLE 2.3

Twelve courtship interaction sequences sorted according to the forms of the first facings-away performed by the two birds and then according to the last facings-away performed by the two birds.

Male–Female	White–white	White–black	Black–white	Black–black
First facing-away	6	3	2	1
Last facing-away	1	3	2	6

the first facings-away of one or both birds were black (median duration: 21·4 seconds). The numbers are too few for statistical treatment, but the difference is in the direction consistent with the idea that white upright facing-away decreases the likelihood of separation. Also consistent with this idea was the fact that the only one of the encounters in which the last facings-away by both birds were white was also the only one that did not end with a departure—the two birds remained standing together, preening, side by side.

However, these observations are just as consistent with the possibility that gulls showing the white form of upright facing-away are likely to stay, and gulls showing the black form are likely to leave. In line with this suggestion was the fact that in the eleven encounters that ended with a departure, the last facing-away by the departing gull was black in all but one, whereas the facing-away by the other gull was black in seven cases and white in four. Furthermore, more than half of all the occurrences of black upright facing-away were immediately succeeded by the displaying bird's turning its whole body away from the other, and in many of these cases the bird moved away also, usually maintaining the black upright posture. Such orientation and movement away less often succeeded performance of the white form of the display (Table 2.4). In the majority of cases a gull

TABLE 2.4

Action immediately following upright facing-away, by the gulls performing the display

	Turning away		No turning away	
	Locomotion	No locomotion	Locomotion	No locomotion
Black facing-away	10	10	7	10
White facing-away	3	1	15	6

performing white upright facing-away walked in front of, in parallel with, or around the other bird.

The gull towards which facing-away was directed remained stationary in most cases, sometimes long-calling, more often standing in an upright posture; and there were several instances of mutual facing-away. The data were too limited and too variable to provide decisive evidence about the species-typical effects of upright facing-away on the behaviour of the other bird, in courtship contexts, or the differences between the white and the black forms of the display as far as such effects are concerned. But, together with my over-all impressions of the courtship interactions I have just considered, and others for which I had less detailed information, they were consistent with the following speculations.

Whatever the immediate consequences of performance of courtship displays by gulls happen to be on particular occasions, formation and maintenance of pair bonds are ends served by these displays, whether one takes one's stand on the basis of available facts or on the basis of definition. The ways in which the displays function in pair formation include the provision of means for coping with the hostility and fear that oppose the coming together and staying together of the two birds of a potential pair. One reason for this belief is that attack and escape behaviour interrupts many courtship sequences, particularly in the early stages of pair formation. In Laughing Gulls, according to my experience, escape interrupts many more such sequences than does attack. There was not a single occurrence of attack in the sample of courtship sequences discussed above, but most of the departing gulls adopted the thin upright posture associated with fear (cf. Tinbergen 1959) just before taking flight, and this posture occurred frequently throughout many of the sequences. Many of the black facings-away in the sample were performed in this thin upright posture, and I have indicated how black upright facing-away appeared to be associated with departure or withdrawal tendency. It is therefore plausible that black upright facing-away expresses mounting fear when performed by a courting Laughing Gull.

The significance of such expression to the other bird could be a warning that unless it adjusts its behaviour appropriately the displaying bird will leave. But then the question arises: since the thin upright posture by itself expresses fear, why should facing-away be superimposed on it? The answer may be two-fold. As far as the displaying bird is concerned facing-away achieves cut-off and may thus serve as a means by which the bird itself can try to reduce its fear. As far as the other bird is concerned the display may signify that its partner is both sexually attracted and afraid—a bird to be courted but not too precipitately. The white form of upright facing-away, on the other hand, appears to express sexual attraction much less trammelled by fear, and hence may signify that the displaying bird can be closely approached and vigorously courted without great risk of its taking flight. Part of the function of facing-away in courtship contexts may thus be to communicate how each bird's behaviour and presence are affecting the disposition of the other, and so mediate the mutual adjustment conducive to their staying together. The intermediates and transitions between the extreme black and the extreme white forms of the display provide a graded signal that may well convey degrees and directions of shift of fear disposition.

But upright facing-away, in interactions between pairing and paired gulls, also occurs when there are no overt signs of fear or hostility immediately attending performance of the displays. When one of the birds of a pair or potential pair alights beside or approaches the other, the two birds usually long-call together, and then one of them adopts the horizontal posture and runs in front of or round the other, bending its head away, then lifts to an upright face-away—the white form as a rule—and ends with one or more head-tosses with the back of the head directed toward the other bird. The regularity or formality of this ritual meeting ceremony increases with the establishment of the pair relationship and as all signs of hostility and fear disappear from the interactions between the birds. It persists in abbreviated forms throughout the laying and the incubation periods, in the displays exchanged at nest-relief. What is signified by facing-away in these ceremonies may not be separable from what is signified by the sequence as a whole, which appears to be an expression of developing or established attachment between the individuals concerned. If hostility or fear are components of this expression at all, they must be so in only a covert form, for attack and escape seldom follow immediately on performance of the meeting ceremony. It could be that this particular juxtaposition of the displays can be regarded as a negative statement as far as their agonistic meanings in other combinations and contexts are concerned. Such qualification of meaning may even carry over to other contexts involving birds of an established pair. At nest-relief the departing bird often shows black upright facing-away just before leaving, apparently as a departure signal rather than as a sign of fear of the other bird.

There is thus reason to believe that both the message and the meaning (see Smith 1968) of upright facing-away vary with differences in details of context within what I have classed as courtship interactions and interactions between the birds of a pair. Comparison of courtship interactions and hostile interactions reveal more marked contrasts. In hostile interactions black upright facing-away was transitional to both attacking and fleeing: in courtship interactions this form of facing-away was frequently transitional to withdrawal and departure accompanied by signs of fear, but not to attack. Attacks do occur during some courtship sequences, and then they sometimes follow immediately on performance of facing-away by the attacker. In all such cases that I have on record (8 in all—there were none in the sample of filmed sequences discussed above) the attack was launched from the white form of the display. White upright facing-away was not associated, either simultaneously or sequentially, with signs of fleeing tendency in my observation of courtship interactions. In contrast, the few birds showing the white form of upright facing-away in my sample of hostile interactions were apparently attempting to avoid being attacked and did not show attack behaviour themselves. To sum up, black upright facing-away appeared to express hostility mixed with fear in hostile interactions, or sexual attraction mixed with fear in courtship interactions; and white upright facing-away appeared to signify appeasement or lack of hostile intent, perhaps coupled with fear, in hostile interactions, or lack of fear coupled with sexual attraction in courtship interactions.

Sequences in which a pair of courting birds interacted with a third provided an interesting combined situation. The order of events in such sequences is variable, and I have too few good records for a representative analysis. But one kind of incident recurred sufficiently often for me to be confident that it is a regular pattern. When, as frequently happened, one of the birds of the pair launched an attack on the third, it typically faced-away in the upright as it did so, but directed the display at its partner, not at the bird being attacked; and the form of the display was white or tending to white. The interpretation that suggests itself here is that by facing-away in this manner the attacking bird counters the possibility that its aggressive behaviour toward the third bird will scare its partner away. But note how the disposition of such a bird apparently differs from the disposition of a bird showing white upright facing-away in purely hostile interactions. In the former case the bird is manifestly aggressive; in the latter it is not.

These observations run counter to the assumption that the same kind of motivational state underlies all occurrences of upright facing-away in Laughing Gulls. The existence of variations in form, with two contrasting extremes, implies at least two kinds of motivational bases or a graded continuum of motivational states. But the behaviour a bird performs along with upright facing-away differs between the various contexts in which the display occurs and thus indicates even more variability in the motivational states underlying occurrences of the display. Even the contrast between the black and the white forms of the display appear to express different contrasts in behavioural tendencies in the different contexts. In other words, there appear to be between-context differences, as well as within-context differences in what the two forms of upright facing-away signify. There may be some continuity of content across contexts. The white form of upright facing-away is associated with lack of hostile intent towards the bird to which it is directed irrespective of context. But the total content of what motivates facing-away appears to incorporate contextual differences and, in consequence, appears to be context specific. It appears as though the gulls use the same displays to frame different messages (see Smith 1968) in different contexts.

The meanings of the displays—their effects on recipients or the information extracted from them by recipients—presumably vary accordingly, although it is also possible that a common thread of meaning may be drawn from some of these different messages. The white form of upright facing-away may be received as signifying 'so-and-so is not going to attack me' on all occasions of its performance, along with context specific information such as 'so-and-so is just passing through', 'so-and-so is attacking someone else', or 'so-and-so is sexually attracted to me'.

A signal that is flexible in its use, in the way I have suggested for facing-away in Laughing Gulls, could be described as an 'open' signal, in contrast to a signal having but one use, such as an alarm call, which could be described as a 'closed' signal. If we follow Smith (1968) and refer what is encoded in a signal to semantics and refer what is decoded from a signal to pragmatics, we can list four possibilities: (1) the signal may be both semantically and pragmatically closed; (2) it may be semantically closed and pragmatically open; (3) it may be semantically open and pragmatically closed; or (4) it may be both semantically and pragmatically open. I do not think these are merely theoretical possibilities. On the contrary I think they should be born in mind when we seek to understand the ways in which animal communication systems work, and their developmental and evolutionary bases. Again, however, the question of what is to count as a signal insinuates itself. I leave discussion of this question until after I have dealt with the long-call.

The long-call

To an ethologist familiar with the behaviour of gulls, the combination of posture and utterance that has been labelled oblique-and-long-call in Laughing Gulls is so obviously a social display, and so obviously homologous to the displays that go by the same name in other species of gulls, that he would probably dismiss a questioner of either point as uninformed. He could be wrong, however, for what has been taken to be a single pattern of behaviour turns out, on closer examination, to be variable in form, in context of occurrence, and in function, in ways that cast doubt on the belief that we are dealing with just one display or display type.

In its typical form the pattern can be divided into three parts: the short-note phase, the long-note phase, and the head-toss phase. In the short-note phase the gull stands in the full oblique posture—neck extended at about 45° to the horizontal, bill pointing in line with the neck or somewhere between that and the horizontal, and carpel joints held away from the sides of the body—and utters a string of more or less uniform,

FIG. 2.4. Full oblique posture of a Laughing Gull uttering the short note part of a long-call.

equally spaced notes which are short in comparison to those of the next phase (Figs 2.4 and 2.7). In the long-note phase the gull maintains the full oblique posture, reduces the extension of its neck while maintaining the angular positions of neck and head, or lowers its head, sometimes all the way down to a position in line with the long axis of the body, while maintaining the neck at full stretch and bill horizontal (Fig. 2.5). The notes

FIG. 2.5. A posture often adopted during the long note part of a Laughing Gull long-call.

FIG. 2.6. A head-toss terminating a long-call by a Laughing Gull. (This and the preceding figures were drawn from cine films.)

uttered in this phase are about twice as long as those in the short-note phase and spaced farther apart. If the gull lowers its head it does so a bit at a time in step with the long notes. The head-toss phase consists of one or more movements in which backward flexing of the neck and upward rotation of the head jerk the head toward the rear, often with so much vigour that the top of the head strikes the feathers of the back (Fig. 2.6). Each such movement is accompanied by a note which is slightly shorter than a long note and which has a dying fall.

Between long-calls there are variations in the numbers of notes in each phase, in the durations of the notes and the intervals between them, and in the frequency and amplitude modulations of the notes and their harmonic spectra. Many long-calls lack a head-toss phase; long notes sometimes precede the short notes, and sometimes occur without short notes; head-tosses, together with head-toss notes, occur frequently by themselves.

A gull does not have to be standing to give a long-call; it can also long-call when sitting, floating, or flying. The range of contexts in which long-calls occur includes hostile interactions, courtship interactions, nest-relief, adult–chick interactions—both between parents and their young and between adults and chicks not their own—and interactions within standing or flying flocks, both inside and outside the gullery area.

Some of the variation in the long-call and its postural accompaniment is associated with differences of context. For example, the longest strings of long notes and the most extreme lowering of the head in the long-note phase occur in courtship interactions, particularly when the calling bird is on the ground and the other bird is flying over or landing nearby. Incubating gulls long-calling to others overhead usually include few long notes, do not lower the head, and often omit the head-toss phase.

However, some features of the long-call vary between individuals rather than between contexts. These are most apparent in the short-note phase. For example, the number of short notes and the rate of repetition are characteristic for an individual, whatever the context, but differ markedly between individuals. In some cases the short-note phase consists of notes that differ in length in such a way as to give a broken rhythm. These individual characteristics of long-calls provide signatures by means of which an observer in the field can identify individuals by ear; and this experience is confirmed to the eye in sonagrams of recordings of the calls (Beer 1970b, 1970c, 1972).

Playback experiments have confirmed the obvious supposition that the gulls can also remark the individual differences in long-calls and so identify one another individually by ear. Playback of the long-call of the mate evoked response in incubating gulls in the field, but playback of long-calls of neighbours or strangers did not (Beer 1970c). Playback of a long-call of one of the parents induced chicks to vocalize and approach the source of the sound, but playback of a long-call of a neighbouring adult, or an adult from a distant part of the gullery, had either no effect or a negative effect (Beer 1973b). These tests were conducted indoors; for descriptions of the experimental procedures see Beer (1970a, 1973b).

The tests from which this last result was obtained were carried out on two groups of chicks: nestlings less than 48 hours post-hatching, and older chicks between 15 days and 21 days post-hatching. The older chicks were much more exclusive in their reactions to the playback long-calls than were the nestlings. Two of the older chicks showed no response to the call of the parent as well as to the call of the neighbour. These two, however, contrasted so markedly with the others, all of which had responded very positively to the parents' calls, that I looked for something special in the circumstances of their testing. I found it in the recordings. Whereas the other chicks had been tested with recordings made on the day of testing or the day before, the two non-responders had been tested with recordings made 2 weeks or so earlier—when they were nestlings. This indication that the time or context of recording might be important was confirmed by another series of tests in which chicks between 2 weeks and 3 weeks of age post-hatching were played two recordings of parental long-calls: one made during the nestling stage and the other made shortly before testing. The chicks responded positively to the latter but not to the former (Beer 1973b).

Although there are other obvious possibilities, the idea that most appealed to me to account for this difference of response to the two calls of the parent was that different versions of the long-call have different meanings, those uttered during the nestling period not meaning approach etc. to a 2-week-old chick, even though the chick may recognize the voice of its parent in such calls. Consistent with this possibility was the impression that long-calls given by parent gulls during the nestling stage are directed at adults—their mates, neighbours, or strangers—and not at their chicks. By the time the chicks are 2 weeks old, however, the parents do direct some of their long-calls at them. Because such calls more often occur in solo than is the case with long-calls directed at adults, I probably selected them, unintentionally, for my test recordings of later calls.

Since a parent with chicks 2 and more weeks old directs some of its long-calls at its chicks, and other long-calls at adults, I could test my idea by selecting recordings of calls from these two classes and doing another playback experiment similar to the last. Chicks in the field generally respond positively only to the calls directed at them by their parents, and ignore the calls directed by their parents at other birds. The behaviour of chicks in the field thus supplied an additional criterion for selection of a class of effective chick-directed calls to compare to a class of adult-directed calls. The results of the tests conducted so far with these recordings are less consistent than those from the tests with early and late recordings; but there are some significant differences that are in the predicted direction. The chicks have shown higher levels of positive response to the recordings of the chick-directed calls than to the adult-directed calls (Beer 1973b).

Examination of sonagrams of the recordings used in these last two series of tests has so far failed to reveal at all clearly the long-call features responsible for the response differences. I have discerned only one feature that generally distinguished the calls to which the chicks responded positively from the calls to which they did not. In amplitude–time displays, 13 out of 15 calls that evoked positive response produced a pattern in which the short notes all peaked at about the same level or the first short note was slightly higher than those following it; all but one of the 15 calls that did not produce positive response (calls recorded during the nestling period and adult-directed calls recorded during the later period) had a pattern in which the first one or two short notes were lower than the rest or amplitude increased progressively throughout the string of short notes (Fig. 2.7). More marked differences between the two test calls of the same

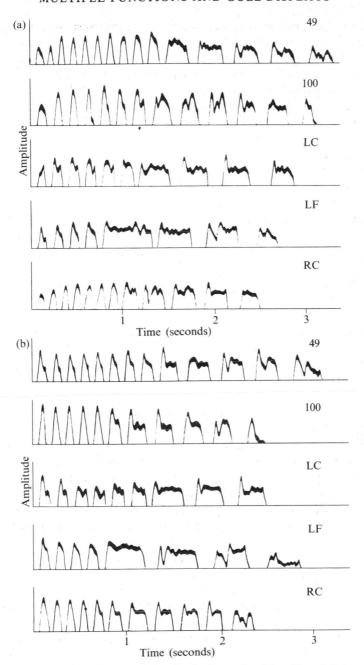

FIG. 2.7. Patterns of amplitude and time in the long-calls of Laughing Gull long-calls. (a) Long-calls directed by parent gulls at other adults. (b) Long-calls directed by parent gulls towards their chicks. The individuals are identified by numbers or letter combinations on the right. The patterns were traced from amplitude–time sonagrams made on a Kay Sonagraph, Model 7029A, using the wide bandpass filter and the 40–4000 Hz frequency range.

individual, but peculiar to it, were present in a number of cases; and, for some of these, examination of sonagrams of calls other than those used in the tests, but assignable to the test categories, showed such differences to be apparently consistent. There is thus a suggestion that, in at least some respects, the feature or features of a parent gull's long-call that encode the command to the chicks to approach, may differ between families. However, I have so far applied only 'eye-balling' comparison to only a small sample of such calls. It could well be that more precise measurement of the many parameters of a larger sample, and even computer analysis of such measurements, will be needed to reveal the kinds of differences that the behaviour of the chicks I have tested has told us must be there. In the meantime the experimental results on selective responsiveness to long-calls raise a number of points for discussion.

In his review of gull displays, Tinbergen (1959) argued that the long-call functions as 'long-distance threat' and to attract unmated females to a male from a distance, and is therefore functionally equivalent to song in songbirds. This can be only part of the story, however, at least as far as Laughing Gulls are concerned. Laughing Gulls use long-calls at close quarters as well as at a distance. For example, long-calls occur in meeting ceremonies and nest-reliefs when the two birds of a pair are side by side. It can be argued that, in such performances, the two birds jointly announce their alliance and defiance to the world at large; but there is also reason to believe that the two birds affect one another as well by these performances. We are dealing with a display which has multiple functions.

My playback experiments indicate two immediate functions for the long-call of Laughing Gulls: individual identification and the conveying of a message having the properties of an imperative or an invitation. Species identification may also be included in the functions of the call—for no other species of gull long-calls quite like a Laughing Gull and no other kind of bird has a call quite like the long-call of a gull—and it is possible that the call conveys indications of the sex, age, and physiological state as well. On the question of physiological state, for example, it has been observed that the long-calls of Laughing Gulls acquire a hoarse tonal quality as the breeding season draws to a close, presumably as a consequence of decline in sex-hormone levels.

The suggestion that all these different kinds of information can be conveyed simultaneously by the one call will not appear far-fetched if it is remembered that we can learn as much from a voice uttering a sentence over the telephone. But just as only part of the information is peculiar to the sentence, the rest being general to the voice, so probably only part of the information in the long-call is peculiar to that call. Perhaps if we could specify the content that is exclusive to the call we should be in a position to come to a definite conclusion about its function. However, although we are, as yet, far from a clear idea of this content, it appears to be so various as to discount the possibility of a short and simple answer to the question, 'What is the function of the long-call of Laughing Gulls?'

Variability in reactions to the call is partly a consequence of the qualifying effects of context, in the broad sense in which Smith (1968) has used that term. One and the same call by an advertising male in the pre-laying period 'means' threat to another male, sexual invitation to an unmated female. But variability in reaction occurs even when the contribution of circumstances external to the call can be discounted. In my playback tests with chicks I found response-differences that implied differences of meaning dependent upon variation in certain features internal to the calls. Here too, however, the information transmitted is textured rather than plain. Some features of long-calls differ between individuals and thus convey individual identity; other features differ between long-calls of the same individual and thus convey different messages. The identification and the message arrive together. Reaction depends upon them both. A call may be ignored either because the individual it identifies is not one of the family, of because the message it conveys is not for the chicks. Failure to take such possibilities into account could frustrate investigation of the communication functions of behaviour and of the means of individual recognition.

But the matter of multiple function of Laughing Gull long-calls is further complicated by temporal changes in both the uses of the call and its consequences. Parent gulls do not appear to use the call to influence the behaviour of their chicks at the nestling stage, although they do use crooning to induce their nestlings to approach and take food. In playback tests nestlings approached and vocalized in response to crooning of any adult, but tended to crouch or withdraw in response to long-calls, although less so to the calls of their own parents than to the calls of other adults (Beer 1970b, 1973b). By the time the chicks are 2 weeks older, however, their parents regularly use the long-call to announce their return to the vicinity of the chicks and to attract them from a distance; crooning occurs only when the chicks draw near and as they are being fed. In playback tests of chicks at this age, crooning, whether by a parent or by a neighbour, failed to elicit any response; long-calls by the parents elicited approach and vocalization, but long-calls by a neighbour elicited either negative response or nothing at all (Beer 1973b). The parents' long-calls thus take over part of what earlier is the function of crooning.

The timing of this transition in function coincides with, or anticipates, the hazards of the new forms of social situations with which a chick has to cope as it extends its range of movement progressively farther from home, and as its parents begin to leave it unattended. Adults other than its parents are usually viciously hostile to such a chick, sometimes with fatal consequences. Hence a 2-week-old chick has to be able to distinguish its own parents from other adults it encounters, if it is to avoid punishment for wrong approaches. Crooning lacks individually identifying features, at least according to the chicks; the long-call possesses such features more distinctly than does any other kind of Laughing Gull call. It therefore makes good functional sense that the long-call should pre-empt the role of crooning in eliciting filial approach behaviour by the time the chicks have reached the age of discretion.

It is also not hard to see the functional point of the chicks' selective responsiveness to differences in the long-calls of their parents. There is a resurgence of agonistic and courtship behaviour during the post-hatching period, and the long-call figures prominently in such behaviour. If a chick approached in response to any and all long-calls of its parents it would probably suffer punishment for presenting itself as a target for redirected hostility. Flexibility in some features of the long-call make it possible for the call to be used for communication both within and without the family.

In summary, the evidence we have at present suggests that the long-call of Laughing Gulls is a form of display by means of which a gull can emphatically identify itself and convey a number of alternative messages with regard to itself, the identification being carried by stable individual characteristics of the calls and the particular message being carried by features of the call that the bird can vary. Anthropomorphically rendered, a long-call might thus signify: 'I am your parent—come and get fed'; or 'I am your mate—let me sit on the eggs'; or 'I am your prospective mate—come and stay close'; or 'I am the occupier of this area—get out'.

According to this view, the long-call of Laughing Gulls is semantically and pragmatically open; but, in contrast to the crooning and facing-away considered earlier, the openness here comes, at least partly, from within the fabric of the display, rather than from the alternative ways of combining the display with other behaviour and contexts. The long-call could thus be described as a display with 'open texture'.

2.3. Conclusion

My rather cumbersome approach to the matter of multiplicity in the communication functions of displays could be avoided if we were to differentiate or designate displays on the basis of communication function. Then we should have unique relationships of displays to functions and no need for notions like open texture. 'Display' itself is a functional term, so it might be argued that functional criteria are the proper means for the demarcation of displays. However, patterns of behaviour are often recognized as displays before their communication functions are understood, the long-calls of gulls being a case in point. If we were to argue, with respect to the long-calls of Laughing Gulls, that what has hitherto been regarded as one category of display should be split up into several on functional grounds, we should not be able to say how many categories or what their defining characteristics are, for as yet the functional grounds have been insufficiently charted. By and large ethologists have used criteria of form rather than function to decide what is to count as a display.

This question about the criteria for demarcation of displays is more than an idle matter as far as ethological theory is concerned (cf. Altmann 1967). In a recent paper, Moynihan (1970) tabulated the numbers of 'major' displays in the repertoires of a variety of vertebrates to make the point that the range in such numbers is surprisingly narrow—between 15 and 35. From this evidence he argued that the display repertoires of diverse species are subject to similar selection pressures on size, as a consequence of which displays are discarded and replaced in the course of evolution. While much of this argument is plausible and ingenious, it contains a number of debatable points, including the criteria for judging whether a pattern of behaviour is a major display.

'A display is considered to be major when it seems to be qualitatively distinct, has some significant characteristic(s) peculiar to itself alone, or is not part of a completely intergrading continuum among other patterns' (Moynihan 1970, p. 90). Are these three independent criteria? If so does a display have to meet each of them, or only one, to be considered major? These questions aside, we are presented with a considerable burden of freedom in making our judgements. It appears that we have to look to our perceptions to decide what forms of behaviour are salient enough to merit major status as displays. A major display will be what counts as such in the opinion of a competent ethologist. But ethologists are just as likely to divide themselves into lumpers and splitters in their judgements about displays, as systematists have been in their judgements about species. I have often been struck by how two equally proficient observers can watch the same behaviour and then argue about what actually went on. Even with film records to go back to they may fail to reach agreement, for though they look at the same thing they perceive it differently; each convinced of the indubitability of his senses, he impugns the veracity of those of his colleague.

Such disagreement between human observers makes one wonder how differently the animals might perceive one another's behaviour compared to the way or ways in which we do. It is a commonplace that speech in a strange language seems to be an unbroken string of sound; we fail to register the pauses that mark off words, phrases, and sentences for someone familiar with the language. A fortiori, naïve human perception will probably be an unreliable guide to the divisions and structure in an animal communication system as it is used and understood by the animals. The consequences could well be detrimental to theorizing about the selection pressures that have shaped the evolution of animal communication systems.

Selection pressures affect displays via their functions. Moynihan's category of major displays is functionally heterogeneous: it contains displays that have but one function, such as the alarm call of Laughing Gulls, and it contains displays that have multiple functions, such as the long-call of Laughing Gulls; it contains displays that function autonomously, again like the alarm calls of Laughing Gulls, and it contains displays that function in concert with other behaviour and context, such as facing-away in Laughing Gulls. The selection pressures on these functionally different categories of displays must be different in ways that Moynihan, with his morphologically defined category of displays, does not take into account. For example, a display that occurs as part of a ritual sequence or interaction chain will be subject to conservative, intrinsic selection pressure (cf. § 2.1), which will slow its rate of evolutionary change or elimination, compared to a display that is functionally autonomous.

Functional autonomy, of the kind implied by the classical ethological concept of social releaser, may well be the exception rather than the rule in animal communication systems, at least those of many species of birds and mammals. Typological thinking may well have been as much a hindrance to ethology as it appears to have been in systematics (cf. Mayr 1963). The word 'system' connotes organization and integration. It probably applies more literally to its denotata in the context of animal communication than has generally been assumed. Preoccupation with aspects of form, part of the inheritance of ethology from the past of zoology, has had a tendency to keep attention away from the full complexities and intricacies of function. For example, the many fine studies of vocal ontogeny in songbirds have, almost without exception, investigated acquisition of a vocal pattern; the semantic and pragmatic aspects of the matter have been neglected.

The semantics and pragmatics of social communication systems, of species such as the Laughing Gull, may well pose problems far more complex than those envisaged by the conception of social communication systems as sets of releasers and innate releasing mechanisms tuned to one another. We may have to deal with an hierarchically ordered repertoire of signal elements and signals (cf. Altmann 1967), together with sequence and combination formulae for the encoding of messages and the decoding of meanings. The multiple functions thus organized at the level of immediate consequences can be thought of, in their turn, as hierarchically related in functional categories assembled at the levels of remote or ultimate consequences, under the orders of factors like food supply and predation. Linguistics is not the only source of analogy for this conception of animal communication as a system of hierarchically related elements and syntactic formulae for their functional combination. In endocrinology we find that some hormones have a variety of effects serving some broad function; for example, sex hormones affect gonadal maturation, secondary sexual characteristics, behaviour, and so forth, all in ways that have to do with reproduction; angiotensin affects drinking, kidney function, adrenal secretion, and blood-pressure in ways that all have to do with the water needs of the body. Some of this variety in hormonal functions is due to synergistic combination of different hormones. Similarly in genetics, the effects of genes can be considered as related hierarchically within the processes into which they enter, and as entering into different processes according to combination alternatives. At all levels we appear to have to deal with functional organization and integration that call for wholistic treatment in contrast to the atomistic or mosaic conceptions that used to prevail.

I began this essay with dissection. I tried to show that if we make certain distinctions in the uses of the word function we can take apparent cases of multiple function apart in such a way as to show that the different functions attach to different aspects of the case or answer to different kinds of question about it. But then I discussed some cases for which this dissection approach was inadequate to disperse the functional multiplicity. Certain displays of Laughing Gulls appear to be semantically and pragmatically versatile by virtue of syntactic and contextual qualifiers on their use and interpretation. The picture of an hierarchically organized system of communication that emerged has suggested to me, finally, that the various senses of function that I pulled apart in the beginning should be drawn back together, or at least considered in relationship to one another rather than in isolation from one another. The study of multiple function at the level of immediate consequences is a meeting point of questions about motivation, ontogeny, survival value, and evolutionary origins. It is important to note that these are different kinds of questions; but I am persuaded also that they eventually have to be brought into connection with one another if more than a fragmented and distorted picture of the biology of a species is to be had. Current developments on the comparative front, in which features of species-typical behaviour are being related to ecological factors, are perhaps pointing the way.

In a paper on the aims and methods of ethology, Professor Tinbergen wrote that ethology is characterized by its concern with four classes of problems about animal behaviour: causation, ontogeny, survival value, and evolution. He went on: 'There is, of course, overlap between the fields covered by these questions, yet...it is useful both to distinguish between them and to insist that a comprehensive coherent science of Ethology has to give equal attention to each of them and to their integration' (Tinbergen 1963, p. 411). I have emphasized the importance of distinguishing these questions, particularly in the nature–nurture issue and in the uses of 'instinct' (Beer 1963–64, 1968a, 1968b, 1973a), to the exclusion of the last part of Tinbergen's view of ethology's task. I now see this error of my ways, and hope that my teacher will be pleased that I have at last learnt the end of his lesson.

Acknowledgements

The work on Laughing Gulls reported here was supported by grants GM 12774 and MH 16727 from the U.S. Public Health Service and a grant from the Research Council of Rutgers University. The U.S. Fish and Wildlife Service granted permission for the research to be carried out in the Brigantine National Wildlife Refuge and made available a building which I and my students have used as a field station. I am grateful to the Refuge Manager and his staff for their hospitality and co-operation. I have enjoyed the co-operation and council of the students and colleagues with whom I have been associated on this project. Dr. Monica Impekoven has been especially helpful to me. For their teaching, encouragement, and example I am profoundly in debt to the late Professor D. S. Lehrman and Professor Tinbergen.

References

ALTMANN, S. A. (1967). The structure of primate social communication. In *Social communication among primates* (ed. S. A. Altmann). University of Chicago Press, Chicago.

BAKER, J. R. (1938). The evolution of breeding seasons. In *Evolution: Essays presented to E. S. Goodrich*. Oxford University Press.

BARFIELD, R. J. (1971). Gonadotrophic hormone secretion in the female Ring Dove in response to visual and auditory stimulation by the male. *J. Endocr.* **49**, 305–10.

BEER, C. G. (1963–4). Ethology—the zoologist's approach to behaviour. *Tuatara* **11**, 170–7; **12**, 16–39.

—— (1968a). Instinct. In *International encyclopedia of the social sciences* (ed. D. L. Sills). Macmillan and The Free Press, New York.

—— (1968b). Ethology on the couch. *Science and psychoanalysis*, Vol. 12, pp. 198–213. Grune and Stratton Inc., New York.

—— (1970a). On the responses of Laughing Gull chicks (*Larus atricilla*) to the calls of adults. I. Recognition of the voices of the parents. *Anim. Behav.* **18**, 652–60.

—— (1970b). On the responses of Laughing Gull chicks (*Larus atricilla*) to the calls of adults. II. Age changes and responses to different types of calls. *Anim. Behav.* **18**, 661–77.

—— (1970c). Individual recognition of voice in the social behaviour of birds. *Adv. Study Behav.* **3**, 27–74.

—— (1972). Individual recognition of voice and its development in birds. *Proc. int. orn. Congr.* **15**, 339–56. Brill, Leiden.

—— (1973a). Species-typical behaviour and ethology. In *Comparative psychology: a modern survey* (eds D. A. Dewsbury and D. A. Rethlingshafer). McGraw-Hill, New York.

—— (1973b). A view of birds. *Minnesota Symposia on Child Psychology*, Vol. 7, pp. 47–86. University of Minnesota Press.

BLEST, A. D. (1957). The function of eyespot patterns in the Lepidoptera. *Behaviour* **11**, 209–56.

BLURTON JONES, N. G. (1968). Observations and experiments on causation of threat displays of the Great Tit (*Parus major*). *Anim. Behav. Monogr.* **1**, 74–158.

BOWLBY, J. (1969). *Attachment and loss*, Vol. I: *Attachment*. The Hogarth Press and The Institute of Psycho-analysis, London.

CHANCE, M. R. A. (1962). An interpretation of some agonistic postures: the role of "cut-off" acts and postures. *Symp. zool. Soc. Lond.* **8**, 71–89.

CHOMSKY, N. (1966). *Cartesian linguistics*. Harper, New York.

CULLEN, E. (1957). Adaptations in the Kittiwake to cliff-nesting. *Ibis* **99**, 275–302.

DARWIN, C. (1872). *The expression of the emotions in man and animals*. John Murray, London.

ERICKSON, C. J. and LEHRMANN, D. S. (1964). Effect of castration of male Ring Doves upon ovarian activity of females. *J. comp. physiol. Psychol.* **58**, 164–166.

EVANS, R. M. (1970a). Parental recognition and the "mew call" in Black-billed Gulls (*Larus bulleri*). *Auk* **87**, 503–13.

—— (1970b). Imprinting and the control of mobility in young Ring-billed Gulls (*Larus delawarensis*). *Anim. Behav. Monogr.* **3**, 193–248.

KIRKMAN, F. B. (1937). *Bird behaviour*. Nelson, London.

MCBRIDE, G. (1966). Society evolution. *Proc. ecol. Soc. Aust.* **1**, 1–13.

—— (1971). The nature–nurture problem in social evolution. In *Man and beast: Comparative social behaviour* (eds J. F. Eisenberg and W. S. Dillon). Smithsonian Institution, Washington, D.C.

MCKINNEY, F. (1973). Ecoethological aspects of reproduction. In *Breeding biology of birds* (ed. D. S. Farner). National Academy of Sciences, Washington, D.C.

MANLEY, G. H. (1960). *The agonistic behaviour of the Black-headed Gull*. Unpublished doctoral thesis. Bodleian Library, Oxford.

MAYR, E. (1963). *Animal species and evolution*. Harvard University Press, Cambridge, Massachusetts.

MILLER, G., GALANTER, E., and PRIBRAM, K. (1960). *Plans and the structure of behaviour*. Holt, Rinehart, and Winston, New York.

MOYNIHAN, M. (1955). Some aspects of reproductive behaviour in the Black-headed Gull (*Larus r. ridibundus*) and related species. *Behaviour, Suppl.* **4**, 1–201.

—— (1970). Control, Suppression, decay, disappearance and replacement of displays. *J. theor. Biol.* **29**, 85–112.

NOBLE, G. K. and WURM, M. (1943). The social behavior of the Laughing Gull. *Ann. N.Y. Acad. Sci.* **45**, 179–220.

RUITER, L. de. (1956). Countershading in caterpillars. *Arch. neerl. Zool.* **11**, 285–341.

SMITH, W. J. (1963). Vocal communication of information in birds. *Am. Nat.* **97**, 117–125.

—— (1965). Message, meaning and context in ethology. *Am. Nat.* **99**, 404–9.

—— (1966). Communication and relationships in the genus *Tyrannus*. *Publs Nuttal orn. Club* **6**.

—— (1968). Message-meaning analysis. In *Animal communication* (ed. T. A. Sebeok). Indiana University Press, Bloomington.

TINBERGEN, N. (1950). The hierarchical organization of nervous mechanisms underlying instinctive behaviour. *Symp. Soc. expl. Biol.* **4**, 305–12.

—— (1951). *The study of instinct*. Clarendon Press, Oxford.

—— (1953). *The herring gull's world*. Collins, London.

—— (1959). Comparative studies of the behaviour of gulls (Laridae): a progress report. *Behaviour* **15**, 1–70.

—— (1963). On aims and methods of ethology. *Z. Tierpsychol.* **20**, 410–33.

—— (1964). On adaptive radiation in gulls (Tribe Larini). *Zoöl. Meded., Leiden* **39**, 209–23.

TINBERGEN, N., BROEKHUYSEN, G. J., FEEKES, F., HOUGHTON, J. C. W., KRUUK, H., and SZULC, E. (1962). Egg shell removal by the Black-headed Gull, *Larus ridibundus*; a behaviour component of camouflage. *Behaviour* **19,** 74–117.

—— and MOYNIHAN, M. (1952). Head flagging in the Black-headed Gull; its function and origin. *Br. Birds* **45,** 19.

TWADDELL, W. F. (1963). On defining the phoneme. In *Readings in linguistics* (ed. M. Joos). American Council of Learned Societies, New York.

WITTGENSTEIN, L. (1953). *Philosophical investigations.* Blackwell, Oxford.

3. Feedback, spontaneous activity, and behaviour

K. D. ROEDER

THE FOURTH SYMPOSIUM of the Society of Experimental Biology, with the title *Physiological mechanisms in animal behaviour* (Cambridge 1950), remains a major landmark in the history of ethology. This symposium focused for the first time on research that attempted to bridge the gap between physiology and behaviour, and it is probable that it first pointed the direction for many now established students of behaviour mechanisms. Among the exciting and sometimes fervent articles in the symposium volume the two shortest and most unassuming are those by E. D. Adrian and N. Tinbergen. Adrian quietly draws attention to spontaneous activity in a variety of nerve cells; Tinbergen firmly points out that progress in his own field—ethology—depends on closer contact with physiology. His hierarchical concepts of behaviour organization can be interpreted as just such an invitation. Each of these articles struck a chord in my own thinking and had a considerable influence on what I have done since that time. The present essay speculates in an area suggested in part by the above-mentioned papers.

3.1. Negative feedback

Negative feedback, regulation, or homeostasis is widely recognized as the principle underlying control in a multitude of body functions. Regulation in the maintenance systems of the body was first recognized by Claude Bernard who realized that 'La fixité du milieu intérieur...' is an important trend in biological evolution that has insulated higher organisms from many of the perturbations present in their external environments. The detailed mechanisms of physiological pathways concerned in regulating body temperature, blood-pressure, osmotic balance, and so on were subsequently worked out, but at first each system was treated as a special case. In his prescient book, *The wisdom of the body*, Cannon (1932) pointed out that these and other autonomic functions operate on a common principle which he called homeostasis. The next major step was a theoretical treatment by Wiener (1948), who introduced the term cybernetics and demonstrated that the principles of regulation are of great power and breadth. They apply to all operations that are self-correcting or steered in relation to a goal or set-point, be it the course travelled by an animal, a certain value of blood-pressure or body temperature, or the desired speed of a man-made engine. Since then there has been a huge and still-mushrooming literature on the application of cybernetic principles to engineering, chemistry, and physics as well as physiology and medicine (reviews by Ashby 1956; Stanley-Jones 1960; Bayliss 1966). One object of this essay is to examine the relevance of certain cybernetic principles to the mechanisms underlying certain forms of animal behaviour.

Negative feedback tends to stabilize a certain value of one parameter of a dynamic system in the presence of conditions that tend to perturb it. It operates by automatically correcting errors or deviations in the chosen value. Since it operates on the basis of deviations negative feedback is inherently incapable of providing perfect stabilization—it can control only within acceptable limits. There have been adequate discussions of control by negative feedback in miscellaneous biological systems by Stanley-Jones (1960), in physiological mechanisms by Bayliss (1966), and in orientation and other forms of behaviour by Mittelstaedt (1964) and Hinde (1970). Therefore, I shall merely list some of the principles and limitations common to control systems.

The significance of the negative sign denoting a control system is evident in the familiar example of a thermostatically controlled heating system or of a person tending a fire so as to maintain an acceptable room temperature. When the temperature drops below the acceptable value fuel is added; when it rises above this point fuel is withheld. Thus, deviations from the acceptable value or set-point must take place before they can be corrected. The greater the delay between deviation detection and deviation correction the greater is the oscillation about the set-point and the tendency for the system to become unstable. Another source of instability lies in imperfect feedback. If the corrective action is insufficient (insufficient fuel added) or excessive (too much fuel added) on a falling temperature there is likely to be an overshoot at either end of the cycle and control fails. Failure due to time-lags in the system or to abrupt external changes (in temperature) can be countered by prediction or anticipation of overshoots or undershoots or of external changes. This has been called feedforward by Mittelstaedt (personal communication). Another way of improving stability of control is to include in the system additional feedback loops operating in a push–pull relation to the first. For instance, the thermostat or fire-tender could also control a damper on the heating system or a means of ventilating the enclosed space.

It is also apparent that the thermostat or fire-tender expends only a relatively small amount of energy to hold or to steer a certain course, in this case a given temperature value, in the presence of external factors tending to perturb it. But the temperature can also be 'steered' through any sequence of values within the operating limits of the system; that is, the set-point can be systematically varied so as, for instance, to maintain different temperatures in the daytime and at night. The situation is analogous to a helmsman holding a ship to various successive bearings in seas that tend to perturb its course. Such arrangements are called servo-systems.

The best known physiological mechanisms operating on these principles are those regulating autonomic functions (Cannon 1932) and body posture (Bayliss 1966). Push–pull control, or two or more negative-feedback loops smoothing and stabilizing the same operation, is exemplified by the sympathetic and parasympathetic control of the heart in regulating blood-pressure, or the parallel mechanisms of heat genesis, dissipation, and conservation in regulating body temperature. The reflex mechanism that sustains the length of the postural extensor muscles at each of a variety of lengths and under different loads is a particularly elegant and interesting example of a servo-system (Bayliss 1966).

The orientating reactions of animals are obvious examples of negative feedback in behaviour (Hinde 1970). An animal follows a faint trail or travels in relation to a distant marker by repeatedly correcting the tendency of its course to deviate due irregularities in its own movements or in the medium through which it is travelling. Its track is subject to the defects implicit in negative feedback mentioned above. Excessive delay or insufficiency in its corrective responses as well as over compensation are likely to cause wide digressions that may lead to loss of the trail or marker.

Orientational homeostasis has been analysed in some depth by Mittel-staedt (1964). Most animals, while awake, are continuously orientating or 'steering a course' with respect to gravity and their visual surroundings. This 'steering' continues irrespective of whether they are at rest or in motion. Continuous visual steering is demonstrated by the well-known optomotor reaction. If an animal is initially at rest in a surround of vertical stripes it will move when the stripes are displaced horizontally. It follows the displacement of the pattern by making nystagmic eye or head move-ments or by turning its body in a direction that reduces or abolishes movement of the pattern relative to its visual field. External displacement of the visual surround is accepted by the animal as a perturbation from its set-point, which is non-displacement of the visual surround while it is at rest. This causes the animal literally to take steps to correct the apparent deviation from this set-point.

The continuous operation of negative feedback with a set-point of non-displacement under these circumstances was vividly demonstrated by Mittelstaedt (1949). The fly *Eristalis* shows a well-marked optometer reaction. Mittelstaedt placed an individual in a striped cylinder after its head had been rotated 180° about its neck and then fixed in this position. The right eye was now directed to the insect's left side. This had an effect equivalent to that which might be produced in an experienced but unsuspecting automobile driver when he discovers that his steering linkage has been reversed. So long as there was no relative movement of the striped pattern (course deviation) the insect remained motionless, but when the striped pattern was displaced slightly the fly with reversed visual fields turned *against* the movement rather than with it. As a result the insect's movement compounded with the relative displacement, producing still greater relative displacement of the stripes over its visual field, and the insect began to whirl madly and continued to do so even after the stripes had ceased to move relative to the platform on which it was standing. Rotation and fixation of the head had reversed the sign of the visual feedback from negative to positive. This caused an uncontrollable reaction to the slightest displacement.

3.2. Positive feedback

Negative feedback tends to preserve the *status quo*. When it is included in a servo-system it enables animals to steer a course from less favourable to more favourable conditions. The principle of homeostasis clearly plays an important part in the evolutionary history of all species as well as in the survival of individual organisms. But it is a conservative principle and alone does not seem able to account for the evolutionary trend towards innovation and complexity or for much of the hour-to-hour and day-to-day behaviour of animals. A simple argument might be that if negative feedback or regulation alone dominated such activities an organism supplied with optimal conditions would reach a dynamic balance and hence would do nothing. Yet it is obvious that animals are usually 'doing something' during most of their waking hours, especially when in good health and under optimal conditions. Much of this activity is probably directed at satisfying some deficiency and hence could be regarded as homeostatic in nature, but abrupt transitions from one behaviour mode to another, wandering or hunting in an apparently sated animal, and so-called play activity find no obvious explanation in terms of homeostasis. It can be argued that this statement is unsupportable since we do not now comprehend all of the interactions within the maze of control systems that compose a living animal. A more heuristic suggestion is that the maze of stabilizing connections must contain within it systems that are inherently unstable.

Positive feedback is the acme of instability. Here a portion of the energy released by a source is fed back to the source in a manner that will increase its output. The result is a chain reaction or run-away to the limits of the energy source. Examples are a fire, an explosion, a nuclear reaction, and an avalanche. Pure positive feedback is unthinkable instability. However, in practice each of these events is limited at its onset and termination by processes that are degenerative or stabilizing in nature—the ignition temperature, resistance, or inertia that must be overcome before the event becomes self-sustaining—and the limits of the fuel or energy available.

Many processes that are clearly controlled by homeostasis may enclose elements of positive feedback that add to the instability of the system as a whole and have the potentiality of 'breaking through' the encompassing control. This is evident in the example of control mentioned earlier where a thermostat or an attendant regulates a furnace or fire with the object of maintaining a steady temperature. The fire provides an element of positive feedback that renders the system metastable and prone to breakdown in three ways. First there is the possibility that the fire becomes extinguished and cannot be re-ignited. A second possibility is that the fire gets out of control and causes a general conflagration. The exterior loop of negative feedback exerts control by containing or limiting these swings in the direction of 'zero' and 'infinity' so that they are evident only as a sequence of oscillations. Even then, they add considerably to the instability of the control operation.

The third source of instability depends on the time course of one cycle of the regenerative process—the flare-up and die-down of the fire, for instance—in relation to the time occupied by the delay in the negative feedback. This can be appreciated if one considers a fire as a moderately slow explosion. Its time of flare-up and die-down is ordinarily long compared with the time needed to sense a drop in the temperature and add fresh fuel. However, the situation gets out of hand in one direction if a long interval elapses before fresh fuel is added, and in the other if the fuel used has a shorter burning time such as dry twigs, gasoline, or gunpowder. In both cases there is increased likelihood of an abrupt transition from regulation to innovation.

With these general characteristics of positive feedback in mind I shall examine its possible role in behaviour by seeking answers to certain questions: (1) Are elements whose operation depends on positive feedback included in the neural mechanisms of behaviour? (2) Does the instability implicit in positive feedback sometimes become manifest in what an animal does? In other words, can examples of behaviour be found where the action in traceable to positive feedback in neural elements? (3) The neural mechanisms responsible for most complex behaviour patterns remain largely or wholly unexplored. Are there indications in such behaviour that positive feedback may play a part in its genesis?

Sources of instability in neural systems

In seeking an answer to the first question one does not have to search far in the nervous system to find events that depend on positive feedback. In fact, responses propagated without decrement fall into this category.

The vertebrate heart is a familiar example. Normally it alternates between a state of complete relaxation (diastole) and one of complete contraction (systole) with no intermediate or graded state. Systole is initiated when increasing instability of a small patch of excitable tissue, the sinoatrial node, reaches critical proportions and gives rise to a self-propagating impulse that travels to neighbouring cardiac muscle fibres as a brief wave of depolarization followed by contraction. This self-excitation spreads in turn over the auricles, bundle of His, and the ventricular muscle. Thus, a minor local instability of the pacemaker triggers a massive response of the whole organ. This total response is accompanied by a refractory period or interval of inexcitability that is characteristic of regenerative events. The muscular 'explosions' of the heart are smoothed or 'muffled' by the elastic properties of blood-vessels and regulated by the push–pull arrangement of negative feedback provided by the autonomic nervous system so that they are externally evident merely as a pulse riding on a relatively steady blood-pressure.

The genesis and propagation of nerve impulses over nerve axons depend on very similar regenerative properties of their plasma membranes. In its resting state the plasma membrane excludes or extrudes positively charged sodium ions, and is in a metastable or poised condition, expressed by its membrane potential of 50–60 millivolts, positive outside. Disturbance of this metastable condition by a suitable stimulus, or by a nerve impulse approaching as it travels along the axon, reduces the membrane potential and favours some inward flow of sodium ions. This influx of positive charge further reduces the membrane potential which permits additional sodium influx, and so a self-promoting 'avalanche' of sodium ions is on its way.

Like other real processes involving positive feedback the genesis of a nerve impulse is 'contained' by degenerative events at either end of its cycle—the tendency for the metastable membrane potential to return to its resting value if a momentary disturbance fails to bring it to a critical level, and inactivation of the mechanism permitting sodium entry as the regenerative influx of sodium approaches its peak. This causes the potential to seek once more its resting value. The transient electrical change that results is the spike potential or nerve impulse. Like each heart beat each impulse is accompanied by a refractory period. The spike potential normally causes neighbouring and hitherto inactive regions of the axon membrane to 'go critical' with the consequence that an impulse spreads to all parts of the neuron capable of sustaining a self-propagating response. An impulse similarly invades and triggers contraction in all parts of each vertebrate striated muscle fibre.

Thus, the resting membrane of a normal axon or striated muscle fibre is in a state that can be likened to that of a fire that has been laid but not lighted. Natural 'ignition' is brought about by graded events at the synaptic region of the neuron or end-plate of the muscle fibre that are themselves degenerative in nature and serve to integrate the effects of stimuli from the outer world or of impulses arriving at synaptic contacts made by other neurons. Certain neurons are spontaneously active, generating a succession of impulses without receiving input from other cells. This property seems to be due to intrinsic instability in the spike-generating regions, a situation similar to that in the vertebrate heart. Signals arriving from other neurons may modulate this discharge, increasing or decreasing its frequency. Adrian (1930) first drew attention to spontaneous activity in nerve cells and later (1950) discussed its significance in certain vertebrate brain-centres. I have reviewed spontaneous activity of of nerve cells and considered its possible role in animal behaviour (Roeder 1955, 1967).

We must conclude that positive feedback is implicit in the mechanism concerned in the genesis and non-decremental transfer of information in the nervous system. However, the regenerative phases of excitation and conduction are largely confined and controlled by processes that are essentially degenerative, such as inhibition, the decremental nature of synaptic events, and sodium inactivation. Thus constrained, neurons may either remain inactive until triggered or they may be spontaneously active, oscillating at a certain frequency that may be subject to external modulation.

These regenerative events in the nervous system are generally not evident in behaviour since their effects are averaged and they are generally enclosed in a multitude of homeostatic loops that smooth their 'explosive' nature and prevent it from becoming manifest in the actions of a normal animal. Yet, under certain pathological conditions positive feedback may 'break through', encompassing degenerative control and become patent as abnormal behaviour in the form of epileptiform seizures, various forms of tetany, and the spasms induced by strychnine and tetanus toxin. The following section examines cases where it has been possible experimentally to identify regenerative events in the central nervous system and to show that they are responsible for sustaining behaviour even after the organism has been deprived of the possibility of sensory feedback derived from its own actions.

Endogenous control of motor rhythms

After they have begun to move most animals generally select a gait and pace that is sustained until circumstances change. As the animal travels it seems inevitable that various degenerative processes are taking place, such as fatigue and adaptation, yet until these become extreme they appear to have little influence on the rate of progress. In counteracting these degenerative effects a steady pace could be sustained by homeostasis from optokinetic and proprioceptive mechanisms operating in much the same manner as those concerned in orientation. The animal might be said to have a speed governor. But there is also evidence that regenerative or self-excitatory processes are involved.

In postulating a spontaneous or self-excitatory biological process one must be able to demonstrate that the process is not due to the action of external stimuli. Since no biological process is completely closed off from the environment the conditions of such a demonstration must be defined. Taking an example of positive feedback mentioned earlier, a nerve impulse transmitted down an axon is normally generated or triggered by integration of the synaptically mediated and decremental effects of impulses arriving from other nerve cells or, if the nerve cell in question is sensory, by decremental generator potentials caused by external physical or chemical changes. But many central neurons as well as sense cells may continue to generate spikes with low frequency when they are prevented from receiving their 'natural' input. They are then considered to be spontaneously active or self-excitatory (Roeder 1955). However, this does not rule out the fact that the instability of self-excitatory nerve cells is influenced by chemical factors in the 'milieu interiéur'. Indeed, nerve cells appear to become inordinately sensitive to such factors when deprived of their normal input (Roeder 1948).

Self-excitation becomes harder to identify when one is dealing with the complex coupled nerve-cell aggregates that compose ganglia and nervous systems. However, there seems no reason to abandon the pacemaker concept that instability arises at some point in the nerve mesh and then spreads by regenerative action accompanied by a refractory period. A classic experiment by the late Donald M. Wilson (1961) demonstrates that an endogenous pacemaker system of this nature provides the motor drive responsible for the rhythmic up-and-down movements made by the wings of locusts during flight.

Locusts flap their wings between 15 and 20 times a second. It was formerly suspected (Weis-Fogh 1956) that the sequence of motor nerve impulses co-ordinating the alternate contractions of the levator and depressor flight muscles is regulated and sustained by feedback from receptors located on various parts of the flight apparatus. Wilson progressively disconnected the thoracic ganglia—the motor centres for the flight muscles—from possible sources of external input by cutting motor and sensory nerves and finally all nerve connections with the exterior. While this was being done he registered the pattern of efferent spikes in the central stumps of the nerves that originally supplied the flight muscles. A rhythmic discharge of motor impulses persisted after the ganglia had been isolated, and Wilson was able to identify this co-ordinated pattern with that driving the flight muscles in the intact insect. In isolated ganglia the endogenous rhythm was slower than normal and tended to die out. However, it showed co-ordination in the form of alternating spike bursts to levator and depresssor muscle groups, and continued without intervention for hundreds of cycles on being triggered by a brief burst of non-specific and unphased electrical stimulation of the nerve cord.

Thus, the phasing and pattern of motor nerve impulses driving the indirect flight muscles of the locust is primarily determined by a self-exciting system of neurons in the central ganglia, called by Wilson a 'central nervous oscillator'. Since this oscillator system operates independently of externally phased (reflex) excitation, it must depend upon elements containing positive feedback. Wilson and Gettrup (1963) investigated the relation between the activity of this central oscillator system and the input from stretch receptors on the wing hinges. These receptors fire one or two impulses at the top of each wing stroke. Wingstroke frequency drops to about one-half normal in the absence of their input, but the precise timing of the stretch receptor input has no effect on the organization and timing of the central oscillator.

The sexual movements made by the male Praying Mantis during coupling enabled a physiological analysis similar to that made by Wilson on locust flight. The intact mature male makes a series of rhythmic movements of his phallomeres after he has mounted and clasped a female. These movements normally lead to copulation and are not observed on any other occasion. If the male is decapitated by a hungry female, or by the experimenter, phallomere movements begin within a few minutes and continue unabated for days irrespective of the presence of a female. If a headless male is placed on the back of a receptive female effective coupling takes place. Normally phallomere movement is inhibited by nerve centres in the brain, but the stimuli that cause disinhibition in the intact male are unknown (Roeder 1935).

It was possible to show that the pattern of motor nerve impulses responsible for these co-ordinated phallomere movements is generated by a central nervous oscillator similar to that postulated by Wilson. The intrinsic or self-excitatory nature of this motor system was revealed by the following experiment. The last abdominal ganglion of the ventral nerve cord contains the motor center supplying the phallomere muscles. All nerves capable of providing sensory input to the last abdominal ganglion were severed, leaving only the central nerve cord connecting it with the inhibitory centre in the brain. The pattern and level of motor nerve impulses were then measured in the central stump of one of the phallomere nerves. When a steady level had been established the nerve cord was severed just above the last abdominal ganglion. This operation had two effects; first, it isolated the last abdominal ganglion from the one remaining route over which it might be reached by sensory (reflex) neural pathways, and second, it disinhibited the centres concerned with phallomere movement. Within 7–15 minutes the level of motor spikes in the phallic nerves increased many-fold. A number of previously inactive motor neurons began to discharge in regular bursts recurring several times a minute—a spontaneous rhythm that continued for several hours and under that normal circumstances would have caused rhythmic phallomere movements (Roeder, Tozian, and Weiant 1960; Milburn, Weiant, and Roeder 1960; reviewed by Roeder 1967).

It must be concluded that the motor patterns responsible for locust flight and for mantis sexual movement originate in a system of nerve cells that is intrinsically unstable and tends to oscillate independently of external input. Normally these motor patterns begin when the respective oscillators are disinhibited or released after higher centres have been triggered by a specific sensory configuration. The action then continues spontaneously so long as inhibition is withheld. In the case of locust flight Wilson and Gettrup (1963) showed that proprioceptive stimulation resulting from the wing movement played no part in shaping the motor pattern generated by the central oscillator.

Positive feedback in behaviour

These experimental studies seem to permit no other interpretation than that given above. This is because the actions are sustained after all possibility of sensory or reflex feedback has been excluded. Spontaneous activity of neuronal pacemakers must be responsible for maintaining the action, and this in turn suggests a regenerative process similar to that maintaining the beat of the heart and the propagation of nerve impulses.

The evidence in these experiments was obtained by eliminating all sensory input that might have been responsible for sustaining the action. This operation entailed extensive transection of nerves and muscles, and it is perhaps surprising that the action of interest could still be identified after this severe surgical insult. At the same time, the extent of the experimental intervention may raise doubts about extending the conclusions to intact animals. However, Davis and Ayers (1972) have provided a suggestive demonstration that positive feedback operating through an opto-kinetic mechanism sustains walking in the lobster and certain other arthropods.

Davis and Ayers placed a lobster in an aquarium having a plexiglass floor, below which was a movable belt marked by transverse parallel stripes. Mirrors at 45° on the sides of the aquarium provided additional lateral visual stimulation from the stripe pattern. The lobster was restrained from making forward progress on the plexiglass floor above the stripe pattern, and the frequency and amplitude of its stepping movements were measured in relation to the velocity of the stripes moving past and below the animal.

Forward stepping activity began when the backward movement of the stripes reached a certain velocity. Stepping frequency then increased in proportion to the stripe speed. Finally the response waned due possibly to sensory and other forms of adaptation. Had the animal been free to progress forward its stepping would have caused similar relative backward movement of the stripe pattern over its ommatidial array, and this experiment suggests that this in turn would have accelerated its forward movement, constituting a system containing positive feedback. Unfortunately, Davis and Ayers do not provide direct evidence on this point. They do, however, discuss various degenerative influences that must counter the run-away tendency always potentially present in such a situation, and suggest that the sustaining effect of positive feedback on locomotion may be widespread in the animal kingdom.

This experiment, when considered beside the physiological evidence already presented, supports positive answers to the first two questions posed earlier, and leads to speculation about the third question. Are regenerative or self-exciting neural systems generally responsible for sustaining specific modes of unrestrained behaviour?

Most action patterns in the higher animals are not now accessible to neurophysiological 'dissection', so that we are limited to what may be called the 'black-box' approach in searching for an answer. In terms of the analogy of the room being maintained at an acceptible temperature by a fire tender we are in the position of an observer located outside the building who attempts to identify the element of positive feedback represented by the fire burning within. Intermittent puffs of smoke from the chimney provide the only possible clue unless the positive feedback 'runs away' and either the fire becomes extinguished or the building burns down. This means that evidence that positive feedback plays a part in shaping the behaviour of intact animals under quasi-natural conditions is inevitably more circumstantial, and conclusions drawn are correspondingly more speculative.

When an animal is presented with two or more stimulus patterns, each of which is known to release a specific response when presented alone, the commonest result is suppression of all but one of the posssible behaviour modes. The animal momentarily acts as though the rejected stimulus configurations were not present. If the conflicting stimulus patterns invite both retreat and attack and are such as would be experienced by a territorial animal on meeting an intruding conspecific at the edge of its territory there may be some degree of postural compromise, but action commonly alternates between brief but distinct episodes of fleeing and aggression, possibly interspersed with displacement behaviour. Hinde (1970) reviews other examples of alternating behaviour in conflict situations, including that shown by small birds when mobbing a predator and by both sexes during the early stages of courtship. In commenting (1970) on his own observations of the courtship of the chaffinch (Hinde 1953) he points out that '...the female may elicit attack and fleeing behaviour from the male as well as sexual behaviour: what he does depends both on the stimuli she presents and on his own internal state'. The presence of these three tendencies was identified and their relative frequency at a given stage of the courtship was measured by how often during a session the male fled, was aggressive, or showed readiness to mount.

The point here is that analysis of this complex situation was possible only because the movements made by the male at a given instant were unequivocally symptomatic of fleeing, or of agression, or of readiness to copulate. For a brief period his behaviour mechanism seemed to be totally committed to a specific behaviour mode. An instant later this mode was abruptly and distinctly supplanted by another pattern if his actions brought him too far from or too close to the female or led to copulation. It seems probable that if each of these three behaviour modes was controlled only by a homeostatic mechanism then the assorted and conflicting stimuli presented by the female would have led not to this concatenation but to a behaviour best described as a compromise or fusion—a 'resultant of vectors', as it were. This may indeed take place to some extent, but a total and simultaneous averaging of fleeing, attacking, and copulating is not only unimaginable but would have made Hinde's analysis impossible.

Another case where simultaneously presented but conflicting stimulus configurations evoke behaviour modes that alternate but fail to compromise has been analysed by J. S. Kennedy. He studied the mutually exclusive activities of flying and settling and feeding in aphids enclosed in a top-lighted arena and presented with a stimulus to settle such as a green leaf. Kennedy says (1967) that 'There is reciprocal inhibition (antagonism) between the two systems, settling and locomotion, resulting in a *singleness of action* [italics mine] which, in the young winged aphids used, shifts spectacularly under natural conditions from settling to persistent flying and back again to settling (if a suitable host plant is found) in a matter of hours or even minutes'. This pattern reminds one of Hinde's male chaffinch.

These behavioural studies suggest that an animal has at its disposal a repertory of separate and distinct behaviour modes. These modes have been likened to musical scores in the repertory of an orchestra; they can be supposed to be represented by specific neuronal networks about which we have no details. At a given moment all but one of these modes is suppressed or inhibited, that is, they are blocked from access to the motor apparatus of the animal. Two or more conflicting stimulus configurations confronting the animal are likely to fluctuate in relative intensity due to movements of both the source or sources and the receiver. When one of the stimulus configurations *momentarily* reaches a critical intensity coincident with a certain internal state of the animal there is disinhibition of an appropriate behaviour mode which then becomes manifest as *continuing* behaviour. It must be emphasized that while the event of disinhibition may be only transient the resulting behaviour mode must be sustained if it is to carry the animal into a situation where the stimulus configuration that originally triggered its disinhibition is relatively weak or altogether absent, or where the stimulus configuration is sufficiently altered in quality so that the system 'flips' to another behaviour mode which then inhibits all other modes and carries the animal into a different action pattern.

This generalization seems to apply to many aspects of behaviour. If it is considered in terms of control theory the principle does not seem accountable in terms only of negative feedback, which would predict an averaging of action in the presence of two conflicting stimulus configurations. But in fact, behaviour is generally either/or in nature. Once triggered by momentary disinhibition a behaviour mode becomes self-exciting much in the manner of a heart systole or the rising phase of an action potential. It then continues independent of the stimulus configuration that triggered or disinhibited it, at the same time suppressing for its duration other stimulus modes in the animal's repertory. Thus, the 'score is played through' or continues at least until the system becomes momentarily susceptible to a new stimulus configuration that disinhibits another self-exciting behaviour mode which in turn inhibits all other behaviour modes. Once released or disinhibited the self-reinforcing quality of a given behaviour mode may be augmented by sensory feedback resulting from the action itself. However, it is difficult to see how this sensory input could account for the onset—the 'rising phase'—of a given action pattern, and the experimental evidence cited above supports the notion that this rising phase as well as maintenance of the action may stem in certain cases from regenerative processes in the central nervous system.

3.3. Conclusion

It is a commonplace that homeostatic or self-correcting systems play a universal part in biological processes. Negative feedback might be thought of as a cloak for survival, shielding and stabilizing that highly improbable stuff, living matter, against external perturbations and steering it to optimal conditions. But homeostasis implies maintenance of the *status quo*, and this principle by itself seems unable to account for the increasing complexity and innovation that has marked evolution. I have speculated about this question as it applies to animal behaviour and its mechanisms, and suggest that an answer may lie with physiological events that are mostly contained within the homeostatic 'cloak' but which are individually capable of being self-sustaining through positive feedback. I have presented evidence that regenerative neural systems may in some cases be directly responsible for sustaining patterns of adaptive behaviour, and have examined certain patterns of action that occur under quasi-natural conditions for evidence that they may be similarly controlled.

In principle positive feedback is the antithesis of negative feedback. The first promotes a tendency to innovation, the second a tendency to stabilization. In the foregoing pages I may have conveyed the impression that these two tendencies are literally struggling for control of the behavioural mechanism. This is far from the relationship I have in mind, which is roughly expressed in the analogy of the fire-tender. Here both events, the regulated refuelling and the explosive fuel consumption, are part and parcel of the same system; both are essential to it and normally operate in a complementary fashion. Yet, there is always the possibility of innovation in the form of an extinguished fire or a general conflagration.

Many of the action patterns that make behaviour interesting to ethologists would seem to be unaccountable if the behavioural mechanism is dominated entirely by a stabilizing tendency. For instance, why are animals while awake generally 'doing something' even under apparently stable and optimal conditions? This suggests that some factor keeps them from reaching complete equilibrium with their environment. A second general observation is that animals seem capable of doing only one thing at a time when presented simultaneously with conflicting stimulus configurations. This also does not find a ready explanation in terms of homeostasis. I am aware that I have examined these questions only from a single viewpoint along a single theoretical axis, as it were, and that there are undoubtedly other axes along which the matter should be explored. But from this viewpoint the element of positive feedback demonstrably present in neural mechanisms seems to offer a possible answer.

A second point that may seem to cast doubt on the tenability of this conclusion relates to the 'explosiveness' implicit in positive feedback. How can this instability be reconciled with the over-all adaptability and stability of behaviour? Why does it not drive action to extreme limits in every case? I have pointed out that real biological events containing an element of positive feedback are always determined at their onset and limited at their termination by degenerative processes, and are thereby constrained from approaching 'zero' and 'infinity'. Thus, they become self-sustaining only between the limits set by these degenerative processes. Reference to the case of nerve-impulse genesis and propagation shows that degenerative processes determining the onset and termination of depolarization are not fixed, and that they dictate whether a neuron shall remain poised and excitable or oscillate spontaneously within these limits. Only under abnormal conditions does a neuron become permanently inexcitable or irreversibly discharged.

It seems to me that the behaviour of neurons shows significant similarities to certain aspects of animal behaviour. Both may remain poised until triggered or they may become oscillatory, depending on conditions, and the action in each is followed by relative or complete inexcitability— the usual aftermath of a regenerative event. This comparison may be discounted as being purely coincidental or notable only because both are biological events. But it also seems possible that the parallel indicates a causal relationship, and that the spontaneous nerve activity that concerns Adrian (1950) lies behind much that makes behaviour interesting to Tinbergen (1950).

References

ADRIAN, E. D. (1930). The activity of the nervous system of a caterpillar. *J. Physiol.* **72,** 132.

—— (1950). The control of nerve-cell activity. *Symp. Soc. exp. Biol.* **4,** 85–91.

ASHBY, W. R. (1956). *An introduction to cybernetics.* Wiley, New York.

BAYLISS, L. E. (1966). *Living control systems.* Freeman, San Francisco.

CANNON, W. B. (1932). *The wisdom of the body.* Norton, New York.

DAVIS, W. J. and AYERS, Jr, J. L. (1972). Locomotion: control by positive-feedback optokinetic responses. *Science, N.Y.* **177,** 183–5.

HINDE, R. A. (1953). The conflict between drives in the courtship and copulation of the chaffinch. *Behaviour* **5,** 1–31.

—— (1970). *Animal behaviour.* McGraw-Hill, New York.

KENNEDY, J. S. (1966). The balance between antagonistic induction and depression of flight activity in *Aphis fabae* Scopoli. *J. exp. Biol.* **45,** 215–28.

—— (1967). Behaviour as physiology. In *Insects and physiology* (eds J. W. L. Beament and J. E. Treherne). Oliver and Boyd, Edinburgh and London.

MILBURN, N. S., WEIANT, E. A., and ROEDER, K. D. (1960). The release of efferent nerve activity in the roach, *Periplaneta americana,* by extracts of the corpus cardiacum. *Biol. Bull.* **118,** 111–19.

MITTELSTAEDT, H. 1949. Telotaxis and Optomotorik von *Eristalis* bei Augeniversion. *Naturwissenschaft* **36.** 90–1.

—— (1964). Basic control patterns of orientational homeostasis. *Symp. Soc. exp. Biol.* **18,** 365–86.

ROEDER, K. D. (1935). An experimental analysis of the sexual behavior of the praying mantis. *Biol. Bull.* **69,** 203–220.

—— (1948). The effect of potassium and calcium on the nervous system of the cockroach, *Periplaneta americana. J. Cell comp. Physiol.* **31,** 327–38.

—— (1955). Spontaneous activity and behavior. *Scient. Mon.* **80,** 362–70.

—— (1967). *Nerve cells and insect behavior.* Harvard University Press, Cambridge, Massachusetts.

——, TOZIAN, L., and WEIANT, E. A. (1960). Endogenous activity and behaviour in the mantis and cockroach. *J. Insect Physiol.* **4,** 45–62.

STANLEY-JONES, D. K. (1960). *The kybernetics of natural systems.* Pergamon Press, New York.

TINBERGEN, N. (1950). The hierarchical organization of neural mechanisms underlying instinctive behaviour. *Symp. Soc. exp. Biol.* **4,** 305–12.

WEIS-FOGH, T. (1956). Biology and physics of locust flight. IV. Notes on sensory mechanisms in locust flight. *Phil. Trans. R. Soc. B.* **239,** 553–84.

WIENER, N. (1948). *Cybernetics.* Wiley, New York.

WILSON, D. M. (1961). The nervous control of flight in a locust. *J. exp. Biol.* **43,** 397–409.

—— GETTRUP, E. (1963). A stretch reflex controlling wing-beat frequency in grasshoppers. *J. exp. Biol.* **40,** 171–85.

4. Behaviour genetics and the study of behavioural evolution

A. MANNING

I FIRST became interested in the genetics of behaviour when I went to Oxford to join Tinbergen's group of research students. I found such work already in progress there—Margaret Bastock's (1956) now classic study on the *yellow* mutant of *Drosophila melanogaster*—although at that time behaviour geneticists could probably be counted on the fingers of two hands. The first review of the field—if it then warranted such a name—appeared that same year (Hall 1951), and it is a measure of Tinbergen's foresight that he recognized so early that the genetic basis of the behaviour was an important area for research. He saw it first as a vital link in the elucidation of behavioural evolution, but also, in an ethological atmosphere which then did not hesitate to label much behaviour 'inherited', as a basic means of studying behavioural mechanisms.

Since then behaviour genetics has, at least in numerical terms, come of age. The latest review (Broadhurst, Fulker, and Wilcock 1974), although following only 3 years after its predecessor in the same journal, cites 283 references, and there are now a society, several textbooks, and a journal devoted to the field.

There is no point in my providing here another review. Rather I want to concentrate upon evolutionary aspects of the achievements of behaviour genetics, but I will also give a personal and selective view of its progress as a whole. Every aspect of this touches upon 'the biological study of behaviour', as Tinbergen (1963) defines ethology.

Tinbergen has always stressed the importance of studying the survival value of behaviour, and some of his own work provides remarkable examples of the detailed perfection of behaviour's adaptation to the environment in which an animal lives. He has also recognized that genes which affect behaviour will often have large selective values and the changes they produce can profoundly affect the course of evolution. This can be clearly seen where genes have fairly direct effects, as when altering a food preference or affecting the selection of a mate—however, genes can also affect the way in which an animal responds to changes in its environment. Such effects may be general and non-specific—affecting learning ability or exploratory activity, for example—but they also influence the direction of evolution. If an animal acquires a new and successful behaviour trait or moves into a new habitat its behaviour may be passed on to its offspring or peers by example. Primates may take to swimming in lakes or even the sea, as has been observed in Japanese macaques. Such a change in habitat might open up new sources of food, and if so will at once impose new selective forces for relevant adaptations—webbed feet and subcutaneous fat are two obvious possibilities. Mayr (1970) discusses further examples and Hardy (1965)—in characteristically lively style—develops this theme of behavioural change preceding and leading evolution.

Such considerations mean that ethologists may see particular relevance in behaviour genetics as an adjunct to evolutionary studies. In the early days they also seemed to hope that genetics would help them when they were in the throes of the nature–nurture controversy. In fact they were rapidly outnumbered by those who came into behaviour genetics from psychology. Fuller and Thompson (1960) relate the extensive history of psychological studies on the genetic component of human intelligence, which extends back to the turn of the century. In spite of this continuing interest, the more biologically minded psychologists of the late 1950's and early 1960's still needed to use genetic analysis as a weapon to attack the deeply entrenched environmentalist schools of behavioural development.

It is often hard for someone reared in an ethological setting to realize the extent to which environmental theories held sway. Certainly some of the new American behaviour geneticists felt like crusaders and some of their papers (for example, Hirsch 1963, 1967; McClearn 1967) have a distinctly evangelistic flavour. Because of the apathetic or even hostile climate then existing, it was important to estabish unequivocal examples of genetically controlled behavioural differences between strains, races and, individuals. Much of the early literature reflects this urge to estabish a phenomenon. Inbred lines of mice were particularly favoured as material and numerous studies on differences in sexual behaviour, aggressiveness, emotionality, exploration, and avoidance conditioning appeared. Some of these studies have been of great importance for removing some of the naïvety of the early ethological hopes for behaviour genetics. I am thinking of such studies as that of McGill (1970) on the inheritance of sexual behaviour differences between strains of mice. He showed that patterns of inheritance depend on the whole genotype one is dealing with; a trait may appear dominant in one cross, but recessive in another. This conclusion was familiar enough to plant geneticists, but less so to students of animal behaviour in search of simple answers. Again, there were a number of studies which revealed subtle interactions between genes and environment during development which obviously precluded a simple nature–nurture classification—Henderson's (1970) elegant work, for example, which showed that different mouse genotypes responded quite differently to environmental enrichment. Their rank order on behavioural scales could change dramatically depending on how they were raised. Such a result was to be noted, not only by ethologists in search of 'inherited behaviour', but also by experimental psychologists who were inclined to make statements about *the* mouse's response to particular treatments.

All the early mouse work, together with its later developments is conveniently reviewed together in Lindzey and Thiessen (1970). This stage was a necessary clearing of the ground to establish the raw material for later work, even though we now find that the cataloguing of strain differences is no longer satisfying unless it is accompanied by some further genetic or behavioural analysis. Behaviour genetics is moving on, and as the field settles down from some of its early insecurities, it is possible to detect three, somewhat-overlapping approaches and aims.

The first concentrates on the nature of gene action, either by studying the effects of single gene mutations, or by looking closely at the behavioural differences between inbred populations or selected lines. Because such studies have as their main aim the elucidation of how genes programme for the development of an animal's behavioural potential, I would call this the 'direct approach' in behaviour genetics.

A second approach employs the biometrical methods originally developed to deal with such continuously varying traits. So much that we measure in behaviour shows similar continuous variation between individuals. The methods of quantitative genetics have been successfully extended to behavioural traits for which, once scaling problems have been overcome, they are admirably suited. The aims of the biometrical approach are to apportion accurately the genetic and environmental components which contribute to the variance of a trait, and to study what Mather has called its 'genetic architecture'.

The third approach in behaviour genetics is more behaviourally orientated for, in effect, it uses the manipulation of genetic variables as a way of studying behaviour. The area of study I have in mind includes that referred to as 'parsing the phenotype' by Ginsburg (1958). By selection or by hybridization we study the distortion or fragmentation and recombination of the behavioural repertoire of an animal. Genetical manipulation allows us to test how far the different elements of an associated group of behaviour patterns are causally connected or whether they appear together fortuitously. In the course of such studies a good deal of information relevant to the genetics of behaviour may be revealed, but this is not often their chief aim.

This classification of work in behaviour genetics should be compared with those of Fuller (1964) and Merrell (1965). Because they use more strictly methodological criteria, they would not recognize my third category—parsing the phenotype—as a distinct one, for it may employ any of several methods. Nevertheless I believe that the three categories outlined above do represent real differences in approach. All of them have become involved in the study of behavioural evolution, but because I shall be most concerned with the direct approach I shall discuss them here in the reverse order.

4.1. Parsing the phenotype

Perhaps the most famous of all experiments in behaviour genetics is Tryon's (1929) selective breeding of maze-bright and maze-dull rats. He achieved a very marked response to selection such that within a few generations there was no overlap between the scores of the two lines on the elevated maze used for measurements. Tryon recognized, and Searle's (1949) later analysis proved, that he was selecting only for performance specific to this particular maze. The selection experiment provided an elegant demonstration that from the complex of behavioural traits which go to make up maze learning certain factors—for example, visual versus kinaesthetic cue responsiveness—clearly separated out in the two lines of rats. Artificial selection, like the natural variety, has little concern for means. The desired phenotypes had to show high and low scores on Tryon's particular maze, and this end was achieved by an appropriate (but certainly not unique) reshuffling of the available genetic variation in the original rat population. Tryon's maze happened to be rich in kinaesthetic cues but poor in visual ones, thus genes predisposing rats to pay especial attention to one or to the other type of cue suddenly became of high selective advantage.

The descendants of Tryon's brights and dulls are still around and they are still proving useful as research material (for example, Rosenzweig, Krech and Bennett 1960; McGaugh, Jennings and Thompson 1962). Thus the first controlled selection experiment for a behavioural character already demonstrated the behavioural gains which can result from employing genotype as an independant variable. A complex phenotype can be prized apart into constituent elements.

The same kind of split was achieved by selection for fast and slow mating speed in *Drosophila* (Manning 1961, 1963). Sexual activity and general locomotor activity which are both high in natural populations proved to be independently manipulable under different regimes of selection. Goy and Jakway (1962) report a similar effect when genes from two inbred lines of guinea pigs were recombined in the F_2 generation of hybrids between them. New combinations of the different elements of sexual behaviour appeared as well as those already found associated in the parental lines. Evidence of this kind enables one to dismiss any hypothesis which implies a simple causal relationship between the cluster of associated elements in the parental lines.

Genetics can be used to great effect in this way when other kinds of test for such an hypothesis might be difficult to make. Perhaps the best example comes from the neurochemical and behavioural work of Rosenzweig and his group on the descendants of Tryon's rat strains. They developed an hypothesis which related the acetylcholinesterase content of the brain directly with learning ability (Rosenzweig et al. 1960). It is very difficult to manipulate levels of this enzyme exogenously for the long periods necessary to test this. Roderick (1960) used selection to change the levels endogenously and could demonstrate unequivocally that the direct association of enzyme level and learning ability must have been fortuitous in the original strains. Rats bred for a raised acetylcholinesterase level in their brain tissue did not show higher learning ability.

If different behavioural elements show largely independant inheritance we may also conclude that these elements do not share any controlling genes of major effect. We are thus provided with some information—albeit of a negative kind—on the relationship between genetic and behavioural 'units' in the population. This is a problem of great importance for unravelling the ontogenetic pathways between genes and behaviour. It is also of considerable interest in the study of behaviour itself. Almost any analysis of an animal's behaviour requires that we abstract from the continuous flow of ongoing behaviour, units which we can identify and count or measure in some manner. It will greatly aid such analyses if we have some confidence that our classification into units has some biological validity. Thus we may be encouraged if we can secure evidence that a unit has a particular set of genes associated with it, whilst other comparable units are affected by different loci.

Certainly we shall not find a one-gene: one-behaviour unit correspondence, but we may find single loci of major effect associated with particular behaviour traits, especially if these are relatively simple. It is difficult to equate all the different levels of organization and types of phenomenon which are involved when measuring behaviour. However, variation in some traits which are easily expressed in qualitative terms can show clear-cut inheritance. A single locus determines the time—early or late after emergence—at which a female mosquito, Aëdes atropalpus, becomes receptive to males (Gwadz 1970) and determines the host preference of the gall-fly, Procecidochares (Huettel and Bush 1972).

Gene action on these traits probably involves a relatively straightforward physiological change. However, even if we look at the motor output side of behaviour—traits which are more complex in form, answering to many of the criteria of 'fixed action patterns' in the ethological usage (Tinbergen 1951)—we may still occasionally find evidence of clear-cut genetic control of performance. A classification of the cell-cleaning behaviour patterns of honey-bees which makes intuitive sense in behavioural terms (that is, first unit—uncapping cell; second unit—removing contents) has been shown by the remarkable work of Rothenbuhler (1964) to have genetic validity also. There is one gene of major effect operating to control the performance threshold of each unit in virtually an all-or-nothing fashion.

This may not be a common situation for so much of behavioural development involves complex interactions between genes, environment and behavioural experience of various types. Sometimes—as with Rothenbuhler's honey-bees—the action of a gene or genes appears to be highly specific, although even here we cannot yet be certain that other behaviour traits are not involved. More often traits will 'share' some gene-controlled processes with others, although for complete development each trait probably depends on its own unique combination of genetic and environmental influences. For this reason we cannot expect to find any regular correspondence between genetic units and behavioural ones. This point is particularly well discussed from rather contrasting viewpoints by Thompson (1967) who deals with the genetical aspects, and Bateson (1975), who concentrates on the developmental side of the problem.

Whatever their genetic basis, the units revealed by parsing the phenotype will continue to provide information of relevance to evolution and to suggest further experiments. Goy and Jakway's (1962) demonstration that the inheritance of the male guinea-pig's behaviour falls into two units—genes affecting general activity and mounting on the one hand, and those affecting intromission frequency and ejaculation on the other— immediately suggest questions of behavioural organization. It also suggests some of the kinds of constraint which will be placed on selection for changes in sexual behaviour during the course of evolution.

4.2. The quantitative genetics approach

The genetic analysis employed in the experiments just described is usually very simple. The nature of the material will usually limit a finer-scale analysis. Further, much of the behaviour we measure—particularly in psychology laboratories—shows continuous variation. Quantitative genetics was developed to deal with just this phenomenon, and for this reason a considerable proportion of behaviour geneticists have been anxious to apply some of the sophisticated biometrical analyses of quantitative genetics to their data. Roberts (1967) provides an excellent critical survey of this approach, discussing both the methods and their behavioural validity.

There has been a great deal of work on the quantitative variation of behavioural characters in *Drosophila*, mice, rats, dogs, and humans. Most of such analyses require measuring the behaviour of a large number of individuals resulting from numerous inter-crossings between the parental strains being investigated. Selection experiments based on behavioural characters are also laborious and involve measuring dozens, if not hundreds of individuals in order to select the parents for each generation.

There can be no doubt that these requirements have influenced the choice of behavioural characters used in such studies. It has got to be something easily and quickly measured, and if measurement can be automated so much the better. The number of squares entered in an open field and faecal boli deposited there (rats), time taken to emerge from a small box, revolutions of a running wheel or speed of acquisition of a conditioned avoidance response (mice), scores of preening, walking or standing still in response to a mechanical stimulus (*Drosophila*)—these are typical measures employed.

Most ethologists would have grave doubts about the relevance of such measures to the behaviour of the animals in their natural habitats. This is not to say that behaviour genetics must always satisfy ethological criteria, but it does make arguments on the evolution of behaviour derived from some quantitative studies much more hazardous. Further, some measures, however easy to make in the laboratory, are subject to all the problems of validating behavioural units touched on earlier.

Thus we can easily quantify the acquisition of a conditioned avoidance response by mice and study its mode of inheritance. It has often been assumed that factors usually labelled as 'learning ability' or 'memory' are primarily involved. But so many other factors may contribute to the behavioural phenotype we measure. Genes affecting avoidance-conditioning may produce their effects via the sensory system (how well do the different genotypes perceive the warning stimulus?) or via some general arousal effect (are there varying tendencies to freeze when shocked?) amongst several other possibilities.

The aims of the quantitative approach to behaviour genetics are sometimes described as the investigation of the 'genetic architecture' of a trait. In a paper largely concerned with the human implications of behaviour genetics, Dobzhansky suggests that the following questions are involved in such an investigation. 'What kinds of genetic systems underlie the individual, racial and species differences in behaviour? Are they mostly monogenic or polygenic? Do they involve additive or epistatic gene actions? Are they organized in supergenes? Do they frequently involve phenomena like heterotic interactions and genetic homeostasis? At present we know next to nothing about these things. Yet they are crucial for understanding and evaluating the alleged genetic differences between social classes and races of man, and the evolutionary future of such differences.' (Dobzhansky 1972, p. 523.)

Dobzhansky thus implies that answers to his questions will yield information on the evolution of behaviour, and similar claims for such genetic analysis have frequently been made (for example, Hay 1972; Wilcock 1972). What kinds of evolutionary evidence can be gleaned from the genetic architecture?

The evidence relates to the past history of behaviour traits in relation to 'fitness', a rough definition of which would be how many descendants one genotype leaves compared with another. If genes at a number of loci affect a particular behaviour trait, then superficially we can classify their action—as by Dobzhansky above—as additive or epistatic. Additive action means that the genes act independently of each other and their effects sum algebraically. Conversely, different loci may interact with each other in a synergistic or antagonistic fashion—the phenomenon of epistasis—so that their effects do not summate in a simple fashion.

It is much easier to alter the level of a trait whose variance is being effected by genes acting additively. Selecting for high or low values will rapidly assemble more homozygous sets of high or low genes. Thus if we can study a behaviour trait in a population and measure the proportion of its variance due to additive effects, we can make an estimate of how much selection has operated on that trait in the past. This variance proportion is often referred to as the heritability of the trait. If natural selection has been consistent in the past, it should have 'used up' most of the additive variation and fixed loci which give the trait its optimum level of expression. Thus we expect traits with high survival value to have low heritabilities. This expectation is certainly borne out by studies on litter size in mice, egg production in *Drosophila*, and other traits which are obviously related to fitness.

Many of the behavioural traits that have been looked at in this way prove to have high heritabilities. For example, artificial selection has been found to produce big changes in the levels of geotaxis, general activity, and mating speed in *Drosophila* (Hirsch 1962; Connolly 1966; Manning 1961, 1963) and open-field activity in rats (Broadhurst 1969). In all these cases heritability proved to be between 30 per cent and 50 per cent or even higher. However, the simple deduction that these traits are not of importance for survival is not really justified. First, the values we observe in our laboratory apparatus may not relate well to the way such traits function in the animal's natural environment where their evolution occurred. Secondly, natural selection often favours an intermediate optimum; for example, it is easy to understand that reactivity should not be set too high or too low for maximum fitness. In such cases a good proportion of additive variance might remain in the population and if it does then changes in the direction or strength of selection can still be responded to. Some of the behavioural selection experiments have shown that natural selection opposes the changes brought about artificially. Relaxing artificial selection for mating speed or geotaxis in *Drosophila* was immediately followed by a return to more normal levels. Clearly a good deal of additive variance remained even after many generations of artificial selection.

Thus far we have considered only the way in which quantitative estimates of additive variance can give us some information relevant to the evolution of behavioural traits. More recently another aspect of the genetic architecture has been increasingly used in arguments about behavioural evolution. This is the degree of 'directional dominance' shown by genes affecting traits.

Directional dominance is taken to represent one manifestation of the end of result of natural selection. As we have seen consistent selection will tend to reduce additive variance; it will also favour the accumulation of genes with dominant effects in the direction of selection. The end result is that we may be able to detect a lot of non-additive variance between populations—particularly artificial populations such as selected lines or inbred strains whose fitness is never put to test under natural conditions. When we cross them together the levels of any behaviour trait may show dominance in direction of one of the parental values. If it does so this is *a priori* evidence that natural selection favoured this direction during the evolution of the original population from which the strains were derived. There is an elegant method—the diallel cross—for inter-crossing a number of strains which enables one to estimate from the F_1 generation alone the proportion of dominant genes in each strain. This greatly facilitates the gathering of data relevant to arguments about directional dominance (see Broadhurst 1967a, 1967b).

From such an analysis Broadhurst and Jinks (1966) could demonstrate that intermediate ambulation scores for rats placed in an open field were controlled by a high proportion of dominant genes. Dominant genes were also responsible for determining a stable level of response to this mildly stressful situation, making it resistant to environmental changes during development. From these results Broadhurst and Jinks argue that developmental stability and an ambulatory response neither too high nor too low contribute to fitness in the rat and both represent the operation of behavioural factors which have been favoured during the course of recent evolution.

These conclusions are valuable as far as they go and suggest further experiments which might identify other types of behaviour which were being affected by the operation of these selected traits. However, as Broadhurst and Jinks recognize, the inferences which can be made from the directional dominance aspect of the genetic architecture are not pariculary strong ones. Some recent claims for this type of quantitative analysis are unrealistic. Thus Wilcock (1972) in advocating its application cites examples where the conclusions, for example, that selection has favoured rapid escape by mice from a water maze, do not seem to justify the rather lengthy procedures required to obtain them. Such a result, it may be argued, gives confidence that dominance does reflect the direction of selection, which is obvious enough in this case. However, such analyses will command more attention when they yield results which go counter to common sense. Only then will they suggest further experiments, as in Rothenbuhler's (1964) example of the nest-cleaning behaviour of honey-bees. Here both the genes involved are recessive and yet there can be little doubt that under modern bee-keeping conditions the trait they control is highly advantageous.

4.3. The direct approach

As an ethologist, one of my dissatisfactions with so much of the quantitative approach is that I do not see it getting to grips with the central issue of behaviour genetics; 'how is behavioural potential itself represented in genetic terms?'

There was certainly a naïve hope in the early days that a direct approach might readily reveal the genetic basis of the fixed action patterns, so characteristic of the behaviour of the insects, birds, and fish, which attracted the bulk of ethological attention. The classic comparative studies by Lorenz, Tinbergen, and their schools had revealed that these patterns could be readily homologized between species and genera. They seem to comprise natural units of behaviour which are comparable to morphological units such as a bristle, a finger, or a metanephric kidney.

To set out to study the genetics of such units is a formidable task, and is perhaps unfair to berate behaviour geneticists for failing to concentrate on the task. In fact it brings us immediately up against the largely unsolved problem of the genetic control of development. Genetics is a powerful tool in the study of some types of developmental variation—why one has blue eyes rather than brown, for example—but it is powerless at present to say much about two eyes versus one, or how eyes themselves develop.

To pursue the morphological analogy for behavioural units used above, we know a great deal about genes affecting bristles in *Drosophila*. We know loci that affect their number, their location and pattern, their length, their shape, and so on. Through it all 'the bristle' remains a recognizable unit, and we know next to nothing about how the genes control the development of the unit itself. Clearly there are many genes that modify the operation of the genes which give the instruction 'build a bristle'. These modifiers are relatively easy to pick up and study, but the genes for bristles themselves are much harder to get at. Exactly the same is true of fixed action patterns. It is not difficult to find variation which affects their performance in a minor fashion, but the unit still appears in clearly recognizable form if it appears at all.

Wilcock (1972) suggests that fixed action patterns could yield to quantitative analyses of the type described above. He points out that, although they vary only within small limits, it should be possible by artificial selection to create strains within a species which differ in the intensity or the completeness with which a particular fixed action pattern is performed. This is undoubtedly true and we already have several examples of selection changing the frequency and intensity with which patterns are performed (for example, Wood-Gush 1960; Manning 1961). However, I doubt if quantitative analysis of the genetic basis of such differences would yield much useful information on the control and development of the patterns themselves. It would be likely to involve only those loci with modifying effects as described above. Brenner and White (1974) call these 'optimizing genes' and contrast them with 'structural genes',† which have major effects determining the development of the specific neural connections. They develop this distinction from work on the relatively simple nervous system and behaviour of the nematode *Caenorhabditis*.

In higher animals the evidence for the role of genes in the development of the potential to perform fixed action patterns is almost entirely circumstantial. Nevertheless, we must surely accept that specific patterns of connection which are built into the developing nervous system must determine the form and behavioural development of most such patterns. Hoyle (1970) provides an excellent and exciting review of the recent remarkable developments in invertebrate neurophysiology which are revealing an extraordinary constancy of structure. Such constancy must be a reflection of extreme consistency and determinacy in development. It is difficult not to believe that this type of development has a high degree of genetic control.

For example, giant cell bodies can be precisely mapped in the ganglion of the mollusc, *Tritonia*. Willows (1967) has shown that stimulation of homologous cells always leads to the same patterns of muscular activity— sometimes a highly elaborate one. Now it is possible to inject a dye or heavy metal ions through a microelectrode which has been stimulating or recording from a single unit (Pitman, Tweedle, and Cohen 1972). Using this technique Hoyle and Burrows (1973a, 1973b, 1973c) are systematically mapping motorneurons in the metathoracic ganglion of the locust, *Schistocerca gregaria*. Again the constancy of structure is remarkable; the dendritic fields of the neurons are of constant form and thus likely to form the same interconnections; the axons emerge along constant paths. Furthermore they (1973c) compare their map of the locust ganglion with the homologous ganglion of the cockroach, *Periplaneta americana*, mapped by Cohen and Jacklet (1967). There are striking similarities which suggest that

† This term is purely descriptive and implies no connection with the structural genes of the Jacob and Monod model of gene operation.

the basic ground plan of the orthopteran ganglion has remained constant since the two sub-groups diverged in the Palaeozoic.

At the moment most of behaviour genetics can only be orthogonal to this anatomical and physiological work. Nevertheless, it is marvellously encouraging to have such direct evidence that our observed constancy of fixed patterns in invertebrates corresponds to a constancy of neural fine structure. It is probable that something comparable holds for vertebrates, although on the sensory side we know that there is considerable developmental plasticity in the central nervous system of young vertebrates. Reviews by Blakemore (1973) and Bateson (1973) describe the effects of visual inputs early in the life of mammals and birds, and there is clear evidence of functional and structural modification as a result of experience. On the motor-output side some of the fixed action patterns of vertebrates are as stereotyped as any seen in insects, and their development is stable and resistant to modification. Thus Dilger (1962) found that hybrid parrots (*Agapornis*) could learn to modify movements of the head used to tuck nest material in the rump feathers, but it took them many months of constant repetition to do so and the birds never eliminated the movements completely.

Such stability must reflect an underlying neural constancy although, because the vertebrate nervous system is organized so differently from that of invertebrates and has so many more units, output stability may be achieved with more flexibility of structure.

By the time an ethologist comes to observe the performance of a fixed action pattern most of the underlying neural development is completed. If we wish to bring genetics to bear on this development we usually have to rely on close examination of the effects of genetic changes on performance. We may then try to deduce the nature of gene action on the nervous system.

There are many advantages to be gained by using single-gene mutations in such studies. At least we shall be dealing with a defined genetic unit, whose ultimate action can probably be related to synthesis of a particular protein. Even allowing for the long chain of intertwined biochemical and physiological events between the gene and any behavioural changes, it ought sometimes to be possible to gain insights from the action of a single genetic unit that would be obscured by a more complex genetic background. With a known mutation we can repeat observations easily and also place the mutant into different genetic backgrounds to study its interactions with other genes.

Having said this it must be recognized that there still remains a difficult task of unravelling causation from consequences. Some clear biochemical or physiological lesion may be detected in mutant animals and it is all too easy to label this the primary site of the gene's action. Bulfield (1972) discusses the stringent criteria that must be met if such a label is to be justified. His own work with hyperphagic mice illustrates the problem. Mice carrying the gene *obese* show much modified feeding behaviour and become grossly fat. There is a clear pattern of abnormal enzyme activity in such mice but this abnormality is not the site of the primary lesion; it is more likely to be a result of the changed behaviour rather than a cause. Bulfield has shown that exactly the same pattern of enzyme abnormality is found in *adipose* mice (this gene is not on the same chromosome as *obese*) and also in normal mice made hyperphagic by an injection of gold thioglucose.

With varying degrees of caution many behaviour geneticists have used the single-gene approach. Wilcock (1969, 1971) provides a thorough review of this literature and is highly critical of many of the studies. In the early days, it was often deemed sufficient simply to establish a behavioural difference between mutant stocks for publication to follow. Wilcock points out that some studies fell into the cause versus consequence pitfall outlined above while others were ill-controlled on the genetic side where, for example, effects were ascribed to single genes that may have been due to several closely linked loci. Further he claims that the behavioural effects reported are often gross, for example, posture and locomotion distortions in mice caused by mutants which affect the inner ear and related auditory centres, or trivial, for example that a mutant which affects temperature preference and nest-building in mice turns out to affect the thickness of the animal's fur. Both categories of result Wilcock considers to be of little behavioural interest. I agree with Wilcock's criticisms and find Thiessen's (1971) reply largely unconvincing. However, I would agree with them both that, if used well, the single-gene approach is potentially of great value.

It was a pity that Wilcock's reviews came just too early to incorporate the most important work of this type, from the groups working with Benzer (for example, Benzer 1971; Hotta and Benzer 1970, 1972), Brenner (Brenner 1973; Brenner and White 1975; Ward 1973), and Kaplan (for example, 1972; Ikeda and Kaplan 1970a, b).

These workers are studying the effects of mutations on the behaviour of two invertebrates—*Drosophila* and a nematode. In all cases they induce most of their mutants artificially by exposing normal individuals to powerful mutagens. Because mutation is always a rare event, this approach has meant developing methods for the mass screening of thousands of individuals and thus it suffers from some of the drawbacks I have already referred to when discussing the measures used by quantitative geneticists. Certainly one has to rely on relatively simple behaviour traits when large numbers of animals must be put through some behavioural test which will readily allow an abnormally responding individual to be picked out for further examination. Benzer's group have been particularly ingenious with such methods, detecting mutants with abnormal responses to light, abnormal circadian rhythms of activity, abnormal responses to mechanical shock, and so on. Brenner's nematode has such a simple behavioural repertoire that the majority of mutants are detected only by their effects on the worm's locomotion. However a large number of quite subtle abnormalities have been isolated.

The next step is to examine mutant individuals in detail, and in many cases a direct effect on the central nervous system can be detected. Thus Hotta and Benzer (1970) describe five different genes (using gene in the functional sense here, that is, cistron) which abolish the normal positive phototaxis of *Drosophila* and whose effects can be traced to their action on the photo-receptor and peripheral visual neurons. Ikeda and Kaplan (1970*a*, *b*) have been able to trace the site of action of one of their 'hyperkinetic' mutants to a group of motorneurons lying laterally in the thoracic ganglion. Hyperkinetic flies were originally detected by rhythmic movements of the legs made under ether anaesthesia, but their behaviour is abnormal in other respects in that they jump and fall over in response to mild shadow stimuli which do not affect normal flies.

The behavioural content of these studies could be regarded as trivial, but they differ from those rightly dismissed by Wilcock in important respects. First, the site and nature of the gene action is clear, and secondly a fine degree of genetic control is possible. Hence it has been possible for Benzer and Kaplan to combine a number of mutants with similar effects in the same individual to study if and how they complement each other's action. It is also possible to obtain mosaic or gynandromorph individuals which could have, for example, a mutant gene affecting the left side of the brain but not the right. Hotta and Benzer (1972) have been using mosaics to plot the site of action of mutants more precisely, both during development and in the performance of behaviour. For some behaviour traits the two halves of the brain act independently; in others one normal half can compensate for a mutant defect in the other.

This type of approach then, enables one to build up a picture of how genes combine in their action to control the development of behaviour's neural substrates. The gaps far outnumber the pieces in position and for the moment perhaps we must be content with analysing one aspect of very simple behaviour elements—the sensory basis of phototaxis, for example— but even this will give insight into the principles upon which more complex patterns are built up by evolution.

It is certain that there must be principles—perhaps they might be better called modules—from which complex patterns are constructed. The evolution of behaviour cannot have involved the gradual superimposition of more and more new gene-controlled processes. This would make no sense in genetic or behavioural terms.

This point has been particularly developed by Brenner and applied to his studies on the nematode *Caenorhabditis*. This worm has only 200 neurons in its entire nervous system and serial electronmicrography has shown that not only are the positions of cells invariant but so are all the synapses between them. Some of the branching of neurons is invariant and some varies, but this latter type is not involved in connections with other neurons. A group of mutants affect this structural constancy and a further group affect behaviour. These two groups do not overlap completely, because behaviour can be affected, by a change in the functional properties of a synapse, for example, without any obvious morphological change. Thus Brenner and White (1975) mention 14 behavioural mutants of which only 6 produce obvious morphological defects.

The nature of the morphological effects of mutation varies. One mutant affects only two neurons at the worm's anterior, but another affects every one of a series of interconnections between motorneurons and the nerve cord, extending the whole length of the body.

Mutants of this second type indicate one 'module' from which the worm's nervous system is constructed, that is, one gene which is available to build a particular type of interconnection wherever this is needed. Brenner suggests that the 'structural genes' which were mentioned earlier will convey instructions of this type, for example, 'build an inhibitory synapse' or 'build an axon growing dorsally'. With such modules the worm is able to construct its whole body with a relatively small number of genes. *Caenorhabditis* has about 2000 lethal targets for mutation in its genome. *Drosophila*, with its far greater complexity of structure and a nervous system at least an order of magnitude larger in cell number than the worm, has 5000 targets—only $2\frac{1}{2}$ times as much. Some estimates give man only 10 000. Sub-routines allowing repetitive gene action are essential and we can directly observe them in morphological terms. In the course of evolution almost all animals have grown larger and more complex by using repeated units in their body construction. Thus the annelids, arthropods, and vertebrates clearly have a gene-controlled sub-routine which instructs 'build a segment'. 'Optimizing genes' can then adjust the form of each segment to particular requirements.

These same principles must apply to the genetic control of elaborate fixed action patterns. Whether feedback control is involved or not, the motor performance of such patterns must involve a neural centre which on appropriate excitation will emit a precisely patterned output to a series of sub-centres and hence eventually to muscle groups. The analogy of a conductor with an orchestra is a useful one. The music is produced by a precise control of the timing, strength, and duration of playing by the different sections and sub-sections of the orchestra. The same routine will, within close limits, produce the same sound again and again. The particular problem in behaviour genetics which we have been discussing may be understanding how the music is originally presented to the conductor. We must suppose he conducts without a score, and we wish to know how during his musical development the score becomes fixed in his mind.

The arguments presented earlier force us to recognize that some 'structural genes' will be shared by different fixed action patterns. The modules they control are certainly not so easy to recognize as those of morphological structures. The units Brenner can detect are of very fine scale but acting on a nervous system with very few cells. The same scale of units acting in a vertebrate nervous system may well go undetected.

Perhaps we can recognize behavioural sub-routines at the sub-centre levels of fixed action patterns. Locomotor patterns obviously have a unity of their own and are accessible to numerous super-ordinated centres. Thus the instructions encoded in all top centres must include variations on 'walk', 'run', 'stand still', etc., but beyond this it is still hard to see.

We have very little information on the genetic basis of fixed action pattern 'structure'. Species which will readily hybridize are not usually well contrasted in their fixed action pattern repertoire—they probably have most structural genes in common. Accordingly most of the evidence from inter-species hybrids indicates that patterns inherit intact in the F_1 generation. It is intriguing that pattern sequences sometimes break down in hybrids even when the units themselves appear normal. This might suggest that there is more inter-species diversity in genes controlling the chaining of patterns. Such abnormal sequences were observed in the courtship displays of duck hybrids studied by Ramsay (1961). The same effect could be detected in the songs of inter-species dove hybrids (Lade and Thorpe 1964), although here it seemed that some of the units were also affected, having abnormal persistence and repetitions. Unfortunately it is usually difficult to take hybrid work further because F_2 and back-cross generations—where one might hope to pick up signs of segregation and recombination—are rarely obtainable. The most promising recent work comes from Bentley (1971) and Bentley and Hoy (1972) working on crickets. They have been able to study the various song patterns of *Teleogryllus* species in great detail and there are a large number of parameters for study—far more than are usually available from fixed action patterns. However they conclude that almost all the parameters are under polygenic control and back-cross individuals reveal smooth intermediate inheritance of even the most fine-scale characteristics. There can be no easy identification between genetic and behavioural units here then, but it remains to be seen if this is always the case.

If we can ever recognize and study structural genes which affect the development of fixed action patterns, we are likely to learn a good deal about the evolution of behaviour. We should be able to relate the major components of a pattern to the much more familiar effects of the optimizing genes whose effects are important in micro-evolution. It has proved easy to relate the effects of genetic changes on behaviour to the differences commonly observed between species. As mentioned above, most selection experiments and inter-strain comparisons have revealed a great deal of genetic variation, much of it additive, affecting behaviour in a quantitative fashion. Strains differ in the frequency with which they perform homologous patterns; mutants and selected lines often show changes of the same type—see Manning (1965) and Ewing and Manning (1967) for full reviews. This kind of difference is also familiar between closely related species and there is no difficulty in understanding how the accumulation of different optimizing genes may be involved in the be-havioural divergence accompanying speciation.

These genes will presumably act by altering response thresholds at some point or points within the chain of events between the stimuli which elicit a pattern and the motor output. Often circumstantial evidence is quite strong that the genes must act on the central nervous system—perhaps on the conductor's own output, on our former orchestral analogy. This conclusion is strengthened when we consider the other types of change which occur during speciation. Close relatives often perform homologous fixed action patterns with subtle differences of emphasis—the angle at which Drosophila hold their wings during courtship vibration varies between species, and the degree of 'throw-back' of a gull's head during the 'long-call' display varies similarly.

Such differences must be mediated by changes to the neural control mechanisms. The conductor's instructions to the orchestra have changed. He brings in the first violins some seconds earlier, and summons greater volume from the woodwind section. It is possible to envisage optimizing genes, each of small quantitative effect, accumulating to change the membrane potential of some elements in the top centre and thereby altering its output in time and pattern, for a given input. This is just one of many possible ways in which genes may act. Whatever the detailed nature of the changes they produce it is reasonable to relate their effects, as those on performance frequency alone, to small quantitative effects. The micro-evolution of behaviour thus fits into a pattern of gene action which can also account for most of the evolutionary changes occurring during the ritualiza-tion of signal movements and postures within a species.

At least, then, we understand the action of genes in the refinement of behaviour patterns for their optimum contribution to survival, even if there is little we can conclude in genetic terms about their origin and develop-ment.

At the conclusion of this brief survey we can certainly acknowledge that Tinbergen was right to encourage behaviour genetics in an ethological group. However, great our present ignorance there can be no doubt that it is concerned with some of the most fundamental problems in all biology. Progress on the central issue—how behaviour is encoded in genetic terms—may well be slow because the right material for tackling it is hard to come by. Further, there are many attractive side-issues which will continue to divert our attention.

Recently the distinguished quantitative geneticist I. M. Lerner borrowed a title from E. M. Forster when he called his address to the Behaviour Genetics Association, 'Two cheers for behaviour genetics!' For the present this is a just portion.

Acknowledgements

I am most grateful to Dr. J. Godfrey, Dr. R. C. Roberts, Dr. M. L. Thompson, and Mr. F. von Schilcher for much helpful discussion and comment on an early draft of this essay. They may well disagree still with some aspects of it, but they helped me greatly to clarify my own thinking.

References

BASTOCK, M. (1956). A gene mutation which changes a behavior pattern. *Evolution* **10**, 421–39.

BATESON, P. P. G. (1973). Internal influences on early learning in birds. In *Constraints on learning* (eds R. A. Hinde and J. Stevenson-Hinde), pp. 101–16. Academic Press, London and New York.

—— (1975). Specificity and the origins of behavior. *Adv. Studies Behav.* **6**. (In press.)

BENTLEY, D. R. (1971). Genetic control of an insect neuronal network. *Science N.Y.* **174**, 1139–41.

—— and HOY, R. R. (1972). Genetic control of the neuronal network generating cricket (*Teleogryllus gryllus*) song patterns. *Anim. Behav.* **20**, 478–92.

BENZER, S. (1971). From the gene to behavior. *J. Am. med. Assoc.* **218**, 1015–22.

BLAKEMORE, C. (1973). Environmental constraints on development in the visual system. In *Constraints on learning* (eds R. A. Hinde and J. Stevenson-Hinde) pp. 51–74. Academic Press, London, New York.

BRENNER, S. (1973). The genetics of behaviour. *Brit. med. Bull.* **29**, 269–71.

—— and WHITE, J. G. (1975). Genetic specification and the nematode nervous system. In *'Simple' nervous systems* (ed. P. N. R. Usherwood). Edward Arnold, London. (In press.)

BROADHURST, P. L. (1967a). An introduction to the diallel cross. In *Behavior-genetic analysis* (ed. J. Hirsch), pp. 287–304. McGraw-Hill, New York.

—— (1967b). The biometrical anlaysis of behavioural inheritance. *Scient. Prog.* **55**, 123–9.

—— (1969). Psychogenetics of emotionality in the rat. *Ann. N.Y. Acad. Sci.* **159**, 806–24.

—— JINKS, J. L. (1966). Stability and change in the inheritance of behaviour in rats: a further analysis of statistics from a diallel cross. *Proc. R. Soc.* B165, 450–72.

—— FULKER, D. W., and WILCOCK, J. (1974). Behavioral genetics. *A. Rev. Psychol.* **25**, 389–415.

BULFIELD, G. (1972). Genetic control of metabolism: enzyme studies of the *obese* and *adipose* mutants in the mouse. *Genet. Res.* **20**, 51–64.

COHEN, M. J. and JACKLETT, J. W. (1967). The functional organization of motor neurons in an insect ganglion. *Phil. Trans. R. Soc.* B252, 561–72.

CONNOLLY, K. (1966). Locomotor activity in *Drosophila*—II. Selection for active and inactive strains. *Anim. Behav.* 14, 444–9.

DILGER, W. C. (1962). The behavior of lovebirds. *Scient. Am.* 206, 88–98.

DOBZHANSKY, T. (1972). Genetics and the diversity of behavior. *Am. Psychol.* 27, 523–530.

EWING, A. W. and MANNING, A. (1967). The evolution and genetics of insect behaviour. *A. Rev. Entomol.* 12, 471–94.

FULLER, J. L. (1964). Physiological and popuation aspects of behavior genetics. *Am. Zool.* 4, 101–9.

—— THOMPSON, W. R. (1960). *Behavior genetics.* Wiley, New York.

GINSBURG, B. E. (1958). Genetics as a tool in the study of behavior. *Perspect. Biol. Med.* 1, 397–424.

GOY, R. W. and JAKWAY, J. S. (1962). Role of inheritance in determination of sexual behavior patterns. In *Roots of behavior* (ed. E. L. Bliss), pp. 96–112. Harper, New York.

GWADZ, R. W. (1970). Monofactorial inheritance of early sexual receptivity in the mosquito, *Aëdes atropalpus. Anim. Behav.* 18, 358–61.

HALL, C. S. (1951). The genetics of behavior. In *Handbook of experimental psychology* (ed. S. S. Stevens), pp. 304–29. Wiley, New York.

HARDY, A. (1965). *The living stream.* Collins, London.

HAY, D. A. (1972). Genetic and maternal determinants of the activity and preening behaviour of *Drosophila melanogaster* reared in different environments. *Heredity* 28, 311–36.

HENDERSON, N. D. (1970). Genetic influences on the behavior of mice can be obscured by laboratory rearing. *J. comp. Physiol. Psychol.* 72, 505–11.

HIRSCH, J. (1962). Individual differences in behavior and their genetic basis. In *Roots of behavior* (ed. E. L. Bliss), pp. 3–23. Harper, New York.

—— (1963). Behavior genetics and individuality understood. *Science, N.Y.* 142, 1436–442.

—— (1967). Behavior-genetic, or "experimental", analysis: the challenge of science versus the lure of technology. *Am. Psychol.* 22, 118–30.

HOTTA, Y. and BENZER, S. (1970). Genetic dissection of the *Drosophila* nervous system by means of mosaics. *Proc. natn. Acad. Sci., U.S.A.* 67, 1156–63.

—— —— (1972). Mapping of behaviour in *Drosophila* mosaics. *Nature, Lond.* 240, 527–35.

HOYLE, G. (1970). Cellular mechanisms underlying behavior—neuroethology. *Adv. Ins. Physiol.* 7, 349–444.

—— BURROWS, M. (1973a). Neural mechanisms underlying behaviour in the locust *Schistocerca gregaria.* I. Physiology of identified motorneurons in the metathoracic ganglion. *J. Neurobiol.* 4, 3–41.

—— —— (1973b). Neural mechanisms underlying behavior in the locust. *Schistocerca gregaria.* II Integrative activity in metathoracic neurons. *J. Neurobiol.* 4, 43–67.

—— —— (1973c). Neural mechanisms underlying behavior in the locust *Schistocerca gregaria.* III Topography of limb motorneurons in the metathoracic ganglion. *J. Neurobiol.* 4, 167–86.

HUETTEL, M. D. and BUSH, G. L. (1972). The genetics of host selection and its bearing on sympatric speciation in *Procecidochares* (Diptera: Tephritidae). *Entomol. exp. appl.* 15, 465–480.

IKEDA, K. and KAPLAN, W. D. (1970a). Patterned neural activity of a mutant *Drosophila melanogaster*. *Proc. natn. Acad. Sci., U.S.A.* **66,** 765–72.

—— —— (1970b). Unilaterally patterned neural activity of gynandromorphs, mosaic for a neurological mutant of *Drosophila melanogaster*. *Proc. natn. Acad. Sci. U.S.A.* **67,** 1480–87.

KAPLAN, W. D. (1972). Genetic and behavioral studies of *Drosophila* neurological mutants. In *The biology of behavior* (ed. J. A. Kiger), pp. 133–57. Oregon State University Press.

LADE, B. I. and THORPE, W. H. (1964). Dove songs as innately coded patterns of specific behaviour. *Nature, Lond.* **202,** 366–8.

LINDZEY, G. and THIESSEN, D. D. (eds.) (1970). *Contributions to behavior-genetic analysis: the mouse as a prototype*. Appelton–Century–Crofts, New York.

McCLEARN, G. E. (1967). Genes, generality, and behavior research. In *Behavior-genetic analysis* (ed. J. Hirsch), pp. 307–21. McGraw-Hill, New York.

McGAUGH, J. L., JENNINGS, R. D., and THOMPSON, C. W. (1962). Effects of distribution of practice on the maze learning of descendants of the Tryon maze bright and maze dull strains. *Psychol. Rep.* **9,** 147–50.

McGILL, T. E. (1970). Genetic analysis of male sexual behavior. In *Contributions to behavior-genetic analysis: the mouse as a prototype* (eds G. Lindzey and D. D. Thiessen), pp. 57–88. Appleton–Century–Crofts, New York.

MANNING, A. (1961). The effects of artificial selection for mating speed in *Drosophila melanogaster*. *Anim. Behav.* **9,** 82–92.

—— (1963). Selection for mating speed in *Drosophila melanogaster* based on the behaviour of one sex. *Anim. Behav.* **11,** 116–20.

—— (1965). Drosophila and the evolution of behaviour. *Viewpoints in biology*, Vol. 4, pp. 125–69. Butterworth, London.

MANOSEVITZ, M., LINDZEY, G., and THIESSEN, D. D. (eds) (1969). *Behavioral genetics: methods and research*. Appleton–Century–Crofts, New York.

MAYR, E. (1970). Evolution und Verhalten. *Verh. d. zool. Ges.* **64,** 322–36.

MERRELL, D. J. (1965). Methodology in behavior genetics. *J. Hered.* **56,** 263–65.

PITMAN, R. M., TWEEDLE, C. D., and COHEN, M. J. (1972). Branching of central neurons: intracellular cobalt injection for light and electron microscopy. *Science, N.Y.* **176,** 412–14.

RAMSAY, A. O. (1961). Behaviour of some hybrids in the mallard group. *Anim. Behav.* **9,** 104–5.

ROBERTS, R. C. (1967). Some evolutionary implications of behavior. *Can. J. Genet. Cytol.* **9,** 419–35.

RODERICK, T. H. (1960). Selection for cholinesterase activity in the cerebral cortex of the rat. *Genetics* **45,** 1123–40.

ROSENZWEIG, M. R., KRECH, D., and BENNETT, E. L. (1960). A search for relations between brain chemistry and behavior. *Psychol. Bull.* **57,** 476–92.

ROTHENBUHLER, W. C. (1964). Behavior genetics of nest cleaning in honey-bees. IV. Responses of F_1 and backcross generations to disease-killed brood. *Am. Zool.* **4,** 111–23.

SEARLE, L. V. (1949). The organization of hereditory maze-brightness and maze-dullness. *Genet. psychol. Monogr.* **39,** 279–325.

THIESSEN, D. D. (1971). Reply to Wilcock on gene action and behavior. *Psychol. Bull.* **75,** 103–5.

THOMPSON, W. R. (1967). Some problems in the genetic study of personality and intelligence. In *Behavior-genetic analysis* (ed. J. Hirsch), pp. 344–65. McGraw-Hill, New York.

TINBERGEN, N. (1951). *The study of instinct.* Clarendon Press, Oxford.
—— (1963). On aims and methods of ethology. *Z. Tierpsychol.* **20,** 410–33.
TRYON, R. C. (1929). The genetics of learning ability in rats. *Univ. Calif. Publ. Psychol.* **4,** 71–89.
WARD, S. (1973). Chemotaxis by the nematode *Caenorhabditis elegans:* identification of attractants and analysis of the response by mutants. *Proc. natn. Acad. Sci., U.S.A.* **70,** 817–21.
WILCOCK, J. (1969). Gene action and behavior: an evaluation of major gene pleiotropism. *Psychol. Bull.* **72,** 1–29.
—— (1971). Gene action and behavior: a clarification. *Psychol. Bull.* **75,** 106–8.
—— (1972). Comparative psychology lives on under an assumed name— psychogenetics! *Am. Psychol.* 531–8.
WILLOWS, A. O. D. (1967). Behavioral acts elicited by stimulation of single, identifiable brain cells. *Science, N.Y.* **157,** 570–4.
WOOD-GUSH, D. G. M. (1960). A study of sex drive of two strains of cockerel through three generations. *Anim. Behav.* **8,** 43–53.

5. Factors affecting the morphology and behaviour of guppies in Trinidad

N. R. LILEY AND B. H. SEGHERS

5.1. Introduction

A PROBLEM of major interest to ethologists is that concerning the survival value of behaviour. Tinbergen (1963, 1965) has argued that apart from its intrinsic interest the study of survival value is essential to an assessment of the part played by natural selection in the evolution of behaviour. Studies of survival value can demonstrate selection pressures operating at the present day and 'suggest (though no more than that) the selection pressures which have in the past molded the species to what it is now' (Tinbergen 1965, p. 523).

Natural populations of the guppy, *Poecilia reticulata*, in Trinidad appear to provide an unusual opportunity to investigate in detail the 'response' of one species to the selection pressures imposed by its environment. C. P. Haskins, E. F. Haskins, McLaughlin, and Hewitt (1961) noted that many of the populations of guppies occupying the streams and rivers of Trinidad are partially or completely isolated from each other. These workers were able to detect differences in the frequencies of genes affecting colour patterns when they compared populations from different streams. In addition their observations suggested that there may be behavioural differences among the populations. Haskins *et al.* (1961) attempted to account for these interpopulation differences largely on the basis of the presence or absence of certain predators.

In 1967 we visited Trinidad and collected extensively in the Northern Range. In 1969 and 1973 one of us (B.H.S.) visited the island again to carry out a more detailed study in the same region. Our objective was to confirm the observations of Haskins *et al.* regarding morphological differences and to look more carefully at any behavioural differences we might detect. We also intended to examine further the relationship between the guppies and their predators. In addition we considered it important to look at other aspects of the biological and physical environment for features which may be related to morphological and behavioural differences among the populations.

It soon became apparent that guppies, particularly the males, collected at different sites differ markedly in a number of characteristics. Males from springs and the upper reaches of mountain streams are larger and (to the human eye) more brightly coloured than those from lowland rivers. A number of behavioural differences were noted: in some areas guppies remain in small schools near the water's edge; in some situations the fish are dispersed over the stream bed; fish from different populations differ in their avoidance responses. In addition, in certain localities there was a marked imbalance in the sex-ratio, with females outnumbering males by as much as 4:1.

These differences, other than that in sex-ratio, persisted through several generations in the laboratory, indicating that the differences in behaviour and morphology have a genetic basis and are presumably the product of natural selection.

These observations provided the basis for a comparative and experimental approach. We believe that such an approach at the level of intra-specific variation provides a powerful tool in the investigation of the nature of micro-evolutionary changes in behaviour and morphology. Briefly, the approach we have adopted is to attempt to correlate the variation in guppies with the physical and biotic variables in the environment. Correlations of this type give rise to hypotheses concerning the nature of the observed relationships. We have attempted to test these hypotheses directly by experiments in the field and in the laboratory.

The main thrust of our work has so far been directed at an understanding of intra-specific variation in relation to selection thought to be due to predation. There has been rapidly increasing interest in anti-predator adaptations, by Blest (1957), Croze (1970), Hagen and Gilbertson (1972), Hoogland, Morris, and Tinbergen (1957), Kruuk (1964), McPhail (1969), Moodie (1972a, 1972b), and Tinbergen, Brockhuysen, Houghton, Kruuk, and Szulc (1962), and many others.

However, although most of our work has been directed to understanding the role of predation, we are well aware that for a full assessment of the evolutionary significance of natural variation in an animal such as the guppy, it is necessary to consider many other features of the environment which result in selection pressures which compete and interact with those due to predation. Indeed in this, and perhaps most other situations, it is unnecessary and perhaps misleading to search for *the* biological function of a piece of behaviour or morphology. It is unlikely that there is a single function or consequence through which natural selection acts to maintain a certain feature of a species or population. Rather, numerous consequences of a behaviour contribute to determining its survival value.

We believe that apart from contributing to our understanding of evolutionary mechanisms this type of study has interest at a more practical level. For example, from the point of view of fisheries management it is important to know something of the impact of predation and other selection pressures upon the subsequent behaviour of a prey population. Raleigh and Chapman (1971) referring to trout-fry migrations stated '...failure to match the innate behaviour of the donor population to the requirements of the recipient environment has led to failure of many fish transplant efforts in the past'. As Calaprice (1969) has warned, the continuous stocking of ill-adapted fish to a natural population can decrease the mean population fitness and possibly culminate in local extinctions.

5.2. Description of the study area

Collection and observation were confined mainly to the north-west quarter of the island of Trinidad, West Indies (see map, Fig. 5.1). Extending along the northern margin of the island is a chain of tree-covered mountains, the Northern Range. The average height of the main chain is 600–700 metres, with several peaks of a little over 1000 metres. The north-facing slopes are dissected by a series of short (<16 kilometres) roughly parallel streams which empty directly into the ocean. Thus there are no freshwater connections between them. A series of similar but longer parallel streams flow off the south-facing slopes. The majority of these are tributaries of two major rivers the Caroni and Oropuche.

Most of the streams flowing off the Northern Range contain populations of guppies as well as other species of fish. Because of the deeply dissected nature of the Northern Range and because several of the streams flow directly into the sea many of the stream populations are virtually isolated from each other. This is particularly true of the north-flowing streams— although it is possible that occasional movements of guppies may occur particularly at times of high water in the Orinoco which results in a movement of water of low salinity along the northern shore of Trinidad and also into the Gulf of Paria (Boeseman 1960). Many of the south-flowing streams are connected through the Caroni or Oropuche but, because of the distances involved and the tendency of guppies to remain in one place (Haskins *et al.* 1961), the stream populations away from the connecting Caroni are probably genetically isolated to a considerable extent.

On several streams there are waterfalls which act as effective barriers to upward migration of fish, serving to increase further the degree of isolation of those populations above the waterfalls. The Paria river is blocked by a major waterfall, 10–14 metres high, approximately 2 kilometres from its mouth. The Aripo river passes over a series of falls 3–5 metres high at a point midway along its length. The Blue Basin stream cascades over a series of falls each 2–4 metres high, about 8 kilometres from its mouth.

Two rivers, the Guayamare and Caparo, to the south of the Caroni were

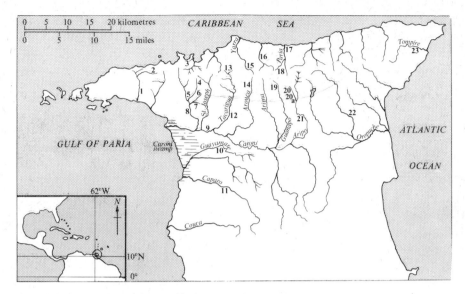

FIG. 5.1. Map of the northern half of the island of Trinidad, West Indies, showing the major river systems. Collection sites are indicated by numbers corresponding to the key below. Inset map indicates the position of the island just off the north-east coast of Venezuela.

Map number	Name of stream	Standard abbreviation
1	Sierra Leone Road	SLR
2	Blue Basin	BB
3	Maracas Village	MV
4	Upper Curumpalo	UCur
5	Lower Curumpalo	LCur
6	Grande Curucaye	GCur
7	Petite Curucaye	PCur
8	Santa Cruz	SC
9	Caroni	Car
10	Guayamare	Guay
11	Caparo	Cap
12	Lower Tacarigua	LTac
13	Upper Tacarigua	UTac
14	Upper Arouca	UArouc
15	Yarra	Yar
16	Marianne	Mar
17	Lower Paria	LPar
18	Upper Paria	Par
19	Upper Guanapo	UGuan
20	Upper Aripo (Naranjo)	UA(N)
20a	Upper Aripo (Crossing)	UA(X)
21	Lower Aripo	LA
22	Oropuche	Oro
23	Tompire Tributary	TT

studied and compared with the streams of the mountain range. Both rivers flow for most of their lengths through low-lying agricultural land devoted chiefly to rice and sugar cane, and empty into the southern half of the Caroni swamp. The Guayamare receives much of its water from the Caroni through a man-made channel some 16 kilometres from the Caroni swamp.

For convenience the streams and rivers have been grouped into four categories: springs, headstreams, midstreams, and lowland rivers. The principal features of each of these categories of collecting site are summarized in Table 5.1. Seghers (1973) provides detailed information on each of the collection sites.

The first group includes small streams close to their source. These are very small bodies of water ranging from a moderately fast-flowing riffled stream, as in the case of the Upper Curumpalo, to a very slow-moving trickle, such as at the Sierra Leone Road site. The Upper Curumpalo and

TABLE 5.1

Classification and physical features of streams in the Northern Range
(July–August 1967; March–June 1969)

Stream type	Stream §	Width (m)	Depth (m)	Velocity (m/s)	Volume of Flow (m³/s)	Temperature (°C)	Shade†	Turbidity‡
Springs	PCur GCur UCur TT SLR MV	0·50–1·0	0·05–0·15	<0·29	<0·0128	24·3–26·2	3–4	0
Headstream	UA(N) UA(X) UTac UArouc UGuan BB Yar Par	1·20–5·0	0·06–0·15	0·32–0·67	0·028–0·267	24·6–27·4	2–3	0–1
Midstream	LA LTac SC Mar Oro LPar LCur	3·0–8·0	0·13–0·20	0·42–1·18	0·150–1·129	24·3–30·0	2	1
Lowland River	Guay Cap Car	2·0–25·0	1·5–3·0	0·33–0·40	0·563–22·50	26·9–29·1	0–1	2

† 0—no shade; 1—small amount of shade restricted mainly to streambank; 2—medium shade (50 per cent cover) 3—medium to dense (75 per cent cover with few exposed parts); 4—very dense cover with virtually complete shading.

‡ 0—always clear; 1—turbid only after heavy rains; 2—turbid throughout year.

§ For a list of stream abbreviations, see Fig. 5.1.

Petite Curucaye rise at over 400 metres altitude, whereas the Sierra Leone Road and Maracas Village streams appear within 130 metres above sea-level.

The headstream and mid-section of the streams flowing off the Northern Range are fast flowing and, in many areas, especially high up, typical tumbling mountain streams. Most of the streams are broken by a series of riffles and pools. The speed of flow over typical sections ranges from 0·32 metres per second to 1·18 metres per second. The lowland rivers are deep and flow smoothly from 0·33 metres per second to 0·40 metres per second. In all sections of the streams and rivers there are slow-moving areas, backwaters, and regions along the edges where fish may be found in still water.

Except for a relatively short period after heavy rain the springs, headstreams, and midsections remain clear and only lightly stained. In contrast the lowland rivers are always very turbid with suspended clay and other material—the water remained an opaque sandy brown throughout the time we spent on the island during both wet and dry seasons.

The substrate in the springs, headstreams, and mid-sections consists of sand, gravel, or bare rock. There is no submerged aquatic vegetation. In pools and backwater areas cover may be provided by leaf litter and fallen branches. During periods of high water the streams may overflow their banks and additional cover is provided by partially submerged vegetation. The lowland rivers are largely devoid of aquatic vegetation in the areas studied. There the substrate is composed of soft mud.

Most streams studied show a marked temperature gradient, almost 6–7 °C, between headwaters and lowland regions (Table 5.1). This gradient persisted through both dry and wet seasons, although of course fluctuations occurred depending upon weather conditions.

The degree of shading from direct sunlight varies considerably along the length of most streams. In general the very smallest streams such as the Curumpalo are almost completely shaded by low as well as high vegetation. For much of their length the headstream and mid-sections flow through citrus, cocoa, or banana plantations and are only partially shaded. In the lowland regions the river passes through open country with little or no cover except close to the banks in some regions.

5.3. Faunal association with particular reference to potential predators of the guppy.

The species of fish collected or observed by us are listed in Table 5.2. There are striking differences in the faunas at the different classes of collection site: only two species, *P. reticulata* and *Rivulus hartii*, commonly occur in springs and isolated headstreams; in headstreams not isolated by waterfall barriers six or seven other species may be present, whereas in the

TABLE 5.2

The families and species of fish collected at 23 sites in the Northern Range region

Family and species	Springs						Isolated				Headstreams					Midstreams					Lowland rivers		
	UCur	PCur	SLR	MV	TT	GCur	UA	BB	Par	Yar	UArouc	UGuan	UTac	LPar	Mar	LCur	SC	Oro	LTac	LA	Cap	Car	Guay
F. Poeciliidae																							
Poecilia reticulata	+	+	+	+	+	+	+	+	+	+	+	+	+	+	+	+	+	+	+	+	+	+	+
F. Cyprinodontidae																							
Rivulus hartii	+	+	+	+	+	+	+	+	+	+	+		+			+							
F. Cichlidae																							
Aequidens pulcher											+	+	+	+	+	+	+	+	+	+			
Cichlasoma bimaculatum																					+	++	+
Crenicichla alta											+	+	+	+	+	+	+	+	+	+	++	‡	+
Tilapia mossambica																			‡	‡		+	+
F. Characidae																							
Astyanax bimaculatus											+	+	+	+	+	+	+	+	+	+	+	+	+
Hemibrycon sp.												+					+		+				
Curimata argentea																				+		+	+

Roboides dayi
Hoplias malabaricus
Corynopoma riisei
Hemigrammus unilineatus
Pristella riddlei or
Aphyocharax axelrodi
F. Loricariidae
Hypostomus robinii
Ancistrus cirrhosus
F. Pimelodidae
Rhamdia sp.
F. Synbranchidae
Synbranchus marmoratus
F. Callichthyidae
Corydoras aeneus
F. Mugilidae
Agonostomus monticola (?)
F. Gobiidae
Sicydium sp. (?)
Unidentified goby

† Species not seen or collected in this study but presence determined through personal communication with local residents.
‡ Species present in river according to Boeseman (1960).
For a list of stream abbreviations, see Fig. 5.1

175884

midstream and lowland sections as many as sixteen species have been definitely identified.

Rivulus hartii is very widely distributed although not marked as present in midstream and lowland rivers in Table 5.2. In these regions *Rivulus* is not found in the main stream but may occur close by in small pools or tributary streams.

The list of species in Table 5.2 is by no means complete. Boeseman (1960, 1964) summarizes information on the fish fauna of Trinidad and indicates a number of species, not recorded by us, which are almost certainly present in the study area, particularly in the midstream and lowland sections. These include freshwater species such as *Gymnotus carapo*, *Hoplerythrinus unitaeniatus*, *Callichthys* sp., *Hoplosternum* sp., and *Polycentrus* sp., as well as a number of euryhaline species which invade estuaries and may occur in completely fresh water some way from the estuarine regions (*Arius* sp., several species of mugilid, eleotrid, and centropomid).

Several species are considered to be important predators of guppies and other fish (Plate 5.1). *Hoplias malabaricus* and *Crenicichla alta* are specialized piscivores, although *Crenicichla* feeds to some extent on insects

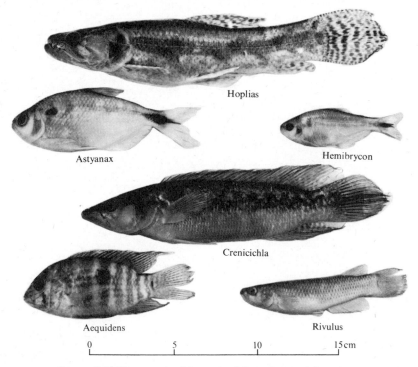

PLATE 5.1. Photograph of the major fish predators of the guppy.

and crustaceans. Gut content analysis of *Crenicichla* caught in the study area confirms that this species does prey upon guppies. It was not possible to identify remains of guppies in *Hoplias* but the readiness with which *Hoplias* captures guppies in the laboratory and the fact that the two species occur sympatrically make it reasonable to assume that this acts as an important predator of the guppy.

Another species we find to be an important predator is the cyprinodontid, *Rivulus hartii*. Haskins *et al.* (1961) considered this to be a 'less severe' predator of the guppy compared with *Crenicichla*, *Aequidens*, and *Astyanax*. However, as we found that 10 per cent of all *Rivulus* sampled had guppy remains in their stomachs there seems little doubt that this species should be regarded as a serious predator.

Several other species are probably facultative predators of the guppy. Chief among these are *Astyanax bimaculatus*, *Hemibrycon* sp., and *Aequidens pulcher*. These species will all readily take small guppies in the laboratory, and both *Astyanax* and *Hemibrycon* in field collections were found to have consumed guppies.

Almost certainly a number of other species prey upon guppies, but there is not sufficient information to determine which should be regarded as 'serious' predators. However, the point we wish to emphasize is that there is a marked contrast between the situation in springs and isolated headstreams on the one hand, and the unisolated headstreams, midstreams, and lowland rivers on the other. In the former situation the principal aquatic predator of the guppy is *Rivulus*, in the latter there are several species of potential predators.

Several species of kingfisher have been reported as occurring in the Northern Range (see for example, Herklots 1961; Beebe 1952; Belcher and Smooker 1936; Junge and Mees 1961). We observed Green Kingfishers (*Chloroceryle americana*) at some of the collection sites. No details of feeding behaviour or diet were given for any of the published reports, so the potential impact on guppy populations of kingfisher predation remains conjectural.

The fish-eating bat, *Noctilio leporinus*, occurs in Trinidad. Although it has been demonstrated experimentally that this species is capable of capturing small fish, including poeciliids (Bloedel 1955), there is no information as to the importance of this predator in our study area.

5.4. Summary of the variation in guppy populations in relation to the physical and biological environment

The following generalizations relate differences in guppy populations to the physical and biotic features of the environment. While recognizing many exceptions we feel that these observations are sufficiently reliable to

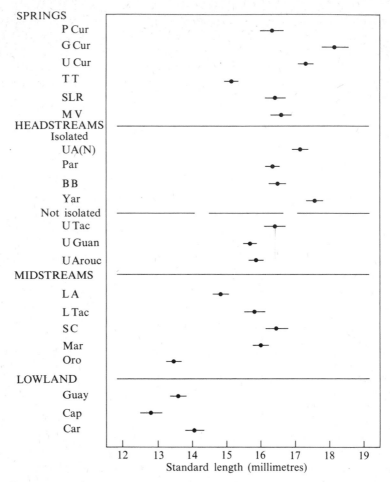

FIG. 5.2. Mean standard lengths (±95 per cent confidence limits) of male guppies taken from 21 collection sites in Trinidad. Sample sizes (40–60) represent the whole collection or a random sample from a larger collection. Names of collection sites abbreviated as in Fig. 5.1.

serve as the basis for developing hypotheses about the nature of the relationships, and subsequent experimental tests.

Guppies taken from springs and isolated headstreams tend to be larger (particularly the males), the males more brightly coloured (to the human eye), and females may outnumber males by as much as 4:1. Fish tend to be dispersed over the stream bed and show relatively poorly developed schooling behaviour and avoidance responses. These features of the *Poecilia* population appear to be associated with clear, fast-running water, relatively low temperatures, and a virtual absence of aquatic predators other than *Rivulus* which may be quite abundant.

In contrast, in the midstream and lowland rivers guppies tend to be small and the males are less brightly coloured and as numerous as females. Guppies show vigorous avoidance responses and may occur in small schools along the edge of a stream or river. In this case guppies are associated with conditions of relatively slow-moving turbid water (lowland rivers), high temperatures, and an absence of shade, Numerous other fish are present including several species known to be predators: *Crenicichla, Hoplias, Astyanax, Aequidens,* and *Hemibrycon.*

One aspect of these relationships will be documented and examined in detail: that of size in relation to predation and temperature. Variation in anti-predator behaviour is considered by Seghers (1973, 1974*a*, 1974*b*). Ballin (1973) has examined differences in sexual and agonistic behaviour in several stocks.

Figs 5.2 and 5.3 indicate the mean lengths of samples of guppies taken

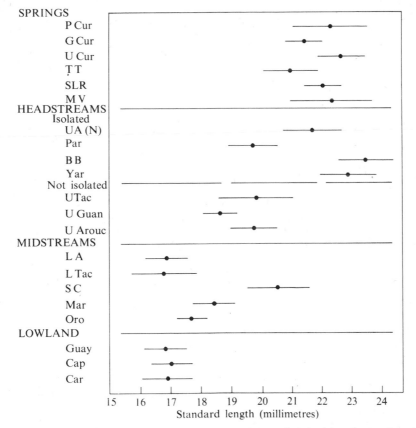

FIG. 5.3. Mean standard lengths (±95 per cent confidence limits) of female guppies taken from 21 collection sites in Trinidad. Samples and abbreviations same as Fig. 5.2.

from different stream populations in 1967 or 1969. The data are based upon standard lengths (S.L.) (tip of snout to base of caudal fin rays) taken from collections preserved in formalin. In both years total length (T.L.) (tip of lower jaw to tip of caudal ray) measurements were taken from fresh collections of many of the populations. Comparison of males of 13 populations for which total length data are available for both years reveals that there is a significant positive correlation between the two sets of measurements ($r_s = +0.90$, $p < 0.001$, using Spearman rank correlation coefficient) indicating that the differences between the populations persisted over that period and are not likely to be transitory phenomena.

In general males and females taken from springs and headstreams tend to be larger than fish from midstreams and lowland rivers. The mean lengths of all but three spring and headstream populations fall above the median value for all populations compared with only one of the midstream populations ($p < 0.01$, using median test). If the populations are divided on the basis of those in 'predator rich' environments (lowland, midstreams, and some headstreams) and those with only *Rivulus* as the chief predator, only one of the latter group falls below the median and only two of the predator-rich streams are above the median ($p < 0.005$, using median test).

There is a significant positive correlation between male and female mean lengths for the different populations ($r_s = 0.81$, $p < 0.01$, using Spearman rank correlation coefficient), however, there are several situations where male length does not appear to be as closely related to female length as might be expected, for example, the females of the three lowland populations are as large as those of the midstreams, Lower Aripo and Lower Tacarigua, whereas the males differ considerably. Similarly Oropuche males are relatively small, whereas the females are comparable in size to those of the Lower Aripo and Lower Tacarigua populations. Upper Blue Basin females are the largest of all populations, whereas the males are only sixth largest. These apparent anomalies in the correlation between the adult sizes of the two sexes suggests that the growth and adult size of the two sexes may be to some extent independent of each other, perhaps reflecting the fact that the two sexes are exposed to different selection pressures or that they respond differently to environmental factors such as food or temperature conditions.

It is well established that temperature has a considerable effect upon metabolism and growth of fish (see Brown 1957; Fry 1971; Ray 1960; Weatherley 1972). A number of studies indicate that each species is adapted to a particular temperature range within which it shows maximum growth: above and below this range growth decreases. Gibson and Hirst (1955) obtained maximum growth up to maturity in guppies at 23 °C and 25 °C with decreased growth at 20 °C and 30 °C. However, Ray (1960) working with another poeciliid, *Xiphophorus maculatus*, found no consistent size–temperature relationship in *Xiphophorus* raised at 20 °C, 25 °C, and 29 °C. Brett, Shelbourn, and Shoop (1969) obtained maximum growth in fingerling salmon at 15 °C; Kinne (1960) found 25 °C and 30 °C to be optimum for growth and adult size in *Cyprinodon macularis*. *Cynolebias adloffi* raised for 5 months at 22 °C and then maintained at either 16 °C or 22 °C lived longer and grew larger at the lower temperature (Liu and Walford 1966).

Our field data for 19 collection sites show significant negative correlations between stream temperatures and standard lengths in both males and females (males: $r_s = -0.5$, $p < 0.05$, females: $r_s = -0.66$, $p < 0.01$, using Spearman rank correlation coefficient). This raises the question, are the differences in size simply a phenotypic response to the different temperature regimes, or is there any evidence that these differences reflect genetic differences which are the product of natural selection?

Laboratory experiments were conducted to determine whether differences in size persist when stocks from different populations are raised under similar conditions. The results of one of these experiments are summarized here; a more detailed treatment will be provided elsewhere (Liley, in preparation).

Fish collected in Trinidad in August 1967 were established in the laboratory. Females descended from the wild-caught fish were used to provide broods of young for growth experiments starting in July 1968. Young of two populations, from Upper Aripo (headstream) and Guayamare (lowland), were raised in batches of 35–40 individuals in 40-litre tanks.

There were two or three replicates of each stock at each of two temperatures, 23 °C and 28 °C, corresponding to headstream and lowland conditions respectively. Apart from this difference in temperature, conditions of feeding, etc. were the same for each population. Each batch was measured at 2-week or 4-week intervals up to 19 weeks of age, when the experiment was terminated.

It was found that males of both populations grew rapidly at first under both temperature regimes, but then growth rate decreased dramatically as males matured at 9–11 weeks. On the other hand, females were still growing rapidly at the end of the experiment.

Consistent differences in the size of fish from the same population held at different temperatures, and between populations raised at the same temperature, became apparent from 9–11 weeks onwards. Males and females of both Guayamare and Upper Aripo populations grew larger under the low-temperature regime. At each temperature Upper Aripo males were larger than Guayamare males: this difference was most marked at the high-temperature regime.

In contrast to the situation in males, Guayamare females were slightly larger than Upper Aripo females at both temperatures, however, in this case the differences were not significant at the 5 per cent level.

We concluded from this experiment that the differences in size of adult guppies, particularly the males, taken from Upper Aripo and Guayamare populations in Trinidad are partly determined by genetic differences, and are in part a phenotypic response to environmental temperature.

The fact that males of the Upper Aripo grew larger than Guayamare fish at both temperatures whereas the females of the two stocks were similar in size at a given age implies that the genetic differences in growth potential are restricted to the males. Other factors such as food availability, water hardness, etc. may also affect the phenotypic growth response under natural conditions and may account for the difference in females taken from headstreams and lowland populations.

Differences in the genetic factors which determine the growth and size are presumably the product of natural selection. We suggest that size-selective predation may be *one* factor involved in the evolution of interpopulation differences in the size of mature fish.

Briefly, we hypothesize that in areas where *Rivulus* is the principal predator (mainly springs and isolated headstreams) selection has favoured larger individuals, particularly among males. *Rivulus* is a relatively small predator and thus is only able to handle relatively small guppy prey. Fig. 5.4 shows that in stomach samples of *Rivulus* collected in Trinidad the maximum size of guppy prey was 23 millimetres T.L. (= approximately 16·8 millimetres S.L.); all except 6 prey were 16 millimetres or less. This suggests that selection due to predation by *Rivulus* will favour guppies which grow large. Once guppies have reached a size of 17 millimetres S.L.

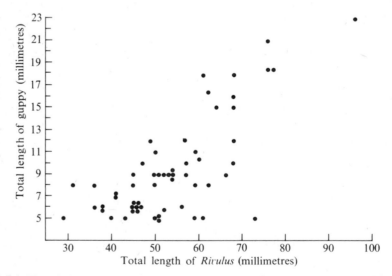

FIG. 5.4. The relationship of *Rivulus* body size to the size of guppies taken as prey. The data are taken from several natural populations.

they are probably relatively invulnerable to predation by *Rivulus*. Size-selective predation will apply more severely to males than to females as growth of males slows or comes to a halt shortly after maturity. Many males of Guayamare and other lowland stocks are mature well within the range vulnerable to *Rivulus* predation, whereas headstream and spring fish do not mature until they reach the size refuge. Obviously fish of either sex or any population are likely to be preyed upon during development but as they grow they become progressively less vulnerable to predation by *Rivulus*.

In contrast, in midstreams or lowland streams there are several species of predators which are large relative to the guppy and are able to handle the largest guppy, male or female. Hence, we hypothesize that in this situation selection by predators is unselective with respect to size, or perhaps directed towards the larger individuals which provide a 'more profitable' kill than smaller individuals. In this situation selection may favour fish which mature and begin their reproductive life at a relatively small size.

There is little field data which bears upon the above suggestions. A number of *Astyanax* were found to have juvenile guppies, ≤13 millimetres T.L. (=9·6 millimetres S.L.) in their stomachs, but as the sample of *Astyanax* was made up of relatively small individuals it is not clear what range of prey size this species may take. There is no doubt that under laboratory conditions average sized *Crenicichla* and *Hoplias* may take the largest guppies available.

An indication that predation and not temperature alone may be important in determining the size of guppies comes from a comparison of headstreams such as Upper Tacarigua, Upper Guanapo, and Upper Aronca where several large predators occur, with isolated headstreams such as Upper Aripo, Yarra, Blue Basin, and Upper Paria, where only *Rivulus* occurs as an important predator. Fish taken in the latter situation tend to be larger than populations in the former category.

It is important to point out that any population sampled in the field has already been subjected to predation and thus differences observed could be due to differential mortality without necessarily reflecting a difference due to temperature or genetic factors. Furthermore, headstream populations are potentially interconnected with populations downstream and differences arising in headstreams may be swamped by gene flow from downstream populations.

Another pointer to the role of size-selective predation is the fact that in certain small springs and headstreams where the *Rivulus* population is particularly dense and forced into close proximity with the guppy population the sex ratio was markedly in favour of females (Seghers 1973). Haskins *et al.* (1961) noted this situation and argued that males were at a disadvantage because of their conspicuous coloration. We agree that this might be an important factor but also suspect that males because of their smaller size may be more vulnerable to predation by *Rivulus*. The sex difference in size may be less important in the areas where the predators prey upon a whole range of sizes.

We attempted to test the hypothesis that selection by fish predators might be size-restricted, by exposing artificial populations to predators in a laboratory situation.

5.5. Experiments to test the size selectivity of predators preying upon guppy populations

A number of experiments were designed to examine both size and sex selectivity in artificial populations of guppies. Table 5.3 lists the stocks and predators used in these tests. In most cases populations consisting of 50 males and 50 females were exposed to predation by *Rivulus* or *Crenicichla*. The prey fish were measured and selected before the start of the experiment and remeasured when about 50 per cent of the prey had disappeared, within 2–4 days.

Experiment 1: *size selection*

Experimental details.   Tests 1(a)–(i) were conducted in 200-litre tanks (length = 92 cm, width = 48 cm, depth = 46·5 cm) in Vancouver. The tanks had a substrate of sand and pebbles and a large flower pot at one side as a refuge, and were without plants. Test 1(j) was carried out in a 1600-litre outdoor concrete pool in Trinidad, 50 centimetres in depth.

Experimental guppy populations were selected at random from stock aquaria and measured 24–48 hours before the experiment began. The predator was resident in the test tank. Guppies were placed in the tank and held in a screen basket for 1 hour before release, normally in late afternoon. Survival was assessed by removing the guppies and counting sexes and sizes. They were immediately returned to the test tank. This procedure did not seem to disturb either predator or prey.

In all experiments the predators were fed an excess of *Tubifex* the day before the start of the experiment; this was to prevent predator 'overkill' on Day 1 as well as to equalize initial hunger level.

The predator, *Hoplias*, and prey in experiment 1(j) were wild-caught fish and had been in captivity for only a few weeks. All other experiments involved laboratory raised stocks of guppies and *Rivulus*. The specimens of *Crenicichla* were caught in Trinidad and held for several months in the laboratory.

Experiments 1(*a*)–(*e*): *size selection by Rivulus.*   Six *Rivulus* (73–90 millimetres T.L.) were used as predators in tests with three populations of Caparo guppies and two populations of Petite Curucaye stock as prey. In each case there were 50 males and 50 females of approximately similar size range (15–23·5 millimetres T.L., = 11–17·3 millimetres S.L.). Males and females were taken equally by the predators and thus the measurements for the two sexes are combined. At approximately 50 per cent survival the mean size of the guppy population remaining had increased by 0·49 millimetres to 1·48 millimetres (Table 5.3), that is, the predators had preyed more heavily upon the smaller fish as predicted.

Experiments 1(f)–(i): size selection by Crenicichla. Two *Crenicichla* predators were tested separately, each with two replicate populations of Caparo guppies. For one replicate with each predator the size of the sexes was matched as closely as possible, while for the other replicate a random sample of adults was taken, that is, the sample showed a greater range in size and females were generally larger than males (Table 5.3).

As females had little if any advantage over males under the experimental conditions the two sexes are not considered separately. There was little evidence of size-selective predation by *Crenicichla*. In 3 out of 4 cases there was a small decrease in mean size indicating that the predators had tended to take more of the larger prey (Table 5.3).

Experiment 1(j): size selection by Hoplias. A population of females was presented to two *Hoplias* (175 millimetres and 205 millimetres) in a large outdoor tank. After 3 days there was a marked decrease in size of the prey remaining, indicating that the predator had taken more of the larger female guppies (Table 5.3).

Discussion. Although not conclusive, these experiments with a limited number of stocks as prey populations do suggest that *Rivulus* may be size

TABLE 5.3
Experiment 1: Total length measurements of prey populations of guppies before and after predation

Test	Guppy stock	Predator	Before predation			After predation			Difference
			n	Mean	Range	n	Mean	Range	
(a)	P. Cur	*Rivulus*	100	20·43	17–23·5	45	20·92	18–24	+0·49
(b)	P. Cur	*Rivulus*	100	20·30	17·5–23·5	51	20·81	18–24	+0·51
(c)	Cap	*Rivulus*	100	19·89	16–23·5	55	20·50	17·5–23	+0·61
(d)	Cap	*Rivulus*	100	19·52	15–23	50	20·05	17–23	+0·53
(e)	Cap	*Rivulus*	100	18·42	15–22	38	19·90	16·5–23·5	+1·48
(f)	Cap	*Crenicichla*	100	20·18	16·5–24	48	20·13	17–24	−0·05
(g)	Cap	*Crenicichla*	100	20·18	16–24·5	54	20·30	17–25	+0·12
(h)	Cap	*Crenicichla*	100	20·81	16–32	50	20·68	16–28	−0·13
(i)	Cap	*Crenicichla*	100	21·83	15·5–32·5	51	21·78	17–33	−0·05
(j)	SC	*Hoplias*	40	27·8	20–38	16	26·4	20–37	−1·4

For a list of stream abbreviations (guppy stock column), see Fig. 5.1.

selective as predicted, whereas larger predators such as *Crenicichla* or *Hoplias* are either unselective or tend to select larger prey items. In all 5 tests involving *Rivulus* there was an increase in mean size of prey remaining, whereas in all except one experiment with *Crenicichla* and *Hoplias* the trend was in the other direction. The probability of this occurring by chance is <0·025 (using Fisher exact probability test).

More convincing evidence that size does affect the vulnerability of guppies comes from observations on the behaviour of predators handling their prey.

Experiment 2: the hunting and attack behaviour of Rivulus

Rivulus was observed hunting and capturing guppies of two different size classes. It was of interest to know how much a difference in length of a few millimetres affects capture efficiency. It was predicted that even a slightly larger fish would be more difficult to capture.

The prey in this experiment were mature males of Guayamare stock, which had been raised at either low or high temperature in the experiment referred to on p. 105 (Liley, in preparation). The fish were accustomed to an intermediate temperature before the experiment began. Fish from the low-temperature environment were larger, mean total length $\bar{x} = 22·8$ millimetres (= 16·8 millimetres S.L.), range 22–24 millimetres; for the high-temperature fish $\bar{x} = 18·1$ millimetres (= 13·3 millimetres S.L.), range 16·5–19·5 millimetres.

Four *Rivulus*, two of each sex, of greater than 70 millimetres T.L. were used as predators in a 40-litre aquarium. For each test two males of each of the two size groups were placed in the aquarium with the predators after a 15 minute adjustment period behind an opaque partition in one corner. Approaches, attacks, and captures were recorded on a Rustrak event recorder until no prey remained. This procedure was repeated 10 times on 10 consecutive days. In most tests each predator consumed one prey, thus they were all at comparable hunger levels at the start of each test.

Results.   Small guppies were caught with greater ease than large ones (Table 5.4). Both the approach and attack scores are significantly different between the size groups (approach; $T = 5$, $n = 10$, $p < 0·01$; attack, $T = 4$, $n = 10$, $p < 0·01$; using Wilcoxon signed ranks test, both tests one-tailed). There was no evidence that the predators were attracted to either size class (the order of captures were random), but once an attack sequence had been initiated, it took more effort (approaches and attacks) to capture the large prey. It might be expected that with increasing experience the predators would preferentially attack the small guppies but this discrimination was not observed over the 10-trial period of this experiment. There was an over-all slight improvement in the ability to handle guppies (compare

TABLE 5.4

Predation efficiency of Rivulus on large and small guppies

Test	Total length of prey (mm) Small (S)		Large (L)		Survival time (s)	Approach Small	Large	Attack Small	Large	Capture sequence
1	18·0	19·0	23·5	22·0	235	27	22	11	10	LLSS
2	19·0	17·0	23·0	22·5	155	14	32	5	17	SLSL
3	16·5	18·5	23·0	23·5	115	13	27	7	13	SLSL
4	18·0	17·0	22·0	22·0	65	16	24	4	9	SLLS
5	18·5	19·0	22·0	24·0	65	15	19	6	5	SSLL
6	18·5	18·5	22·0	24·0	55	17	16	6	7	LLSS
7	19·0	19·5	24·0	22·0	90	15	26	4	8	SSLL
8	17·0	18·5	22·5	23·0	100	16	34	4	12	LSSL
9	16·5	18·0	22·0	22·0	60	9	18	2	10	SSLL
10	17·0	19·0	23·0	23·0	35	10	14	2	4	SLSL
Mean	18·1		22·8		97·5	15·2	23·2	5·1	9·5	

approaches and attacks, tests 1–5 v. tests 6–10), but this was not size-dependent. The survival time (time to consume all four prey) also decreased over the course of the experiment, due to (a) the increased proficiency in handling guppies, and (b) a decrease in the latent period following the release of the prey from behind the partition.

Experiment 3: handling efficiency of Crenicichla

Although it was known that this predator could handle guppies larger and smaller than those offered as prey in this experiment the object in this case was to see if a mean size difference of 6–7 millimetres (typical of adult fish in nature) might affect the handling efficiency of a large predator.

Five large (24–27 millimetres T.L., $\bar{x} = 26·1$ millimetres (19·2 millimetres S.L.)) and 5 small (18–20·5 millimetres T.L., $\bar{x} = 19·4$ millimetres (14·3 millimetres S.L.)) female guppies of the same stock were placed in a 400-litre aquarium with a single *Crenicichla* (250 millimetres T.L.). The predator had not been fed for 24 hours. Approach, attack, and capture scores were recorded on a Rustrak event recorder for 60 minutes after introduction of the prey.

Results. Although this test was not replicated, it is clear that it took many more approaches (pursuits without prey contact) to capture a small guppy than it did to capture a large one (Table 5.5). Even when the small fish were attacked, they could sometimes escape the predator's attempted grasp. Attacks on large fish were 100 per cent successful under these experimental conditions (no prey refuge, very clear water, bright illumination).

TABLE 5.5

Predation efficiency of Crenicichla on large and small guppies

Response	Small guppies	Large guppies
Approaches per hour	28	13
Attack	5	4
Capture	2	4
Approach/capture ratio	14:1	3·25:1
Attack/capture ratio	2·5:1	1:1

Although this experiment was conducted in a large aquarium, the size differences did not appear to affect the detection of the prey from a distance (the predator reacted to even the smallest guppy from the maximum available distance—1·2 metres), but it is possible that the smaller guppies were less easily tracked during the crucial milliseconds prior to the opening of the jaw.

Conclusions and discussion

In general the pattern of micro-geographic variation in body size conforms to Bergmann's rule as applied to poikilotherms by Ray (1960), namely, that fish taken from populations living in cool water are larger than those residing in warmer waters. Ray (1960) discusses the parallel between phenotypic changes in size in relation to temperature, as observed in laboratory experiments on a variety of poikilotherms, and the difference in size, assumed to have a genetic basis, observed in natural populations adapted to different temperature regimes. The 'Baldwin effect' is suggested as a means by which environmentally induced changes might be incorporated into the genotype.

In a laboratory experiment (Liley, in preparation) males of both a headstream and a lowland population were found to mature at a greater size when raised at the lower of two temperature regimes. Females also tended to grow larger at the lower temperature, but possibly because growth of females continues after maturity, the difference in size achieved by the time the experiment was terminated was less marked. At both temperatures the males from the headstream population tended to grow larger than the lowland fish, suggesting that the difference observed in field populations is not only a phenotypic response to differences in environmental temperature, but also reflects a genetic difference between the populations. In contrast, females of the two populations did not differ significantly at either temperature—Guayamare females tended to be larger than Upper Aripo females at a given temperature. This suggests that selection has not favoured a size difference as an adaptation to different temperature regimes as such (at least in females), and that a selection pressure specific to males is responsible for the differences in size observed in natural populations.

The question now arises as to why natural selection has favoured large-bodied males in some environments and smaller males in others.

Two studies of size-selective predation on fish by piscivores have suggested that heritable changes in growth rates and body size may have occurred in the evolutionary history of the species. Jackson (1961, 1965) argued that the impact of the tiger-fish, *Hydrocynus vittatus*, on small fish (less than 20 centimetres long) has selected for fish species that are large as adults (but see Fryer (1965) for critique). More recently Parker (1971) has speculated that, in the Bella Coola River, chum salmon fry have 'evolved a strategy' to outgrow a small size-selective predator (coho salmon parr). For evolution in the opposite direction, Roberts (1972, p. 134) has commented that minute fish in the Amazon and Congo River systems may be less vulnerable to predaceous fish because they are 'below the size threshold for attack...'

While predation may be an important selection mechanism in some cases, body size must inevitably be a compromise between numerous complementary and conflicting selective forces. Hamilton (1961) has tabulated seven selective forces which might interact to cause intraspecific size trends in birds; most of these are peculiar to homeotherms or concern problems of flight.

For numerous animals natural selection may favour an optimal body size for: (1) food-getting (Brooks and Dodson 1965; Estes and Goddard 1967); (2) resisting abiotic stresses such as high water velocity (Hubbs 1940; Hartman 1969), or wave action (Struhsaker 1968; Berry and Crothers 1968); (3) avoiding non-predatory inter-specific interactions (Hamilton 1961; Soulé 1966); (4) securing some form of mating advantage (Hanson and Smith 1967; Hartman 1969).

It seems unlikely that in the guppy feeding specialization or inter-specific competition would affect only males. Sexual selection might favour larger males in the dispersed populations in clear headstreams, but as yet there is no experimental evidence available. In this paper we have described a preliminary attempt to test one of the many alternate explanations for the size trends—the hypothesis that variation in body size may reflect an adaptation to size-selective predation. This idea arose out of the observation that guppies in the Lower Aripo River are much smaller than in the Upper Aripo River, though apart from predation, the ecological conditions (including temperature) are very similar.

Field and laboratory evidence shows that large guppies do enjoy an advantage with respect to *Rivulus* predation, but are more vulnerable to

Crenicichla or *Hoplias*. It has been shown that there is a good association of small body size in guppies with the presence of characids and cichlids, and large guppies with *Rivulus*. However, temperature and predator distribution are also associated so it is impossible except in isolated cases such as the Aripo River, to determine which mechanism is operative, or whether both factors are important.

Predation might also act on body size indirectly. Where the large predators are present, guppies are found in very shallow water at the stream edge. This restricted environment presumably exposes the fish population to new selection pressures, favouring efficient feeding, reproduction, etc. in very confined surroundings. Thus an initial behavioural response to escape characid and cichlid predation may have indirectly resulted in the evolution of small fish.

Though the association of large body size with size-selective predation by *Rivulus* is an attractive hypothesis, there are several anomalies in the field collections (for example, the fish appear to be too large or small for a particular temperature regime). It is possible that food availability might result in retarded or accelerated growth rates, thereby determining the size of males at maturity (cf. Svärdson in Alm (1959, p. 97)). This might be an important factor in the Lower Tacarigua and Santa Cruz streams which are both enriched through human activity. The resultant increase in primary productivity appears to have favoured the growth of guppies to a larger size than in comparable 'natural' streams.

If *Rivulus* does exert directional selection on body size, the largest guppies should be in populations exposed to the densest populations of *Rivulus*. This clearly is not always the case since the mean body size of males at Tompire Tributary and Petite Curucaye (both with dense *Rivulus* populations) is well below the maximum of 18·1 millimetres S.L. (Grande Curucaye).

For Tompire Tributary, the small size may be partly attributable to a relatively high temperature (26·2 °C), and a different genetic history (it is isolated from the Caroni system). But this does not account for the small males at Petite Curucaye. Petite Curucaye is a tributary of Grande Curucaye and appears to be identical to it in temperature, pH, hardness, substrate, etc.; both fish populations in these streams are undoubtedly closely related historically and are probably connected at present by some gene flow (male colour patterns are very similar). However, there are two conspicuous differences between them: Petite Curucaye is smaller and has a greater abundance of *Rivulus*.

Size (and hence velocity) of the stream might be an important selective factor. There are at least two ways this could operate. The first is the mechanism alluded to earlier for populations of guppies 'forced' to live in a very restricted environment—small size may have definite advantages, partly counteracting the greater vulnerability to *Rivulus*. In the Petite

Curucaye River there is literally no place for the fish to go. This presents unique survival problems which cannot be solved by large body size and swimming speed alone.

The second factor is water velocity. Though guppies generally select micro-environments where water velocity is well below the stream maximum, large body size (with its concomitant effect on swimming speed) might allow a fish to manoeuvre more easily up and down or across a fast-flowing stream. This may have advantages in intraspecific behaviour and escape from *Rivulus* and terrestrial predators. This might explain the large body size of Grande Curucaye males.

Finally, it should be recalled that many of the populations in the Caroni system are isolated only by distance. Geographic trends may simply reflect the degree to which local populations are able to preserve adaptive features in the face of gene flow from other populations.

This study demonstrates something of the complexity of the selection pressures imposed upon guppy populations in the natural environment. We believe that predation is an important selection pressure affecting morphology. However, we have already emphasized that morphological differences, size of adults in particular, cannot be thought of as being a direct response to predation alone: larger size is likely to affect numerous physiological parameters, swimming speed, perhaps ability to survive in the more rapidly flowing waters of headstreams, capacity of the male in competitive situations, etc. Any or all of the above consequences of size differences may contribute to their survival value, making it impossible to isolate a single primary function.

5.6. Summary

Guppies collected at different sites in Trinidad, West Indies, differ in a number of characteristics of morphology and behaviour. In general guppies taken from springs and isolated headstreams tend to be large and the males brightly coloured, and females may outnumber males by as much as 4:1. Fish tend to be dispersed over the stream bed and show relatively poorly developed avoidance responses. These features appear to be associated with clear, fast-running water, relatively low temperatures, and a virtual absence of aquatic predators other than the small cyprinodontid, *Rivulus hartii*.

Guppies in the lower sections of streams and rivers tend to be small and the males less brightly coloured and as numerous as females. They show vigorous avoidance responses and may occur in small schools along the edge of a stream or river. In this case guppies are associated with conditions of relatively slow-moving—often turbid—water, high temperatures, and an absence of shade. Numerous other fish are present including several relatively large species known to be predators.

One aspect of the relationship between the variation in the guppy populations and environmental factors is examined in detail: that of size in relation to predation and temperature. Experiments in which guppies of two different populations were raised at different temperatures indicate that the interpopulation differences in size of adult guppies, particularly the males, are partly determined by genetic differences and are in part a phenotypic response to environmental temperature conditions.

Differences in the genetic factors that determine growth and size are presumably the product of natural selection. It is suggested that size-selective predation may be one factor involved in the evolution of interpopulation differences in the size of mature fish. Experiments in which 'populations' of guppies were exposed to predation by a variety of natural predators provided evidence that predation may be size-selective. In particular it was found that *Rivulus* preyed selectively on smaller guppies, whereas larger predators (*Crenicichla, Hoplias*) were unselective or took larger guppies as prey.

It is concluded that size-selective predation is one important selection pressure affecting morphology. However, it is emphasized that size is likely to affect numerous physiological parameters, swimming speed, perhaps ability to survive in the more rapidly flowing waters of headstreams, capacity of the male in competitive situations, etc. Any or all of the above consequences of size differences may contribute to their survival value, making it impossible to isolate a single primary function.

Acknowledgements

This work was supported by a National Research Council of Canada Operating Grant to N. R. Liley and a National Research Council Postgraduate Scholarship awarded to B. H. Seghers. For hospitality in the Department of Biological Sciences, University of the West Indies, Trinidad, we are indebted to Dr. B. D. Ainscough, Dr. J. S. Kenny, and Mr. R. L. Loregnard.

Bibliography

ALM, G. (1959). *Connection between maturity, size, and age in fishes.* Report of the Institute of Freshwater Research, Drottningholm, Vol. 40, pp. 5–145.

BALLIN, P. J. (1973). M.Sc. Thesis. University of British Columbia.

BEEBE, W. (1952). Introduction to the ecology of the Arima Valley, Trinidad, B.W.I. *Zoologica* **37**, 157–83.

BELCHER, C. and SMOOKER, G. D. (1936). On the birds of the Colony of Trinidad, and Tobago. *Ibis* **6**, 792–813.

BERRY, R. J. and CROTHERS, J. H. (1968). Stabilizing selection in the dog-whelk (*Nucella lapillus*). *J. Zool. Lond.* **155**, 5–17.

BLEST, A. D. (1957). The function of eyespot patterns in the Lepidoptera. *Behaviour* **11**, 210–56.

BLOEDEL, P. (1955). Hunting methods of fish-eating bats, particularly *Noctilio leporinus. J. Mammal.* **36**, 390–9.

BOESEMAN, M. (1960). The fresh-water fishes of the island of Trinidad. *Stud. Fauna Curaçao* **10**, 72–153.

—— (1964). The fresh-water fishes of the island of Trinidad: addenda, errata, et corrigenda. *Stud. Fauna Curaçao* **20**, 52–7.

BRETT, J. R., SHELBOURN, J. E., and SHOOP, C. T. (1969). Growth rate and body composition of fingerling sockeye salmon, *Oncorhynchus nerka*, in relation to temperature and ration size. *J. Fish. Res. Bd Can.* **26**, 2363–94.

BROOKS, J. L. and DODSON, S. I. (1965). Predation, body size, and composition of plankton. *Science* **150**, 28–35.

BROWN, M. E. (1957) (ed.) Experimental studies of growth. *The physiology of fishes*, Vol. 1, pp. 361–400. Academic Press, New York.

CALAPRICE, J. R. (1969). *Production and genetic factors in managed salmonid populations*. In Symposium on Salmon and Trout in Streams (ed. T. G. Northcote), pp. 377–88. Institute of Fisheries, University of British Columbia.

CROZE, H. (1970). Searching image in Carrion crows. *Z. Tierpsychol., Suppl.* **5**, 1–86.

ESTES, R. D. and GODDARD, J. (1967). Prey selection and hunting behavior of the African wild dog. *J. Wildl. Mgmt* **31**, 52–70.

FRY, F. E. J. (1971). The effect of environmental factors on the physiology of fish. In *Fish physiology* (eds. W. S. Hoar and D. J. Randall), Vol. 4, pp. 1–98. Academic Press, New York.

FRYER, G. (1965). Predation and its effects on migration and speciation in African fishes: a comment. *Proc. zool. Soc. Lond.* **144**, 301–22.

GIBSON, M. B. and HIRST, B. (1955). The effect of salinity and temperature on the pre-adult growth of guppies. *Copeia* **1955**, 241–3.

HAGEN, D. W. and GILBERTSON, L. G. (1972). Geographic variation and environmental selection in *Gasterosteus aculeatus* L. in the Pacific Northwest, America. *Evolution* **26**, 32–51.

HAMILTON, T. H. (1961). The adaptive significance of intraspecific trends of variation in wing length and body size among bird species. *Evolution* **15**, 180–95.

HANSON, A. J. and SMITH, H. D. (1967). Mate selection in a population of sockeye salmon (*Oncorhynchus nerka*) of mixed age-groups. *J. Fish. Res. Bd Can.* **24**, 1955–1977.

HARTMAN, G. F. (1969). *Reproductive biology of the Gerrard stock rainbow trout.* In Symposium on Salmon and Trout in Streams (ed. T. G. Northcote), pp. 53–67. Institute of Fisheries, Univ. British Columbia.

HASKINS, C. P., HASKINS, E. F., MCLAUGHLIN, J. J. A., and HEWITT, R. E. (1961). Polymorphism and population structure in *Lebistes reticulatus*. In *Vertebrate speciation* (ed. W. F. Blair), pp. 320–95. University of Texas Press, Austin.

HERKLOTS, G. A. C. (1961). *The birds of Trinidad and Tobago.* Collins, London.

HOOGLAND, R., MORRIS, D., and TINBERGEN, N. (1957). The spines of sticklebacks (*Gasterosteus* and *Pygosteus*) as means of defence against predators (*Perca* and *Esox*). *Behaviour* **10**, 205–36.

HUBBS, C. L. (1940). Speciation of fishes. *Am. Nat.* **74**, 198–211.

JACKSON, P. B. N. (1961). The impact of predation, especially by the tigerfish (*Hydrocyon vittatus*) on African freshwater fishes. *Proc. zool. Soc. Lond.* **136**, 603–22.

—— (1965). Reply to G. Fryer and P. H. Greenwood. *Proc. zool. Soc. Lond.* **144**, 313–21.

JUNGE, G. C. A., and MEES, G. F. (1961). *The avifauna of Trinidad and Tobago.* Rijksmuseum van Natuurlijke Historie, Leiden.

KINNE, O. (1960). Growth, food intake, and food conversion in a euryplastic fish exposed to different temperatures and salinities. *Physiol. Zool.* **33**, 288–317.

KRUUK, H. (1964). Predators and anti-predator behaviour of the black-headed gull (*Larus ridibundus* L.). *Behaviour, Suppl.* **11**, 1–130.

LIU, R. K. and WALFORD, R. L. (1966). Increased growth and lifespan with lowered ambient temperature in the annual fish *Cynolebias adloffi. Nature, Lond.* **212**, 1277–8.

MCPHAIL, J. D. (1969). Predation and the evolution of a stickleback (*Gasterosteus*). *J. Fish. Res. Bd Can.* **26**, 3183–208.

MOODIE, G. E. E. (1972a). Predation, natural selection and adaptation in an unusual threespine stickleback. *Heredity* **28**, 155–68.

—— (1972b). Morphology, life-history, and ecology of an unusual stickleback. (*Gasterosteus aculeatus*) in the Queen Charlotte Islands, Canada. *Can. J. Zool.* **50**, 721–32.

PARKER, R. R. (1971). Size selective predation among juvenile salmonid fishes in a British Columbia inlet. *J. Fish Res.Bd Can.* **28**, 1503–10.

RALEIGH, R. F. and CHAPMAN, D. W. (1971). Genetic control in lakeward migrations of cutthroat trout fry. *Trans. Am. Fish. Soc.* **100**, 33–40.

RAY, C. (1960). The application of Bergmann's and Allen's rules to the poikilotherms. *J. Morph.* **106**, 85–108.

ROBERTS, T. R. (1972). Ecology of fishes in the Amazon and Congo basins. *Bull. Mus. comp. Zool. Harv.* **143**, 117–147.

SEGHERS, B. H. (1973). *An analysis of geographic variation in the antipredator adaptations of the guppy*, Poecilia reticulata. Ph.D. Thesis. University of British Columbia.

—— (1974a). Geographic variation in the responses of guppies (*Poecilia reticulata*) to aerial predators. *Oecologia* **14**, 93–8.

—— (1974b). Schooling behaviour in the guppy (*Poecilia reticulata*): an evolutionary response to predation. *Evolution* **28**, 486–9.

SOULÉ, M. (1966). Trends in the insular radiation of a lizard. *Am. Nat.* **100**, 47–64.

STRUHSAKER, J. W. (1968). Selection mechanisms associated with intraspecific shell variation in *Littorina picta* (Prosobranchia: Mesogastropoda). *Evolution* **22**, 459–80.

TINBERGEN, N. (1963). On aims and methods of ethology. *Z. Tierpsychol.* **20**, 410–33.

—— (1965). Behaviour and natural selection. In Ideas in modern biology (ed. J. A. Moore). *Proc. int. Zool. Congr.* **16**, 521–42.

—— BROEKHUYSEN, G. J., HOUGHTON, J. C. W., KRUUK, H., and SZULC, E. (1962). Egg shell removal by the black-headed gull *Larus ridibundus* L.: a behaviour component of camouflage. *Behaviour* **19**, 74–118.

WEATHERLEY, A. H. (1972). *Growth and ecology of fish populations.* Academic Press, London.

6. Functional aspects of social hunting by carnivores

H. KRUUK

6.1. Introduction

THAT ANIMALS defend themselves against predators by various means has long been taken for granted. But it was not until after experimental studies such as those of Tinbergen, Broekhuysen, Feekes, Houghton, Kruuk, and Szulc (1962) that it was realized to what intricate detail this defence against a species' enemies went. This was confirmed by further investigations in the series of gull studies under Niko Tinbergen's supervision (Kruuk 1964; Tinbergen 1967; Tinbergen, Impekoven, and Franck 1967; Croze 1970). It was studies such as these that showed the close adjustment between anti-predator behaviour of an animal and hunting strategies of its predators; they gave a first indication of the complexity of the defence system which a predator must penetrate in order to eat.

Overcoming this defence of a potential prey is the biological function of *hunting* behaviour with its many corrollaries, and in the Carnivora this is well developed along many different specialized lines. Hunting, the pattern of activities adapted to the capturing of other animals, is virtually absent in other orders of mammals such as primates (except in man), ungulates, and rodents, with the exception of the insectivores.

One of the striking aspects of hunting behaviour of Carnivora is the close integration with other aspects of the social behaviour, such as territory and group structure. Several carnivores hunt communally, which appears to increase their effectiveness (see below). Furthermore, it has been shown that hunting and foraging patterns of carnivores are variable in different environments and also vary towards different prey species. It is these characteristics which make hunting and social behaviour in the Carnivora worthwhile studying in considerable detail, for the following reasons.

First, recent comparative studies in birds, primates, and antelopes suggested the existence of a close relationship between differences in species' social behaviour and organization, and differences in the distribution of their food supply (Crook and Gartlan 1966; Crook 1964, 1965, 1970; Clutton-Brock 1972; Jarman 1974). However, functional aspects of interactions between individual animals in the process of food acquisition in the wild have rarely been studied. As it is there that social organization and food supply actually come together such a study would be important; it would be difficult to carry out in vegetarian monkeys and ungulates, but easier in carnivores.

Secondly, the special adaptations which are necessary for a predatory species to overcome the anti-predator behaviour of different prey under varying environmental conditions provide a good opportunity to study the range of a species' adaptability, which is essential for further discussion of the effects of environment on behaviour. Because in studying hunting behaviour we are dealing with a type of activity which the carnivores have in common with man, this may eventually give us an opportunity to clarify ways and means of studying some of man's original functional relationships with his environment. Tinbergen (1968, 1972) argued that in man's rapidly changing environment his adaptability is being severely tested; this adaptability will, in fact, determine to what extent a stable relationship with the new, self-created environment can be developed.

Thirdly, more inter-specific comparisons of social behaviour and ecology are still badly needed, to study evolutionary processes in these particular aspects of the animals' biology. In this respect the Carnivora should be particularly rewarding because of the large variation in this order: species vary from completely solitary to highly gregarious, from entirely vegetarian to exclusively carnivorous, and they are found in many different forms and sizes (for instance, the largest species is in the order of 50 000 times heavier than the smallest).

Lastly, it has been repeatedly shown that results of studies of this kind can be useful for conservation-management purposes, and carnivores are in need of conservation as much as (if not more than) any other large mammals.

In this essay I shall be mostly concerned with the Hyaenidae, but it will be shown that several evolutionary trends in social organization and hunting in that family have their parallels elsewhere. First I will discuss different social hunting techniques in one species, then differences between populations. After that, comparisons will be made with closely related species and carnivores of different families. Finally, I will discuss evolutionary trends, and draw some general conclusions about the effect of technique of food acquisition on the behaviour of Carnivora and some primates, including man.

6.2. The Spotted Hyaena (*Crocuta crocuta*) in Ngorongoro

Most of the data on which the following account on Spotted Hyaenas is based have been obtained from my own studies (Kruuk 1966, 1970, 1972).

The main food of the species in areas of Northern Tanzania where it was studied consists of ungulates such as wildebeest, and zebra, which are considerably larger than the hyaenas; the prey is mostly killed by the hyaenas themselves. Less frequently, the hyaenas eat smaller prey, such as gazelles, or they scavenge parts of animals which died of other causes. In one area, the Ngorongoro Crater, Spotted Hyaenas live in societies ('clans') of between 30 and 80 individuals, each clan occupying an area of between 10 and 40 square kilometres. These clans do not operate as a single unit; within a clan range each hyaena can be solitary or in a group with others, and it frequently changes its 'gregariousness'. All clan members tolerate each other on kills, and they join in the defence of the area against neighbouring clans. Clan ranges have a well-defined and marked boundary. In the Crater, wildebeest and zebra are by far the most important prey of hyaenas; they are hunted on the clan territory, and all the clan ranges have large numbers of wildebeest and zebra on them most of the time.

When Spotted Hyaenas hunt adult wildebeest the prey is selected and a chase is usually started by just one animal (74 per cent of observations), which is then later joined by others for the actual kill. Characteristically, a hyaena will walk close to a herd of wildebeest, then run into it, causing the wildebeest to scatter and run off in small groups. Out of this turmoil the hyaena may chase an individual wildebeest (though more often no chase results at all). After a chase of approximately a kilometre (at maximum 5 kilometres), in which speeds of over 50 kilometres per hour are reached, the wildebeest is captured (37 per cent of 49 chases). Often the hunting hyaena has been joined by this time by several others; the wildebeest is killed by an average of 7·5 hyaenas, and eaten by an average of 23·6 hyaenas and up to as many as 52 (see Plate 6.1).

Before a zebra hunt, Spotted Hyaenas gather in a pack which has an average of 10·8 members (maximum 25). When they approach a family or a herd of zebra, a stallion will usually attack them, though sometimes the zebra flee without this aggression. If the hyaenas continue their hunt the group of zebra will run close together, not always very fast, followed by the pack, with a stallion keeping in between whilst attacking the hyaenas. If one of the hyaenas manages to grab one of the zebra the other hyaenas will immediately concentrate their attack on this victim, slowing it down, then tearing it apart, starting at the loins and anal region, in the same way as a wildebeest is killed. Whether the victim is a mare, a foal, or the stallion himself, the attacks of the stallion stop soon after the quarry has been grabbed. As in the case of the wildebeest the zebra chase goes on over approximately a kilometre (maximum 3 kilometres) at speeds of up to 50 kilometres per hour, but usually much more slowly; the hyaenas managed to kill in 34 per cent of the 47 observations. There is little increase in the number of hyaenas actually killing the victim (mean 11·5) from the number chasing, but several more hyaenas eat the carcass (average 22).

PLATE 6.1. Six Spotted Hyaenas just before pulling down a wildebeest bull, after the chase.

Thus, there are some important differences between the ways in which wildebeest and zebra are being hunted by hyaenas (for further statistical details, see Kruuk 1972); most striking is the variation in the numbers of hyaenas setting out to hunt. I observed several times how a pack of hyaenas would walk through large herds of wildebeest without paying a great deal of attention to them, until it reached some zebra up to several kilometres away, which were subsequently chased. Pack formation took place well before actual confrontation with different prey species.

An important question in considering causality of this association between pack size and prey is: what other activities are related to gregariousness? Apart from hunting particular prey species the temporary groupings also perform another group of typical activities, related to the defence of the clan territory. A group of spotted hyaenas may walk over the range, repeatedly 'social sniffing'—a behaviour which may have a bonding function (Kruuk 1972). When they reach the territorial boundary they will usually perform a number of scent-marking activities on special marking sites ('lavatories'): communal defaecating, pawing the ground (that is, marking with interdigital glands), and pasting anal secretions on grass stalks. If hyaenas of a neighbouring clan are encountered in the boundary area this may result in chases and sometimes a fight. Such territorial activities may take place before a chase of prey, or after the animals have eaten, or indeed at any time that hyaenas are active; they are mostly performed by animals in packs, with the exception of pasting.

Thus it appears that at least on some occasions, after hyaenas have fed and do not respond to the presence of potential prey, a pack shows only territorial behaviour; on other occasions one sees only hunting. In both kinds of situation one can argue that group formation is adaptive: for zebra-hunting the function would be to overcome the anti-predator defence of the prey (see below); for territorial defence it would be to combine individual strengths against the social neighbours.

So far I have considered the interactions between Spotted Hyaenas and only two kinds of prey; wildebeest and zebra. However, different hunting techniques are used to obtain other prey species and food supplies as well. Hunting for Thomson's Gazelle is a relatively solitary occupation; if a chase occurs the number of hyaenas increases little or not at all (mean number of hyaenas starting a chase is 1·2, finishing it 1·3, and those finally eating the carcass 3·2). Hyaenas scavenging around human inhabitation, around lion kills, or near carcasses of animals which died of other causes are mostly alone or with one or two companions only, picking up small bits. A rhinoceros, on the other hand, is always attacked by many hyaenas simultaneously.

6.3. Advantages and disadvantages of hunting in packs

If we want to discuss the biological function of different degrees of gregariousness of hyaenas hunting various prey species we will have to look at some of the social characteristics and anti-predator behaviour of these ungulates.

Wildebeest live in large herds of unstable membership, and some of the bulls defend individual territories (Estes 1966). Wildebeest cows defend their calves (see below), but otherwise wildebeest do not defend each other against predators. In fact, they hardly defend themselves at all. Their chances of survival when hunted are determined almost entirely by their means of escape. Possibly the large herds in which they occur make it difficult for any predator to concentrate on catching individuals; the 'bunching up' of wildebeest in reaction to a predator may be of selective advantage in this context.

The initial part of the wildebeest hunt by Spotted Hyaenas, in which the hunter scatters the herd, is probably an important stage. It may well be that if this initial chase (during which selection takes place) were carried out by a number of hyaenas simultaneously, chaos would result which would make selecting more difficult for the hunters. During the following fast chase more than one hyaena would be of little use, since wildebeest do not attempt to shake off pursuers by fast cornering. But once the wildebeest is stopped, it is more efficient for the hunter to kill it with many hyaenas than alone; although the quarry's defence is ineffective it can still often keep one hyaena away. The hunters' co-operation in quickly dispatching and eating the prey is important also as protection against inter-specific competition; often a lion is attracted to the commotion and chases hyaenas away from their food (for example, in Ngorongoro, September 1972, 20 of the 26 hyaena kills observed were stolen by lions). Thus, in the hunting of wildebeest by hyaenas it may be more efficient if the hunt is started by one; finished by many.

The basic social unit of the zebra is the family (Klingel 1967), consisting of up to 6 adult mares and their foals, and a stallion; the stallions without families join in stallion herds. Many families may join into large herds. One of the striking aspects of any interaction between hyaenas and zebra families is the usually very active defence of the family by the stallion, who will attack hyaenas with his feet and his teeth. Against this kind of defence the co-operation between several predators is an excellent answer since the stallion cannot be in many places at once. Consequently, one might expect then a large hyaena pack to be more successful in capturing zebra than a small one, but in my observations of hyaena–zebra interactions this was not the case. It may be that any existing relationship between pack size and hunting success is obscured by the fact that not all pack members always take an active part in the hunt, or there may be some other explanation.

At least at times when hyaenas are hunting wildebeest calves, it has been possible to demonstrate clearly the usefulness of collaboration. Wildebeest calves stay with their mother from birth, and they are able to run when approximately half an hour old. If a Spotted Hyaena attempts to grab a calf the mother will often defend it with her horns, chasing the hyaena and sometimes hitting it hard; but before the hyaena gets close mother and calf have first tried to outrun the hyaena. In 15 per cent of the 74 attempts in which a single hyaena attacked it was successful and killed the calf; in contrast, 74 per cent of the 34 attempts carried out by two or more hyaenas were successful. This is because two or more hyaenas can always dodge the attacks of the wildebeest mother once they have overtaken the calf, but the single hyaena was prevented from grabbing the calf in all observations in which the mother attacked. Thus, co-operation between two hyaenas can make a hunt much more than twice as successful, and there is obviously a considerable saving of energy here. However, why hyaenas do not co-operate *more often* in catching wildebeest calves is still an open question.

For comparison, it is interesting to look at some observations of co-operative hunting by lions. Schaller (1972) found an increase in the success of hunting lions with increasing numbers of participants: where a single lion was 15 per cent successful in capturing Thomson's Gazelle, two or more lions together caught their prey in 32 per cent of the hunts; similar figures obtain for lions hunting wildebeest and zebra. In the case of lions hunting gazelle the joining of their activities probably does not pay, because the relatively small gazelle has to be shared amongst several hunters which together are only twice as successful as alone. When hunting larger ungulates their collaboration becomes lucrative, since one lion's portion is small relative to prey size.

The increase of hunting success of each individual hyaena with the increase in the number of participating hyaenas hunting wildebeest calves is an effect of actual co-operation. If the hunting success of a pack increases only proportional (or even less than proportional) to the number of participating hunters, as in the case of hunting lions, the over-all result could be explained by entirely independent hunting of the participants in each others' proximity.

As with most hunting techniques employed by the predator, prey species show an adaptive response to the increase in pack size of their enemies. Walther (1969) and Kruuk (1972) mentioned the considerable increase in reaction distance of Thomson's Gazelle and other ungulates in response to an increase in pack size of such predators as hyaenas and wild dogs, a phenomenon also known in reactions of birds to their enemies (Kruuk 1964). This may be adapted to the predator's greater likelihood to hunt, or to greater effectiveness of co-operative hunting, or to both. However, the pack's increase in success is in itself an indication that the prey's adaptations are not fully effective in counteracting the predator's strategy.

6.4. The Spotted Hyaena in the Serengeti

In this section I will compare the Ngorongoro hyaenas' social system and foraging with that in a neighbouring area, the Serengeti, where the Spotted Hyaenas feed on largely the same kinds of prey as in the Crater but with some differences. In the Serengeti the ungulates migrate over large distances, and huge areas are often devoid of potential prey for long periods of the year. The prey species occur in different proportions in the two areas, there being relatively more Thomson's Gazelle in the Serengeti; finally, there is much more carrion available in this last area (Kruuk 1970).

In the Serengeti there are no hyaena clans with more or less fixed membership which defend a specific range (as in the Crater). Animals originating from areas more than 100 kilometres apart may form temporary associations, and only when the herbivore migrations concentrate in an area for a long period do Serengeti hyaenas form 'clans' of a size comparable to those of Ngorongoro. Even then the area which is defended is much less well defined than in the Crater, and a border clash between two packs may take the participants over distances of up to 3 kilometres. On the whole, the reaction against strange hyaenas is less aggressive in the Serengeti. There, too, dens show large seasonal fluctuations in numbers of cubs and location, whilst they are rather constant in Ngorongoro; the fluctuations in Serengeti are clearly related to the distribution of the hyaenas' prey species (there is no particular breeding season, but cubs are

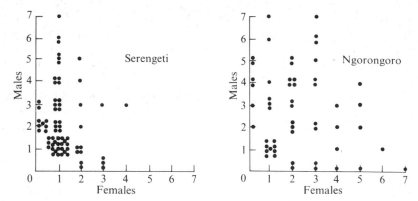

FIG. 6.1. Spotted Hyaena packs in Serengeti and Ngorongoro. Each dot represents one observation of a pack consisting of x females and y males.

distributed over fewer or more dens). Scent-marking, which in Ngorongoro is performed largely on the boundaries of clan ranges, shows a different distribution in Serengeti: there the 'lavatories' are positioned along the main pathways of hyaenas, often alongside car-tracks, with little or no reference to any boundaries.

Related to these differences between the Ngorongoro and Serengeti hyaena communities is the difference in composition and occurrence of packs. It is much rarer to see Serengeti hyaenas in packs, and packs that do occur are considerably smaller. Typically also, a Serengeti pack consists of one female and several males, while in Ngorongoro it usually contains several females and males, although in both areas the sex-ratio is approximately 1:1 (Fig. 6.1).

Almost all these differences in social organization of Spotted Hyaenas appear to be advantageous with respect to differences in food supply. Clearly, with almost all areas of the Serengeti showing seasonal absences and abundances of food, the Ngorongoro clan system would be impossible there. The Serengeti organization of dens, boundaries, etc. makes it possible to keep up with the moving herds of herbivores. Live zebra is a less frequent prey, and the greater abundance in the Serengeti of carrion and Thomson's Gazelle is associated with a more solitary existence of the hyaenas there (which would be predicted on the base of hunting observations).

The difference in female participation in packs, too, could possibly be related to a different exploitation of the food supply. In Ngorongoro, hyaena and prey populations are in an approximate balance with each other. There is just enough food for the number of hyaenas, and consequently there is a great deal of intra-specific competition for food. In the Serengeti, the migratory nature of the herds of ungulates makes it impossible for most of the carnivores to increase in numbers so as to exploit fully all the available food; consequently, the amount of food consumed per day by each individual carnivore is far greater than in Ngorongoro (Kruuk 1970). A carcass in Ngorongoro disappears in a short time, usually well within the hour, and if a female hyaena is to compete effectively with the others she probably should be in on the kill right from the start. In the Serengeti it is so much easier to obtain food that females can 'afford' to arrive late on a carcass (for example, staying longer in the den) rather than having to join the hunting packs.

It is possible therefore, to show that within one social system of the Spotted Hyaena the tendency to join in packs may be closely related to the kind and distribution of food. Furthermore, different social systems exist in other hyaena populations and some of the differences are probably related to variations in groupings used when hunting different prey species.

6.5. Other Hyaenidae

I will now draw some comparisons with a closely related carnivore, the Striped Hyaena, *Hyaena vulgaris*, which was observed mostly in the Serengeti and in Israel (Kruuk 1975). Faecal analysis of the two species occurring in the same area indicated some clear-cut qualitative and quantitative differences in diet; where Spotted Hyaenas ate the large mammals, the striped species lived on reptiles, birds, insects, and fruits, but also mostly on mammals, very small as well as large. Direct observations showed that most of the mammals eaten by Striped Hyaenas had been dead before the hyaena found them: Striped Hyaenas are largely scavengers (Plate 6.2).

The difference in diet of the two hyaenas is probably related to a variety of factors, including their geographical range, and habitat selection. It is also related to a difference in social behaviour: the Striped Hyaena is a solitary species which usually forages alone. Whilst in the Serengeti the Spotted Hyaena shows a flexible social organization adapted to a migratory existence, the Striped Hyaena appears to use the same home range throughout the seasons. This home range is comparatively large (approximately 44 square kilometres and 72 square kilometres in the two individuals which were followed by radio-location). Striped Hyaenas spent more time searching for food than did Spotted Hyaenas; data from the radio-collared animals of both species showed that Striped Hyaenas were active (as opposed to lying down) during 26 per cent of the time, Spotted Hyaenas during only 16 per cent. This may be related to the feeding regimes; Striped Hyaenas eat smaller food items, and it has been shown at least for Spotted Hyaenas that the total effort per individual hyaena for living off large prey items is less than for living off small things (Kruuk 1972).

PLATE 6.2. Striped Hyaena with a Leopard Tortoise, one of its usual prey items in Serengeti.

The variation between the social organizations of the two species is reflected also in differences in their intra-specific communication behaviour. The loud 'whoo-oop' call with its advertising effect in the Spotted Hyaena has no equivalent in the striped species. In addition, while many of the agonistic and juvenile calls are similar (to the human ear) there is a striking difference between the two species in volume and occurrence of these sounds: the Striped Hyaena is usually much less noisy than the Spotted Hyaena. However, there are geographical variations in this aspect of the behaviour of Striped Hyaenas; I noted that some of the calls in Israel were considerably louder than those in Serengeti. This, too, may be indicative of an environmental effect on social behaviour; in the Serengeti the Striped Hyaenas are often chased by their dominant spotted relatives or other carnivores, or kept away from food, and it may be advantageous to be quiet there, whilst in Israel such advantage does not exist.

There are some differences between the two hyaenas in their visual displays; short-range social bonding mechanisms appear to be much reduced in the Striped Hyaena compared with those of the spotted relative. Striped Hyaenas have a well-developed mane which can be erected in agonistic situations, but they appear to be less inclined to use the many different tail positions as signals as do Spotted Hyaenas. A striking difference is the fact that Striped Hyaenas do not use the penis or clitoris as a signalling device in social contexts, as do Spotted Hyaenas, nor has the clitoris evolved into a penis-like structure, nor do they show the long drawn-out meeting ceremony which in Spotted Hyaenas must be associated with this genital development.

The Brown Hyaena, *Hyaena brunnea*, from southern Africa lives in arid country, mostly off small mammals, fruits, and scavenged animal remains (G. Mills, personal communication). Its behaviour and social organization resembles that of the Striped Hyaena in all aspects mentioned above, only its territory tends to be larger.

A more distant member of the Hyaenidae, the aardwolf, *Proteles cristatus*, which is considerably smaller, is a highly specialized termite feeder (Kruuk and Sands 1972). It lives mostly off harvester termites of the genus *Trinervitermes*, which graze in columns on the surface in African savannas. A column is licked up in about 20 seconds, and a foraging aardwolf finds about one per minute, scattered throughout its range—clearly, gregarious hunting would be disadvantageous for the utilization of this resource. The aardwolf shows many of the social characteristics of the Striped Hyaena: it lives a solitary life on its territory (about 1 square kilometre), and it is a very silent animal. Its use of the mane, tail, and other aspects of its morphology are similar to that of the Striped Hyaena.

Comparing these Hyaenidae it is clear that there is a close relationship between foraging behaviour on the one hand, and social behaviour on the other; (see Table 6.1). In the studies of the Spotted Hyaena there are at least some clues for the causality in this relationship. It was possible to show that food intake is closely related to food availability. Theoretically it would be possible that this food availability would be affected by hyaena predation itself, but it has been suggested (Kruuk 1970, 1972) that this is not the case—herbivore densities in the areas concerned are determined by other environmental factors, probably related to the herbivore's own food supply (Sinclair 1974). Thus, hyaena food intake and selection follow whatever determines the availability of prey species and carrion, although clearly this selection remains within a species-specific range. The hyaenas' food supply, affecting population density and age-composition, is likely to exercise strong pressures, too, on the social structure of the population. This is likely especially after the observations of differential effectiveness of different groupings in hunting. What evidence there is points, therefore, to the utilization of a particular set of food resources as an important shaping force in the evolution of social organization of the hyaenids.

6.6. Other carnivores: Felidae

Similar evolutionary processes appear to have taken place in the behaviour of several other carnivore families, such as the Felidae. The

TABLE 6.1

Some aspects of behaviour of the four members of the Hyaenidae

	Spotted Hyaena	Striped Hyaena	Brown Hyaena	Aardwolf
Main food	Large or small ungulates	Carrion; insects; fruits	Carrion; small mammals; fruits	Termites
Foraging strategy	Solitary or social hunting	Solitary searching	Solitary searching	Solitary searching
Gregariousness	Solitary, or temporary groups within 'clan'	Solitary	Solitary	Solitary
Land tenure	Clan territory (group defence)	Large individual territory	Very large individual territory	Small individual territory
Parental organization	Many females in one den; 2 cubs per litter; no food brought to cubs; male not involved	One female per den; 3–4 cubs per litter; male and female bring food to cubs	One female per den; 3–4 cubs per litter; male and female bring food to cubs	One female per den; 3–4 cubs per litter; ? ; ?
Vocalization	Varied, loud, frequent; has long-range call	Rare; no long-range call	Rare; no long-range call	Rare; no long-range call
Displays	Many tail and genital displays. Small mane	Few tail and no genital displays. Large mane	Few tail and no genital displays. Large mane	Few tail and no genital displays. Large mane

great majority of members of this family are solitary species, living off prey considerably smaller than they are themselves. The only species which lives in large groups is the lion, *Panthera leo*, studied intensively by Schaller (1972) in the Serengeti National Park. A lion pride there may number 4–37 individuals; this includes 2–11 adult females and 2–4 adult males. Prides occupy well-defended territories, in which the members operate singly or in groups; the latter may number up to 25 individuals, but the mean group size is about 4. There is large overlap between neighbouring territories. Schaller found that in an 'average' habitat in the Serengeti (plains–woodland edge) the lions' prey consisted of wildebeest (37 per cent), zebra (24 per cent), gazelle (12 per cent), buffalo (15 per cent), and a few other species in those size classes—mostly ungulates considerably larger than the lions themselves, and a diet similar to that of the Spotted Hyaena. About the size of lion groups encountered in the pride territories Schaller remarks: 'An average group consists of four to six animals, just the right number to enable each individual to gorge itself on a wildebeest, topi or a similar prey. But during the dry season when Thomson's gazelle was the principal prey around Seronera, lions often hunted alone and average group size was only about half as large as at other times of the year'. In fact in 51 per cent of all the wildebeest and zebra hunts observed by Schaller at least three lions participated, whereas this was so in only 27 per cent of all the gazelle hunts. Thus in this species, too, food appears to be closely related to the size of groups encountered within the pride; it is not known, however, to what extent this affects the over-all pride composition in different areas.

The increase in hunting success of lions in hunts with several participants (to which I referred earlier) is not yet sufficient evidence to show that a lion actually promotes another's chances of prey capture by co-operating; a lion may merely benefit with a meal from a kill made in its proximity. There is also no good evidence yet to show that lions actively co-operate when hunting, that is, that the reactions of an individual lion to potential prey are different depending on whether other lions are present. Nevertheless, the sharing of large prey items could in itself be a considerable incentive for hunting in groups.

In simultaneous studies in the same area on cheetah, *Acinonyx jubatus*, Schaller (1972) found this species to be solitary in 52 per cent of his observations, with 31 per cent of cheetah encountered in groups of two, 14 per cent in threes, and 3 per cent in fours. There is hardly any indication of territorial defence in this species; ranges of individuals overlap extensively, and they seem merely to keep out of each others' way (MacLaughlin 1970; Schaller 1972). In the Serengeti the cheetah's prey consists mostly of gazelle (91 per cent), although sometimes larger prey species are taken (Foster and Kearney 1967; Kruuk and Turner 1967; Schaller 1968). There is evidence that these larger prey items are usually killed by several cheetahs together (Foster and Kearney 1967); the most common prey, Thomson's Gazelle, is considerably smaller than the predator and hunted solitarily.

The third species of the Felidae studied in the Serengeti is the leopard, *Panthera pardus*, which is almost invariably solitary and prefers the denser vegetation of watercourses, rocks, and woodlands above the more open savannah habitat of lion and cheetah. Each individual appears to have its own home range, the ranges of females being more or less exclusive of each other, and those of males encompassing ranges of several females (Eisenberg and Lockhart 1972; Schaller 1972). The leopard diet in the Serengeti is more varied than that of other cats in the same area; it contains gazelle (31 per cent), reedbuck (11 per cent), and impala (16 per cent), but also birds and many smaller mammals. These figures are biased in favour of the larger prey species, however (Kruuk and Turner 1967).

The example of these three cats exploiting the same areas gives support to the hypothesis concerning the relation between prey size and social organization. And at least some aspects of the communication behaviour appears to vary between these species in a manner similar to that indicated in the hyaenas. Vocalizations of lions are strikingly diverse, frequent, and loud (Schaller 1972), whilst leopards and cheetah are silent. Although these latter two species use, for instance, an advertising call, they do this much less often and much softer than the lion with its frequent thundering roar. But visual displays appear to be similar and almost equally well developed in most species (Leyhausen 1956; Schaller 1972).

There are other species amongst the Felidae, however, which if considered together with the previous three would rather change the picture of a simple relation between prey size and social behaviour. For instance, the tiger, *Panthera tigris*, is clearly a solitary hunter, although the social and territorial organization is by no means clear (Schaller 1967). About half of its prey, in several areas, consists of species of the same size or larger than the tiger itself (deer, gaur, buffalo). Similarly, the puma, *Felis concolor* has a territorial organization as described above for leopard (Hornocker 1969)—but its average prey size is considerably larger, and the majority of its victims (various species of deer) exceed the predator in weight (Robinette, Gashwiler, and Morris 1959; Hornocker 1970). It may well be that the tiger and the puma have overcome the difficulties of exploiting the large-ungulate resource in a way different from that of the lion, possibly because they live in a habitat where visibility is poor. Tiger, puma, leopard, jaguar, and many other species of cat hunt in often very dense vegetation, where prey is difficult to find and the number of prey encounters may be at a premium. In that case dispersed hunting would give an increased chance of prey encounter, whilst for species like the lion the easy detectability of prey in the open savannah habitat would ensure a sufficient number of prey sightings. Predators using hunting techniques which rely on the co-operation between hunters to overcome the aggressive defence of prey species (such as the social hyaenids and canids) might not be able to exploit the ungulates in forests because the paucity of encounters would make it impossible to feed a pack; but the felids, with their hunting technique relying on surprise to overcome the prey's anti-predator mechanisms, would still be able to utilize this food resource solitarily. They would thereby forgo some of the advantages of group hunting.

6.7. Canidae

There are several gregarious species of Canidae, notably the wolf, *Canis lupus*, the African Wild Dog, *Lycaon pictus*, and the Indian dhole *Cuon alpinus*; but most members of the dog family are solitary, exploiting populations of small mammals and insects; they may also be partly vegetarian. Wolf packs may consist of up to 36 individuals (Rausch 1967), although the species is also often seen solitary; in spring and summer the activities centre on the den where several adults may attend to what is usually only one litter of pups (Murie 1944; Mech 1970). It is supposed that families and individuals often join into packs in wintertime. There are several observations of aggression between wolves related to the defence of a territory (Murie 1944; Cowan 1947; Mech 1966, 1970). Amongst the Canidae it is this species which takes on the largest prey—often large herbivores such as the moose (which weighs 500 kilograms) (Mech 1966), but also other smaller animals, even as small as voles (Murie 1944). Although no single study has been made of the number of wolves involved in hunting different kinds of prey it is clear that there is a close relationship between prey species and the number of hunters, when accounts from different sources are compared. For instance, Dall Sheep and caribou are usually run down by wolves alone (Murie 1944; Crisler 1956), whilst the moose with its formidable defence is habitually taken on only by large packs (Burkholder 1959; Mech 1966). Mice, not surprisingly, are hunted by solitary wolves with the characteristic pounce (Murie 1944).

In contrast with this flexible arrangement as found in wolves is the behaviour and food ecology of species such as the various foxes (Scott and Klimstra 1955; Englund 1965; Burrows 1968), jackals (van der Merwe 1958, and personal observation) and many others, which all have a diet of small rodents, insects, fruits, carrion, etc. Interestingly, when jackals, *Canis mesomelas* and *C. aureus*, attempt to take young Thomson's Gazelle (which are actively defended by the mother) it was observed that their hunting success was only 16 per cent when hunting alone, but 67 per cent when male and female were operating as a team (Wyman 1967). All these species of fox, jackal, and other canids are solitary or live in pairs, and they appear to defend a relatively large but vaguely defined territory.

Several studies have been carried out on intra-specific communication in the canids. The classic work on displays in wolves (Schenkel 1948) indicates the enormous versatility of expression in this species; whilst comparison with communication in foxes shows that the wolf has 1·4 times as many distinct display patterns as has the red fox (Tembrock 1954, 1957). This relation between social organization and differentiation of communication patterns in the Canidae has been the subject of several studies since then (Kleiman 1967; Fox 1970). It runs closely parallel to that described in the Hyaenidae, but there is a striking exception in the differentiation of the communication behaviour of a highly gregarious species of canid—the African Wild Dog. In this species one is struck by an exceptional lack of variety in the visual display patterns (Kühme 1965; Kleiman 1967), and these dogs have remarkably few different sounds (mostly rather soft). They are gregarious to such a degree that they rarely move out of sight of each other, roaming around in tight packs of 2–40 individuals (average 11; Kruuk 1972), with very large and completely overlapping home ranges. Also, African Wild Dogs are mostly active and hunting in daylight, in open savannah country (Kühme 1965). The packs have a rather constant membership over long periods; there is little evidence for the existence of dominance order (Kühme 1965; Estes and Goddard 1967), except for a not very strict ranking order during the hunt (Kruuk and Turner 1967) and temporarily around a den site (van Lawick-Goodall 1970). There is little overt aggression between pack members (although there are exceptions to this (van Lawick 1973)); young pups have right of way near the kill, and all members of the pack participate in regurgitating food for pups and pack members which did not reach a kill before it was completely consumed. In short, Wild Dog packs are tightly knit, 'altruistic' communities, where members are in almost continuous close proximity to each other with little intra-pack strife; there is less need for an elaborate, unambiguous signalling system. Probably the social organization where a real need for conspicuous and unambiguous means of communication exists is the society in which transitions between individual and group are frequent, not the 'simple' non-gregarious territorial system or the only-gregarious permanent group system.

The hunting habits of Wild Dogs have been the subject of several studies (Estes and Goddard 1967; Kruuk and Turner 1967; Pienaar 1969; Schaller 1972). In the Serengeti their prey consists largely of gazelle and wildebeest, which they run down over long distances with a very high success rate; their numbers per pack do not appear to vary with the kind of prey that is being hunted. Wild Dogs are clearly very dependent on each other during the hunt; I have observed individual dogs chase and catch gazelle, but they were not able to kill the victim until others arrived on the scene.

6.8. Viverridae and Mustelidae

From the preceding sections it is clear that in the Hyaenidae, Felidae, and Canidae the foraging and social organization complex has gone through several convergent evolutionary developments. There are other carnivore families, however, where a quite different phenomenon must have affected group organization. Among the Viverridae, for instance, there are some highly gregarious species of mongoose which, superficially at least, appear to have foraging techniques and a food selection not very different from those of related solitary species.

Thus, in the Serengeti National Park there are six species of mongoose, of which two, the Banded Mongoose, *Mungos mungo*, and the Dwarf Mongoose, *Helogale undulata*, live in troups of up to respectively 32 (Neal 1970; Rood 1973) and 15 individuals (Hendrichs 1972), within a definite home range, and troup members almost continuously in contact with each other through vocalization or by sight. Both species are diurnal, both obtain most of their food (arthropods, molluscs, bulbs, fruits, etc.) by rummaging through the vegetation, scratching the soil, and digging. There is a difference in habitat choice between the two species: the Banded Mongoose occurs frequently in open grassland, and the Dwarf Mongoose mostly in or near dense cover. Another species, the Slender Mongoose, *Herpestes sanguines*, occurs in the same habitat as the Dwarf Mongoose; this species is also diurnal, and entirely solitary, feeding mostly on insects and small vertebrates, even fairly large birds. In the Banded Mongoose's habitat the Egyptian Mongoose, *Herpestes ichneumon*, occurs, much larger but with very similar habits to the slender one, and the White-tailed Mongoose, *Ichneumia albicauda*, which is nocturnal, solitary, and also appears to feed on small vertebrate prey, insects, fruit, etc.

After some of the initial studies such as those carried out by Zannier (1965), Taylor (1970), and Rasa (1972) clearly more information is needed before more definitive statements about the relationship between environment and social behaviour can be made; but whatever these relationships are, they are almost certainly different in principle from those found amongst the carnivores discussed earlier. It may be suggestive that the Banded Mongoose (perhaps also the Dwarf Mongoose) shows a spectacular social response towards predators (eagles and jackals): the troup members join in a very tight bunch with mouths pointing in all directions, giving the appearance of one large organism defending itself (Rood 1973, and personal observation). The defence against predators may be a much stronger socially organizing force in these small creatures than in the carnivores discussed previously; a diurnal, relatively small species feeding in open grassland may benefit greatly from social protection against enemies. An anti-predator function has also been ascribed to schooling fish behaviour (Cushing and Harden Jones 1968; Neill 1970), and to the formation of large groups in open-country primates (Crook and Gartlan 1966).

Similar problems are found in the Mustelidae, where solitary species such as stoats, *Mustela erminea*, and weasels, *M. nivalis*, usually live on voles, mice, etc. but may take rabbits which are many times larger than they are themselves (Day 1968; Hewson and Healing 1971). Perhaps they are able to exploit these larger prey species, as it was suggested for some of the Felidae, by virtue of their special killing technique (Wüstehube 1960). Badgers, *Meles meles*, however, exist largely on earthworms, insects, vegetable matter, a few small rodents, and young rabbits—and they are a relatively social species, occurring in groups of up to 12 individuals with an obviously highly developed but little-known organization (Neal 1948). The North-American Badger, *Taxidea taxus*, is solitary (Sargeant and Warner 1972) with a more typical carnivore diet, such as ground-squirrels, mice, birds, eggs, and insects (Cowan and Guiguet 1960).

It is not impossible that also in these carnivore families the necessity for most efficient exploitation of particular food resources exerts an important influence on the tendency to socialize (there are interesting parallels between the foraging of badgers and that of social mongooses). But even if this were the case this kind of pressure would not necessarily be the same as that described above for the hyaenas, cats, and dogs. For instance, food distribution might be the important factor, rather than prey size. A study is in progress at present in an attempt to shed more light on the ecological significance of social organization of the carnivores discussed in this last section.

6.9. Discussion

In the previous section I suggested that in at least some carnivores there is a close relationship between the success of exploiting the important resource of 'relatively large and difficult prey', and the degree to which the predators co-operate with each other. Carnivores which exploit such large prey resources tend to show a social organization different from those of the same species which live off smaller-sized prey which is more efficiently acquired solitarily. This trend is visible also when comparing various related species of carnivores which differ in their food preferences.

Thus, foraging appears to be closely related to group size and structure, and it was indicated also that several other aspects of social behaviour, such as communication, are involved in this relationship. Possibly, this holds true also for aggression, parental care, and other behaviour patterns. I have argued that it is the prey availability and acquisition through which important selection pressures operate on the social organization and behaviour of the carnivores. The same selection pressures appear to have been effective in several different carnivore families, in which one or two species have evolved a flexible group organization whilst most of the others are solitary, with large overlapping feeding territories.

The complex of relations between habitat, food exploitation, and social organization was first investigated in birds (Crook 1964, 1965) and in primates (Crook and Gartlan 1966; Crook 1970). Several comparisons between these primate studies and those of carnivores suggest themselves. In general, primates show little active co-operation when foraging and it is therefore difficult to see the selective advantage of large group size as an adaptation to the handling of food. It has been argued that group size is affected by the degree of heterogeneity of the food distribution. Spatial and seasonal variation in the availability of food necessitates a large feeding range, which can then be inhabited by many individuals (for example, Clutton-Brock 1972, on *Colobus* species). There are many possible advantages to all these individuals for joining in a group, rather than exploit the large range on their own; for instance, they may learn new foods and feeding techniques from each other (Kawai 1965).

There are a few primates, which have a social organization such as that found in the social cats, dogs, and hyaenas, within which individuals have considerable flexibility as to their gregariousness. These are, for instance, the chimpanzee (van Lawick-Goodall 1968; Izawa 1970) and primitive man; they allow for a solitary existence and various degrees of gregariousness within 'societies', and these societies are separated from each other by more or less pronounced group territorial behaviour, as in, for example, the Bushmen (Tobias 1964; Eibl-Eibesfeldt 1972).

Having seen how (in the Carnivora) this type of social organization has arisen several times in evolution one cannot help wondering if the same selective forces could not have been operating also on the social organization of man, especially since the diet of primitive man was and is to an important extent carnivorous, and since his prey animals are often considerably larger and faster than man himself. There is evidence that also in primitive hunting man the group size is related to the species of prey hunted, and to the means of hunting (for example, Balikci 1968; Woodburn 1968). Many aspects of man's behavioural plasticity itself (for example, in community organization, adaptation to different environments, and variation in the occurrence of aggression) may have a similar evolutionary origin as the behavioural plasticity found in carnivores. There are several other aspects of man's organization and ecology which appear to be reminiscent of carnivores rather than of the other primates—foodsharing, occupation of a central living and rearing site, and other behaviour patterns (Tiger and Fox 1966; Schaller and Lowther 1969). But appropriate studies are still badly needed; there is an enormous gap in our knowledge of man as a hunting animal—a gap much greater in the case of man than for many of the carnivores.

If the original biological function of some aspects of man's social organization is, indeed, similar to that of some of the gregarious carnivores one could extend this inter-specific comparison to some of the details of the respective societies. It may be, for instance, that the original biological function of our present day 'gangs' (or other male groupings) is similar to that of packs of hyaenas or wolves: communal hunting, and/or group territorial defence. If this is so, it would be useful to understand the former 'natural' situation in which this kind of grouping occurred, in order to find an appropriate stimulus situation which could provide a proper feedback to the participating individuals—in order to prevent frustrations, or to direct aggression.

An argument is often put forward that in order to understand the present-day behavioural problems of cultural man one has to study man's psychology as it is now, and that it is merely wasting time to draw comparisons with the behavioural organization of other animals. I would suggest that, on the contrary, it is probably in just these kinds of comparisons between animal and human behaviour that ethology can contribute something to the management of the human species. The study of biological function may provide human psychology with clues about the stimuli which would constitute the most appropriate feedback to regulate ('satisfy') our own behaviour.

6.10. Summary

A comparison has been made of hunting by several Carnivora, with special reference to aspects of co-operation and gregariousness. Solitary or social hunting has been related to diet and food selection, and to aspects of social behaviour such as communication and social organization. It is concluded that a carnivore's food is often related to its social behaviour; more specifically, social behaviour is probably in several ways adapted to the exploitation of the species' resources. The conclusions have some importance also for the study of human behaviour. It is one of the very important contributions of Niko Tinbergen to have pointed the way towards this approach.

Acknowledgements

I am very grateful to Dr. G. Barlow, Dr. B. Bertram, and Dr. T. H. Clutton-Brock for their criticisms of the manuscript for this essay.

References

BALIKCI, A. (1968). The Netsilik eskimos: adaptive processes. In *Man the hunter* (eds R. B. Lee and I. DeVore), pp. 78–82. Aldine, Chicago.

BURKHOLDER, B. L. (1959). Movements and behavior of a wolf pack in Alaska. *J. Wildl. Mgmt* **23**, 1–11.

BURROWS, R. (1968). *Wild fox.* David and Charles, Newton Abbot, England.

CLUTTON-BROCK, T. H. (1972). *Feeding and ranging behaviour in the red colobus monkey,* Colobus badius. Ph. D. Thesis, Cambridge University.

COWAN, I. MacT. (1947). The timberwolf in the Rocky Mountain national parks of Canada. *Can. J. Res.* **25**, 139–74.

—— GUIGUET, C. J. (1960). *The mammals of British Columbia.* British Columbia Provincial Museum, Victoria, B.C.

CRISLER, L. (1956). Observations of wolves hunting caribou. *J. Mammal.* **37**, 337–46.

CROOK, J. H. (1964). The evolution of social organisation and visual communication in the Weaver Birds (Ploceinae). *Behaviour, Suppl.* **10**, 1–178.

—— (1965). The adaptive significance of avian social organisations. *Symp. zool. Soc., Lond.* **14**, 181–218.

—— (1970). The socio-ecology of primates. In *Social behaviour in birds and mammals* (ed. J. H. Crook), pp. 103–68. Academic Press, London and New York.

—— GARTLAN, J. S. (1966). Evolution of primate societies. *Nature, Lond.* **210**, 1200–3.

CROZE, H. (1970). Searching image in carrion crows. *Z. Tierpsychol., Suppl.* **5**, 1–85.

CUSHING, D. K. and HARDEN JONES, F. R. (1968). Why do fish school? *Nature, Lond.* **218**, 918–20.

DAY, M. G. (1968). Food habits of British stoats (*Mustela erminea*) and weasels (*Mustela nivalis*). *J. Zool. Lond.* **155**, 485–97.

EIBL-EIBESFELDT, I. (1972). *Die !Ko-Buschmanngesellschaft. Gruppenbildung und Aggressionskontrolle bei einem Jäger und Sammlervolk.* R. Piper Verlag, München.

EISENBERG, J. F. and LOCKHART, M. (1972). An ecological reconnaissance of Wilpattu National Park, Ceylon. *Smithson. Contr. Zool.* **101**, 1–118.

ENGLUND, J. (1965). The diet of foxes (*Vulpes vulpes*) on the island of Gotland since Myxamatosis. *Viltrevy* **3**, 507–30.

ESTES, R. D. (1966). Behaviour and life history of the Wildebeest (*Connochaetes taurinus* Burchell). *Nature, Lond.* **212**, 999–1000.

—— GODDARD, J. (1967). Prey selection and hunting behavior of the African Wild Dog. *J. Wildl. Mgmt* **31**, 52–70.

FOSTER, J. B. and KEARNEY, D. (1967). Nairobi National Park game census, 1966. *E. Afr. Wildl. J.* **5**, 112–20.

FOX, M. W. (1970). A comparative study of the development of facial expression in Canids: Wolf, Coyote and Foxes. *Behaviour* **36**, 49–73.

HENDRICHS, H. (1972). Beobachtungen und Untersuchungen zur Ökologie und Ethologie, ins besondere zur soziale Organisation ostafrikanischer Säugetiere. *Z. Tierpsychol.* **30**, 146–89.

HEWSON, R. and HEALING, T. D. (1971). The stoat *Mustela erminea* and its prey. *J. Zool. Lond.* **164**, 239–44.

HORNOCKER, M. (1969). Winter territoriality in mountain lions. *J. Wildl. Mgmt* **33**, 457–64.

—— (1970). An analysis of mountain lion predation upon mule deer and elk in the Idaho Primitive Area. *Wildl. Monogr.* **21**,

Izawa, K. (1970). Unit groups of chimpansees and their nomadism in the savanna woodland. *Primates* **11**, 1–46.

Jarman, P. J. (1974). The social organisation of antelope in relation to their ecology. *Behaviour* **48**, 215–67.

Kawai, M. (1965). Newly acquired precultural behavior of the natural troop of Japanese monkeys on Koshima Isles. *Primates* **6**, 1–30.

Kleiman, D. G. (1967). Some aspects of social behavior in the Canidae. *Am. Zool.* **7**, 365–72.

Klingel, H. (1967). Soziale Organisation und Verhalten freilebender Steppenzebras. *Z. Tierpsychol.* **24**, 580–624.

Kruuk, H. (1964). Predators and anti-predator behaviour of the black-headed gull (*Larus ridibundus* L.). *Behaviour, Suppl.* **11**, 1–129.

—— (1966). Clan-system and feeding habits of spotted hyaenas (*Crocuta crocuta* Erxleben). *Nature, Lond.* **209**, 1257–8.

—— (1970). Interactions between populations of spotted hyaenas (*Crocuta crocuta* Erxleben) and their prey species. In *Animal populations in relation to their food resources* (ed. A. Watson), pp. 359–374. Blackwell, Oxford.

—— (1972). *The Spotted Hyena.* University of Chicago Press, Chicago.

—— (1975). Social behaviour and foraging of the Striped Hyaena (*Hyaena vulgaris*). *E. Afr. Wildl. J.* **13** (in press).

—— Sands, W. A. (1972). The aardwolf (*Proteles cristatus* Sparrman, 1783) as predator of termites. *E. Afr. Wildl. J.* **10**, 211–28.

—— Turner, M. (1967). Comparative notes on predation by lion, leopard, cheetah and wild dog in the Serengeti area, East Africa. *Mammalia* **31**, 1–27.

Kühme, W. (1965). Freilandstudien zur Soziologie des Hyänenhundes (*Lycaon pictus lupinus* Thomas 1902). *Z. Tierpsychol.* **22**, 495–541.

van Lawick, H. (1973). *Solo.* Collins, London.

—— and van Lawick-Goodall, J. (1970). *Innocent killers.* Collins, London.

van Lawick-Goodall, J. (1968). The behaviour of free-living chimpanzees in the Gombe Stream Reserve. *Anim. Behav. Monogr.* **1**, 165–311.

Leyhausen, P. (1956). *Verhaltensstudien an Katzen.* Paul Parey, Berlin.

MacLaughlin, R. (1970). *Aspects of the biology of cheetahs* Acinonyx jubatus (*Schreber*) *in Nairobi National Park.* M. Sc. Thesis, University of Nairobi.

Mech, L. D. (1966). The wolves of Isle Royale. *U.S. natn. Park Serv. Fauna* **7**, 1–210.

—— (1970). *The wolf.* National History Press, New York.

van der Merwe, N. J. (1958). The jackal, fauna and flora. *Z. Tierpsychol.* **15**, 121–23.

Murie, A. (1944). The wolves of Mt. McKinley. *U.S. natn. Park Serv. Fauna* **5**, 1–238.

Neal, E. (1948). *The badger.* Collins, London.

—— (1970). The banded mongoose, *Mungos mungo* Gmelin. *E. Afr. Wildl. J.* **8**, 63–71.

Neill, S. R. St. J. (1970). *A study of anti-predator adaptations in fish, with special reference to silvery camouflage shoaling.* D. Phil. Thesis. Oxford University.

Pienaar, U. de V. (1969). Predator-prey relationships amongst the larger mammals of the Kruger National Park. *Koedoe* **12**, 108–76.

Rasa, O. A. E. (1972). Aspects of social organisation in captive dwarf mongooses. *J. Mammal.* **53**, 181–5.

Rausch, R. A. (1967). Some aspects of the population ecology of wolves, Alaska. *Am. Zool.* **7**, 253–65.

ROBINETTE, W. L., GASHWILER, J. S., and MORRIS, O. W. (1959). Food habits of the cougar in Utah and Nevada. *J. Wildl. Mgmt* **23**, 261–73.

ROOD, J. P. (1973). The banded mongoose. *Africana (Nairobi)* **5**, (No. 2), 14.

SARGEANT, A. B. and WARNER, D. W. (1972). Movements and denning habits of a badger. *J. Mammal.* **53**, 207–10.

SCHALLER, G. B. (1967). *The deer and the tiger.* University of Chicago Press, Chicago.

—— (1968). Hunting behaviour of the cheetah in the Serengeti National Park, Tanzania. *E. Afr. Wildl. J.* **6**, 95–100.

—— (1972). *The Serengeti lion.* University of Chicago Press, Chicago.

—— LOWTHER, G. R. (1969). The relevance of carnivore behaviour to the study of early hominids. *Southwest. J. Anthropol.* **25**, 307–41.

SCHENKEL, R. (1948). Ausdrucksstudien an Wölfen. *Behaviour* **1**, 81–129.

SCOTT, T. G. and KLIMSTRA, W. D. (1955). *Red Foxes and a declining prey population.* Monograph Series Vol. 1, pp. 1–123. Southern Illinois University.

SINCLAIR, A. R. E. (1974). The natural regulation of buffalo populations in East Africa. *E. Afr. Wildl. J.* **12**, 291–311.

TAYLOR, M. E. (1970). Locomotion in some East African Viverrids. *J. Mammal.* **51**, 42–51.

TEMBROCK, G. (1954). Rotfuchs und Wolf, ein Verhaltensvergleich. *Z. Säugetierek* **19**, 152–9.

—— (1957). Zur Ethologie des Rotfuchses. *Zool. Gart., Lpz.* **23**, 289–532.

TIGER, L. and FOX, R. (1966). The Zoological Perspective in social science. *Man* **1**, 75–81.

TINBERGEN, N. (1967). Adaptive features of the Black-headed Gull (*Larus ridibundus* L.). *Proc. int. orn. Congr.* **14**, 43–59.

—— (1968). On war and peace in animals and man. *Science, N.Y.* **160**, 1411–18.

—— (1972). Functional ethology and the human sciences. The Croonian Lecture, 1972. *Proc. R. Soc.* B**182**, 385–410.

—— BROEKHUYSEN, G. J., FEEKES, F., HOUGHTON, J. C. W., KRUUK, H., and SZULC, E. (1962). Egg shell removal by the black-headed gull, *Larus ridibundus* L., a behaviour component of camouflage. *Behaviour* **19**, 74–117.

—— IMPEKOVEN, M., and FRANCK, D. (1967). An experiment on spacing out as a defence against predation. *Behaviour* **28**, 307–21.

TOBIAS, P. V. (1964). Bushman hunter-gatherers: a study in human ecology. In *Ecological studies in Southern Africa* (ed. D. H. S. Davis). W. Junk, The Hague.

WALTHER, F. R. (1969). Flight behaviour and avoidance of predators in Thomson's gazelle (*Gazella thomsoni* Guenther 1884). *Behaviour* **34**, 184–221.

WOODBURN, J. (1968). An introduction to Hadza ecology. In *Man the hunter*, (eds R. B. Lee and I. DeVore), pp. 49–55. Aldine, Chicago.

WÜSTEHUBE, C. (1960). Beiträge zur Kenntnis besonders des Spiel- und Beuteverhaltens einheimischer Musteliden. *Z. Tierpsychol.* **17**, 579–613.

WYMAN, J. (1967). The jackals of the Serengeti. *Animals (Lond.)* **10**, 79–83.

ZANNIER, F. (1965). Verhaltensuntersuchungen an der Zwerg-manguste *Helogale undulata rufula* im Zoologischen Garten Frankfurt am Main. *Z. Tierpsychol.* **22**, 672–95.

7. The extension of the orientation system of *Bembix rostrata* as used in the vicinity of its nest

J. J. A. VAN IERSEL

7.1. Introduction

IT IS about thirty years ago that Dr. Tinbergen and I were watching the orientation behaviour of Digger Wasps in a beautiful part of the sand-dunes at Hulshorst (Veluwe, the Netherlands). He was studying the bee-wolf (*Philanthus triangulum*), and I had fallen in love with *Bembix rostrata*, a species catching flies as prey for the larva. Since that time I have continued the study of this fascinating wasp species and it seemed appropriate for me to contribute to this book dedicated to my highly esteemed friend Niko Tinbergen an essay on some aspects of the orientation behaviour of *Bembix rostrata*.

The point I want to stress in this essay is that in evaluating the orientation behaviour of the wasp while it is flying around in the vicinity of the nest we have to take into account the influence of landmarks located far away, even as far as what, for a given wasp, may be called its horizon. The old idea that a wasp returning from far away uses a set of successive beacon systems, narrowing down the nearer the wasp comes to its nest, can not be true. Van Beusekom (1946, 1948) has already shown that in *Philanthus* optical points (big branches) at a distance of about one metre from the nest play a role in the location of the nest-entrance. Also Baerends (1941, p. 260) describes an experiment suggesting that landmarks farther away than the direct vicinity of the nest of *Ammophila campestris* (but in his opinion not too far) are important in locating the nest. In *Bembix rostrata*, I think, we have to accept that all characteristics of the wide surroundings, if possessing potential beacon value, are used by the wasp even when very close to landing. It uses all of what it is able to see.

This does not deny, of course, that marks in the direct vicinity do play a role. In fact, Tinbergen 1932 and Tinbergen and Kruyt (1938) have shown the importance of these marks. After training *Philanthus* on a system of two types of objects differing in only one respect and arranged in a circle close around the nest, they were able to determine the orientational value of factors such as the height, size, or contrast of objects by carrying out choice experiments consisting of two circles, one at each side of the nest and composed of either one of the object types. In such a situation the wasp could show its preference.

It became soon clear to me that this training method could be applied to *Bembix* only with great difficulty. Though the wasp did follow objects to which it had been trained, and though disorientation of the wasp occurred when the contrast pattern of large parts of the nest surroundings was disturbed by covering the bottom with a thin layer of sand, *Bembix* very soon stopped this behaviour and landed time after time on the site of the closed nest-entrance without any hesitation.

Fortunately I found another method suitable to draw conclusions about the importance of various object-characteristics for orientation. When a wasp lands without any hesitation, it will leave its nest, when the time comes, also without hesitation. However, when it is disturbed by an unknown object in the nest vicinity, it will show a reorientation flight before leaving, even when the disturbing object is taken away during the wasp's stay in the nest. Measurements of the disorientation time (DOT, in seconds) and the reorientation time (ROT, in seconds) were correlated together and also with characteristics of the disturbing objects (van Iersel and van den Assem 1965). We shall not deal with the question of whether 'training method' and 'disturbance method' lead to the same conclusions about the orientational value of object-characteristics, but concentrate on factors that are responsible for the fact that the disturbing effect of objects is dependent on their location. At this point the reader is referred to Fig. 7.1 which illustrates the terminology adopted in classifying the environment around a *Bembix* nest. Using a number of vertical flat black sheets, it very soon became clear that a frontal object, that is one placed at right-angles to the wasp's direction of approach, and beyond the nest-entrance, resulted in a longer DOT than the same object placed caudally, i.e. at right-angles to the direction of approach, but behind the entrance (see Fig. 7.1). As to the sides of the nest, it was very regular to find a side of 'heavier disturbance' and it is especially this latter phenomenon, which we shall call 'sidedness', that we are interested in. What are the factors determining this 'sidedness'?

In the 1965 paper of van Iersel and van den Assem, an experiment was described showing that *Bembix* uses optical characteristics of the far-distant horizon to locate the nest-entrance. When the frontal view was blocked by a large vertical sheet at a distance of 60 cm a strong disorientation was found, but it was reduced dramatically by opening a window in the sheet through which part of the horizon could be seen. In studying the

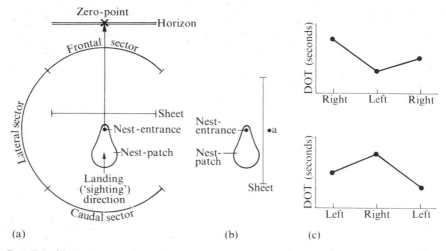

Fig. 7.1. (a) The system of notation used in describing a nest and its environment for testing, showing the nest-patch with the entrance and the normal landing or sighting direction of the wasp which cuts the apparent horizon-line at the 'zero-point'. The sheet is a vertical flat sheet as used in tests, presented frontally in (a) and laterally in (b). (b) Position of 'false sitting place'—marked a—(see text, §7.4), when the vertical sheet is presented in a lateral position. (c) Typical results from a pair of triplet tests, for a right-sided wasp. Successive tests are shown on abcissa. The time of disturbance on the ordinate is greatest when objects are placed on the right side.

phenomenon of 'sidedness' a possible influence of far-distant cues has to be taken into account.

7.2. Material and methods

The observations were made in two colonies. Data are given from a big inland colony at Hulshorst (Hu-colony; years 1958–62) and from a smaller one situated in the dunes of one of the islands (the Schouwen-Duiveland, or S.D.-colony; years 1963–72) of the province of Zeeland, the Netherlands. The situations in the two colonies were very different. The Hu-colony was situated on a flat, horizontal, sandy area, blown out by the wind, and covered by a sparse vegetation of short grasses and patches of dark moss; the S.D.-colony was situated on a sandy slope of the dunes, bare except for some moss. The axes of the nest were directed towards all sides in Hulshorst, but more or less parallel to each other and facing slope-upwards in S.D. At a rather great distance (about 60 metres from the centre of the plain) the Hu-colony was surrounded by a wood of pine and spruce; solitary birches were standing on the plain itself. The horizon-line, that is the line between the dark wood and the clear sky above it, was rich in detail. The tops of bigger trees or groups of trees as well as the silhouettes of solitary birches projected above the rather horizontal flat sections of other parts of the horizon. So from every place in the colony 'horizon-objects' stood clearly out against the sky as seen from ground level. The horizon at S.D. is of a very dull type as compared with the Hu-colony. The edge of the pine wood is much nearer to the centre of the colony (about 6 metres) and, besides that, the horizon-line has scarcely any distinct breaks, for the trees are all of similar height.

The disturbing objects used were all flat hardboard sheets, painted dull black (or white) with the same board paint and were all presented vertically in a small device made of iron wire. In order to find out differences in disturbing value (DOT) as related to position, a triplet of presentations is given in quick succession (e.g. left–right–left), without allowing the wasp to enter the nest between presentations. It is necessary to carry out three presentations in one trial, since in many cases repetition of presentations results in a decrease of DOT (Fig. 7.1 (c)). We shall not deal with the question how the wasp 'adapts' to the object disturbing its expected pattern of orientation points, but simply state that a side of heavier disturbance can be established only when the middle presentation of a triplet either yields a lower or a higher DOT than found with the object in the other position.

During a reproductive season (July–August) a *Bembix* female constructs a number of nests. When we speak, therefore, of a number of wasps showing an effect, in fact the number of nests tested is meant. However, many different individuals are present in any comparison made in the following pages. In its good years the Hu-colony certainly consisted of more than 300 nests, the S.D.-colony of more than 100. One- and two-sample χ^2 tests have been used to calculate p-values (0·05 or less is taken to be significant). Fisher's exact probability test was used when appropiate (Siegel 1956).

7.3. Factors not affecting sidedness

In trying to find an explanation for the 'sidedness' phenomenon revealed by a triplet of presentations, we will first consider a number of factors which appear not to have any great importance.

The position with which a triplet is started

Irrespective of a left or right starting position, the sidedness for a given nest is the same in 85·1 per cent of the cases in the Hu-colony ($n = 137$; $p < 0·001$) and in 90·3 per cent in S.D. ($n = 604$; $p < 0·001$). Interestingly enough, in a number of wasps the sidedness in not independent of the starting position, but we shall not analyse these cases in this essay.

The influence of time of testing

Sidedness is also largely independent of the time at which testing is done during the period the wasp occupies a particular nest. When tests are taken on two consecutive days, no changes in sidedness were found in the Hu-colony ($n = 17$), but a few wasps changed in the S.D.-colony (6·5 per cent instability, $n = 107$). Many wasps were tested on a variable number of days over a variable time-span. Taking the time-spans together in groups of 2–6, 7–11, and 12–20 days, changes in sidedness were found in respectively 7·9 per cent ($n = 63$), 12·5 per cent ($n = 24$), and 24·4 per cent ($n = 41$) for the Hu-colony; 8·3 per cent ($n = 397$), 16·0 per cent ($n = 175$), and 17·2 per cent ($n = 70$) for the S.D.-colony. In all groups of time-spans a significant proportion of the wasps showed stability in sidedness over time (Hu-colony, $p < 0·001$; S.D.-colony, $p < 0·001$). With longer time-spans there is a tendency for instability to increase (Hu-colony; $p < 0·05$; S.D.-colony: $p < 0·01$). In a number of cases we were able to determine the cause for such changes (see §7.4: *Changes in horizon-line*).

The presence and position of the sun (shadow)

Three arguments can be given against a possible role of the sun in determining sidedness.

1. The probability of scoring an inconclusive triplet (i.e. one with a continuous decreasing DOT) is exactly the same when wasps are active under sunny conditions or remain active under complete thick overcast when air-temperature is high (Hu-colony: respectively 23·9 per cent, $n = 825$ and 23·4 per cent $n = 111$). Furthermore, individual wasps tested under clear and overcast conditions appear to have the same sidedness (20 out of 21 cases). Though the presence or absence of the sun appears to have no influence on the persistence of sidedness, it could still be true that the position of the sun and shadows could have an effect on the sidedness— left or right—actually found.

2. In many cases tests with a particular wasp were carried out with the sun either shining from the left or from the right. If the sun's position was determining, one would expect a preponderance of type of sidedness with a particular position of the sun. Table 7.1 shows that no such preponderance exists whatsoever: given a certain position of the sun the probability of scoring a left or a right animal is exactly the same.

3. According to Table 7.1 in 216 cases (Hu-colony) stable sidedness is found with stable sun positions. In 21 cases (9 per cent) changing sidedness may also occur when the position of the sun (while taking triplets) is not the same. It occurs in 11 per cent of the 45 cases of variable sun position. Apart from the fact that stability of sidedness is present under such variable sun positions ($p < 0·001$), the equal percentage of exceptions under both conditions allows us to conclude that instability of sidedness is also independent of the position of the sun. It must depend on other factors.

TABLE 7.1

Sidedness of wasps as related to the position of the sun

	Sidedness		Sun position and sidedness	
	Left	Right	Same	Opposite
Hu-colony (1958–62)				
sun from left	58	47	116	100
sun from right	53	58		
S.D.-colony (1963–72)				
sun from left	58	42	63	46
sun from right	4	5		

Note: The small number of nests in the S.D.-colony tested with the sun from the right is a consequence of three compounding factors: (a) the very similar orientation of all the nests; (b) the apparent course of the sun in relation to the location of the colony; (c) the period of day when the wasps are most active.

From the arguments given it is concluded that sidedness is not determined by the sun at the moment of testing. Of course, it is still possible that the sun had an influence in the period the wasp learned the orientational system on which to rely later on and during which the asymmetry in this system was established. Unfortunately, we did not systematically record the sun's position during the phase of digging of the new nest, but data from the S.D.-colony may suggest an answer. Since digging of new nests starts at about noon and is completed, in the majority of cases the same afternoon and, because the axes of the nests in this colony are mostly parallel, it can be deduced that the sun must have been to the left during the phase the wasps learned their nest locations. If the sun played a role, one would expect a high preponderance of either left or right animals. Though there are significantly more left animals in the S.D.-colony ($n = 1402$, 63·6 per cent left, $p < 0.001$), the percentage of right animals is far too high to suggest a determining effect of the sun: the departure from an absolute preponderance for either left or right is very significant ($p < 0.001$).

Summarizing this section we may conclude that neither the presence nor the position of the sun are relevant to the occurrence of the direction of sidedness, nor does the sun determine the asymmetry in the wasp's orientation pattern during the phase when it was learning which environmental features to rely upon.

Characteristics of the nest surrounding

As already mentioned, some aspects of the immediate surroundings of the nest have orientational value. The position of these local characteristics may, therefore, (co-)determine the side of heavier disturbance. In order to investigate this possibility the ground to the left and right of the nest (up to about 50 cm) was classified according to four criteria (Hu-colony: $n = 100$ for each of them): (a) degree of darkness in general: a black sheet placed on ground covered by dark mosses was expected to be less disturbing than when placed by a nest whose background was pale greyish sand; (b) the degree of 'patchiness'; a dark sheet on an area of irregular contrast was predicted to be more disturbing than when placed at the other side of a more uniform appearance; (c) the degree of asymmetry of the whitish nest-patch; especially when placed close to the nest the vertical sheet may cut off a great portion (side of heavier disturbance) of such a conspicuous nest-patch; (d) the degree of density of vegetation (grasses); in many cases the ground was bare, but in others a clear difference existed between sides; the more densely overgrown side was expected to be the side of heavier disturbance.

Apart from the fact that in many cases no clear differences existed between sides (ranging from 39 per cent to 82 per cent for the types of qualifications), in which cases sidedness still was recorded, no significant correlations whatsoever were found between qualification and sidedness in case such differences did exist. For the S.D.-colony no other conclusions could be drawn.

Characteristics of the horizon-line as seen over the vertical sheet

Since many of the triplet tests were carried out using a low vertical (4 cm) sheet at a distance of 10 cm from the nest, the wasps could still have seen parts of the horizon-line projecting above the upper rim of the disturbing object while landing and even when sitting on the nest-patch. One could hypothesize that a difference between the horizon characteristics on the two sides determines the sidedness of the wasp; the more of the horizon remains to be seen, the less disturbing is the vertical sheet. What could be seen was judged by the observer viewing with one eye (the other closed) from about 2 cm above the nest entrance and looking over the 4-cm sheet at a distance of 10 cm parallel to the nest-axis ($n = 84$, Hu-colony). Again, the two sides were classified according to the following four criteria: (a) the length of the visible horizon-line; (b) the number of tree-tops projecting; (c) the position of such high objects—whether they are situated more to the frontal or caudal section of the horizon; and (d) the presence and position of a particular set of horizon elements to which I shall return below.

Apart from the fact that sidedness was found to exist in cases when there were no differences between the sides, using the above criteria (these varied from 11 per cent to 21 per cent of nests), there were also no significant correlations between sidedness and horizon characteristics even when differences were present (e.g. a clear sky at one side contrasting with a large section of visible horizon-line at the other). It is concluded, therefore, that the sidedness of a wasp cannot be a direct result of asymmetrical blocking of its view of the horizon. This conclusion is supported by the fact that testing with higher sheets, which always block the horizon completely, also reveals sidedness.

Characteristics and positions of the test object

With many wasps triplet tests were carried out with a number of different vertical test objects. They differed in dimensions, colour, and position. The sidedness of a wasp appears to be the same irrespective of the type of testing applied. For example, 98 per cent of the wasps (S.D.-colony, $n = 149$) showed the same sidedness when tested with a black or a white sheet of similar dimensions placed in the same lateral position. Similarly, using a black sheet 30 cm long and, 4 cm high, at a distance of 2 cm parallel to the nest-axis revealed the same sidedness as that found using a black sheet only 4 cm long, and 4 cm high, at a distance of 10 cm (Hu-colony: 96 per cent correspondence, $n = 125$; S.D.-colony: 95 per cent correspondence, $n = 562$). At the disturbed side the free frontal (and caudal) angle increases from $7\frac{1}{2}°$ in the former to $79°$ in the latter position. Again, a corresponding sidedness is found when a wasp is tested laterally and frontally or caudally. In such a frontal (or caudal) test a vertical sheet is presented in the frontal or caudal sector (see Fig. 7.1(a)) left and right from and at right-angles to the sighting (landing) direction. (Hu-colony: correspondence of sidedness 98 per cent, $n = 51$; S.D.-colony: 99·4 per cent correspondence, $n = 187$.)

Apparently, symmetrical disturbance of any part of a wasp's horizon reveals the same sidedness. Many positions of the disturbing object certainly do not block from view objects conspicuous from an objective point of view. We must conclude that by some kind of central integrative process the informational value of all parts of one side of the far-distant environment is given more weight than that from all corresponding parts of the other side.

What then are the characteristics of the far-distant environment which determine the sidedness of a wasp?

7.4. Relation between sidedness and conspicuous characteristics of the frontal horizon-line

General remarks

As mentioned already, at some points the rather flat and featureless horizon-line around the Hu-colony is interrupted by birches and higher trees standing out clearly against the sky. Fifteen of these elements could be distinguished and the frontal section of each nest (that is, that between 30° left and 30° right from the zero-point, see Fig. 7.1) was classified as to the presence, number, and size of these elements visible and their deviation in angle from the zero-point, which was found by extrapolating from the landing direction of the wasp. Wasps hovering above the nest-entrance are also nearly always viewing in this direction. The classification of the frontal horizon was said to be 'left' when one, two, or more of the conspicuous horizon elements were present at the left whereas the other side was free of such elements, or in case of competition between two objects (of equal size) when the left one was nearer to the sighting-point than the one at the right. Applying a set of rules, which will be discussed in some detail below, there appears to be a high correlation between the side classification of the frontal horizon and the sidedness of the wasp. For the Hu-colony there was correspondence in 82·6 per cent of nests ($n = 230$, $p < 0·001$).

If one classifies the frontal horizons, applying the same set of rules but now considering the conspicuous horizon elements in so far they are situated within the sections of 30–60° left and right from the zero-point, then no correlation between sidedness of a wasp and side classification of the horizon is found (Hu-colony, 53·7 per cent correspondence). It is clear that horizon elements determining the sidedness of a wasp must be situated not too far away from the sighting-point.

In a number of cases the sidedness of wasps could not be established because of inconclusiveness of the test triplet (ever-decreasing DOT). In many of these cases, however, the wasps gave some behavioural indications as to their possible side of heavier disturbance. These consisted of (a) asymmetry in flying pattern—a wasp spending more time searching at one side than the other; (b) asymmetry in the positions at which wasps would rest on the ground within one metre of the nest (we exclude those positions where the wasp suns itself after alighting, recognizable by the characteristic sunning posture); (c) the direction the wasp faced when resting. In some cases this deviated from the normal frontal facing direction; the side to which the sitting deviated was classified as the heavier-disturbance side.

Analysis showed that these behavioural criteria also correlated well with the horizon classification. The side at which wasps searched most, alighted more often, and showed more deviations from the normal orientation when resting, was that with the strongest horizon elements. The correlation was as good as when standard DOT data from triplet tests were available (Hu-colony, $n = 164$, 86·6 per cent, $p < 0·001$). Behavioural criteria seem, therefore, as good as DOT measurements. In order to be more certain about this point one special so-called 'false sitting place' which was often observed even during conclusive triplet tests was analysed a little further.

Often a disoriented wasp lands exactly on the 'east–west' axis through the nest-entrance but at the far side of the vertical sheet (position a in Fig. 7.1 (b)) as close to it as possible (2 cm). The probability of choosing this position to sit for a short time increases as the sheet is moved closer to the nest: we have called this position 'the false spot'. In the Hu-colony sitting in this specific position occurred in 137 out of 277 tests when the vertical sheet was 2 cm from the nest. This proportion had dropped to only 10 cases out of 219 tests with the sheet at 5 cm from the nest and to 11 out of 456 cases at 10 cm distance (difference 2–5 cm, $p<0.001$ difference 5–10 cm, $p<0.01$). Furthermore, sitting at the false spot is very asymmetrical. In only 14 cases (out of 137 at 2 cm distance) was a wasp observed to sit on both sides. In all other cases the spot was only chosen when the sheet was on one side ($p<0.001$) and in no less than 91·1 per cent of cases this side was that of heavier disturbance as measured by the DOT ($p<0.001$). As with the DOT-measurements the side chosen was independent of the side used first in a triplet test. From these facts it is clear that (a) sitting on this false spot (and in general asymmetrical alighting) is a good indicator of sidedness; (b) the wasp is tempted to alight more often at a place which gives a view of the horizon on the important side most similar to that of the nest-entrance itself; and (c) that we are justified in combining all the cases of sidedness whether it can be determined by a clear DOT-measurement or not (correspondence between sidedness and horizon classification: Hu-colony: 84·3 per cent, $n = 394$).

Properties of horizon elements as related to sidedness

In Fig. 7.2 a schematical representation is given of the various types into which the frontal horizons have been classified. Not all possible situations are illustrated, for example competition may also occur between one object at one side as opposed to two at the other. In all competition cases, however, angle a is smaller than b.

Effect of frontal position.   The degree of correlation between sidedness and horizon classification as shown in Fig. 7.2 (a) is given in Table 7.2 (a). Apart from the fact that the presence of an object projecting above the horizon-line on one side remains significantly correlated with the wasp's sidedness (see rows), it can be concluded that this correlation significantly decreases as the angle between zero-point and the object increases (comparison between rows). This trend is also apparent when one analyses two particular objects (Table 7.2 (b)), for which sufficient data is available. It could be that this trend, suggesting an influence of an frontal position of an object as such, correlates with a decreasing width (in degrees) under which it is seen from the nest-entrance when its position shifts from a more frontal to a more lateral one. However, when one calculates the average width of objects 1 and 2, no difference in width appears to exist between the frontal and more lateral positions, nor does one find a difference in width when one compares lateral positions ($>15°$) which do and do not correlate with sidedness. 'Frontality' of an object as such seems, therefore, to play a role; the more frontal the position of an object the greater the

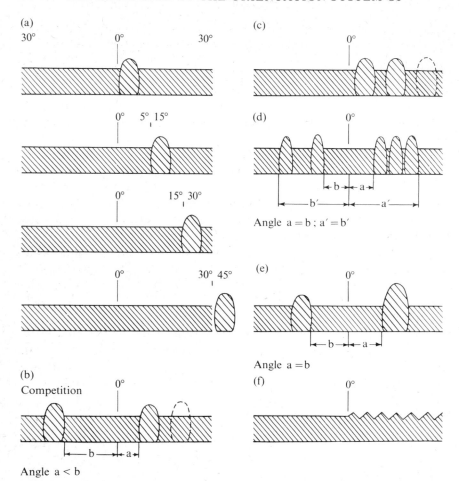

FIG. 7.2. Schematical representation of frontal horizons. In all cases the horizon has to be classified as 'right' (see text).

probability that its position relative to the zero-point corresponds to the sidedness of the wasp.

The effect of 'frontality' is also clear when one studies competition cases (Fig. 7.2 (b)) where one horizon element projects at each side of the zero-point and both are about equal in size. The prediction that sidedness would correlate with the more frontally placed element is fullfilled in 38 out of 41 cases ($p<0.001$).

The effect of number of horizon elements. Cases in which more than one object projects at one side ($n=27$; Fig. 7.2(c)) do show a 100 per cent correlation with the wasp's sidedness. There is therefore a trend (non-significant on our data) that multiple objects to one side 'determine' the

TABLE 7.2

(a) *Correlation between sidedness and horizon qualification; cases of only one conspicuous object in the frontal section at different angles from the sighting-point (Hu-colony)*

Angle between sighting-point and horizon element	Side of object correlated with sidedness	
	Yes	No
(a) 0°	78	11 (12·3%) $p<0·001$
(b) 2–15°	31	4 (11·4%) $p<0·001$
(c) 16–30°	33	15 (31·3%) $p<0·01$
(d) 30–45°	39	15 (28·3%) $p<0·01$

Difference between (a)+(b) and (c)+(d): $p<0·001$.

(*b*) *The same correlation for objects* 1 *and* 2

Angle between sighting-point and horizon element *Object* 1	Side of object correlated with sidedness		Angle between sighting-point and horizon element *Object* 2	Side of object correlated with sidedness	
	Yes	No		Yes	No
(a) 0°	27	1 (3·5%)	(a) 0°	13	0 (0%)
(b) 2–15°	13	0 (0%)	(b) 1–15°	12	0 (0%)
(c) 16–30°	31	7 (70%)	(c) 16–30°	8	4 (33·3%)
(d) 30–45°	5	7 (58%)	(d) 30–45°	5	1 (16·7%)

Difference between (a)+(b) and (c)+(d): object 1, $p<0·001$; object 2, $0·02<p<0·05$.

sidedness better than when only one object is present (the first two rows of Table 7.2(a). Another indication of the same trend is given by those cases of competition in which more tree-tops project at one side than at the other, although both groups of objects lie within similar angles from the zero-point (Fig. 7.2(d)). In 12 out of 12 cases the expected side scored a higher DOT. Whether this effect of number has to be described in terms of total length of the line of contrast between wood and clear sky or in terms of number of interruptions of a nearly horizontal horizon-line, cannot be decided without further standardized experimentation (see discussion, §7.6).

The effect of size.   Fig. 7.2(e) represents the situation from which we may draw a conclusion about the influence of size. In all of 17 available cases the side with the larger object corresponds with the sidedness of the wasp. Likewise, when frontality and size are cooperating (9 cases) no exceptions as to the prediction are found.

Knowing the rules by which 'frontality', size, and number affect sidedness, the importance of the fact that elements project above the horizon-line can be assessed. Nests with 3 (projecting) elements were selected. If (situation A) these elements are close to the nest then, from the wasp's sighting position, they will stand out clearly against the sky. However, when at a greater distance (situation B) they disappear below the horizon, from the wasp's viewpoint. As expected, in situation A in 38 out of 40 cases sidedness was correlated with the positions of the 3 elements with respect to the sighting points ($p<0.001$). However, under condition B the wasp's sidedness coincided in only 6 out of 26 cases (difference between situation A and B: $p<0.001$). On the other hand, classifying the latter 26 horizons using other elements farther away but still projecting above the wasp's horizon then a good correlation with sidedness is found (22 out of 26). No correlation whatsover exists when an equivalent classification is applied to the nests of situation A (18 out of 40). This data proves beyond doubt that horizon elements play a role only when they project and that their degree of 'frontality' is very important.

The effect of stronger indentation.   In a number of cases the wasp's sighting points were positioned on a rather flat horizontal part of the horizon extending 30° to the left and the right. In a proportion of these cases, however, a clear difference between sides existed as a consequence of the composition of the nearby woodland (Fig. 7.2(f)). Taller spruce or pine at somewhat larger distances from each other gave a more indentated horizon-line than more closely packed thinner pines of similar height. Again, a group of broad-leafed trees gave a much smoother top-line than the spruce adjacent to it. In 17 of such cases a clear difference between sides was present, and in all of them the sidedness of the wasp corresponded with the more indentated side of the frontal horizon. There was no correlation between sidedness of these 17 wasps and the position of one (or more) of the 15 conspicuous objects lying more laterally within 30–60° degrees left or right from their sighting points. In one case there was no such object, randomness between site of object and sidedness existed in the remaining 16 (8 versus 8). Therefore, the difference in indentation between the two frontal portions of the horizon seems to be the crucial factor determining sidedness in these cases.

The effect of width (apparent size) of horizon elements.   From the first row of Table 7.2 it can be seen that the correlation between sidedness and horizon qualification can be absent, in spite of a horizon element having a good frontal position. It is clear that with a decreasing distance between the nest and the object concerned, the latter's apparent width (taken in degrees of the frontal section) and its apparent height (taken in vertical degrees) both increase. This increase in apparent size is even stronger relative to the

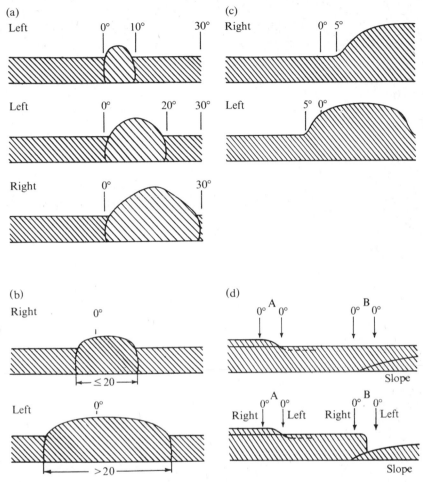

Fig. 7.3. (a) Increase of 'massiveness' on nearing an horizon-object. (b) Sighting-points falling within a frontal horizon element. (c) Hu-colony: effect of position of dip. (d) S.D.-colony: effect of position of dip; horizon-line before and after cutting of part of the wood. Point A is not always conspicuous. The (dominant) sidedness of wasps having the respective horizons is given in the figures.

background of woods when the effective object is not part of the wood itself, but stands in front of it (for example solitary birches). In order to analyse a possible effect of apparent size, the data from Table 7.2 were classified again taking into account the apparent width of the objects as seen from the nest-entrance (see Fig. 7.3(a)). Table 7.3 shows the result for 3 different positions of sighting-points. In order to get more data, cases of 'competing' frontal elements were used as well in Table 7.3 all classified according to the rule of frontality.

TABLE 7.3

Correlation between sidedness and horizon classification (Hu-colony), as related to different angles between the sighting direction and the object and the width of the horizon element

Angle between sighting point and object	Width of horizon element	Side of object correlated with sidedness		
		Yes	No	
(a) 0°	(*i*) 0–10°	30	1	
	(*ii*) 0–20°	45	0	(*i*)+(*ii*) versus (*iii*): p<0·001
	(*iii*) 0–40°	10	13	
(b) 5–15°	(*i*) 0–10°	34	2	
	(*ii*) 0–20°	34	1	(*i*)+(*ii*) versus (*iii*): p>0·30
	(*iii*) 0–40°	13	2	
(c) 16–45°	(*i*) 0–10°	26	16	
	(*ii*) 0–20°	28	9	(*i*)+(*ii*) versus (*iii*): p>0·30
	(*iii*) 0–40°	10	2	

The difference between (a) (*i*)+(*ii*) and (c) (*i*)+(*ii*) is significant at a level of $p<0·001$. The difference between (a) (*iii*) and (b) (*iii*), and between (a) (*iii*) and (c) (*iii*) is significant at a level of $p<0·0001$.

The table shows the following points.

(a) When the sighting-point coincides with one side of the horizon object, the correlation between sidedness and horizon side breaks down with very broad (massive) frontal objects.

(b) When, however, the zero-point is at some distance from the frontal element, increase of width of the frontal object has no effect (Table 7.3(b) row (*iii*) and (c) row (*iii*)).

(c) On the other hand, when there is a large angle between the zero-point and the conspicuous object the correlation is significantly reduced if objects are apparently small (compare Table 7.3(a), row (*i*)+(*ii*) with (c), row (*i*)+(*ii*), another demonstration of the rule of frontality.

Conclusions. The decrease in correlation of sidedness with horizon elements when there is an increase in the angle between sighting-point and horizon object, as shown in Table 7.2, has to be explained by the fact that small objects are less important the more lateral their position. On the other hand the absence in correlation in some cases of frontally positioned objects has to be understood on the basis of an increase in apparent size when they are viewed by the wasp from a nest close to them. The question of why this visual situation induces an opposite sidedness is difficult to answer. It could be that the amount of dark (wood) is favoured so much over light (sky), that the side of the massive dark frontal object appears to the wasp as just 'dark' and details are not responded to.

Examples where the sighting-point is included within the horizon element. If we exclude cases in which the wasp's sighting-point exactly bisects a horizon object, 34 cases of sightings falling within the angle of an element are available. When the object is less than 20° wide (see Fig. 7.3(b)) the side including the larger portion corresponds with the sidedness of the wasp (17 out of 18 cases, $p < 0.001$). Apparently the size of that part or the total length of the line of contrast is more important than the sudden vertical interruption of the horizon-line, which is more lateral at the preferred side in such cases. When, however, frontal elements exceed a width of about 20° (Fig. 7.3(b)), there is no longer a preponderance of sidedness coinciding with the larger portion of the element (5 out of 16 cases, $p > 0.10$). The difference between the two groups is significant ($p < 0.001$). Either large apparent size plays a role, or the more frontal position of the vertical interruption (probably as a consequence of the large apparent size of the portion at the non-preferred side) is of importance in these cases. That the position of a sudden dip in an otherwise rather featureless horizon is correlated with sidedness is shown by the last type of horizon we shall examine.

The effect of sudden dips in the horizon-line. This horizon type is shown schematically in Figs 7.3(c) and (d). It is characterized by a smooth upper line extending over more than 30° to either side of the zero-point, but with a sudden vertical dip at one side of this point. The Hu-colony had only a few nests with this type (Fig. 7.3(c)), of which in 11 out of 12 cases the side including the interruption coincided with the sidedness of the wasp. The S.D.-colony provided many more cases, especially after the autumn of 1970. In that year part of the surrounding wood was cut down and as a consequence a clear vertical dip was created, visible from anywhere in the colony area (see Fig. 7.3(d); also under point B in Table 7.4). From some parts of the colony another conspicuous drop could also be seen (see under point A in Table 7.4). However, it was invisible from many nests because of their location. From Table 7.4 it can be seen that, when visible, this point A does indeed exert a sidedness-determining influence, which decreases with distance (according to the rule of frontality); the same conclusions can be drawn with respect to Point B.

Horizons difficult to classify. Two types of horizons are difficult to classify. Sometimes the characteristics of the horizon lead to a contradictory expectation about the sidedness of a wasp, in other cases a clear difference is absent between the left and right frontal portions of a wasp's horizon. When, for example in a one to one object situation, the position of the larger of the two is more lateral than that of the smaller, it cannot be predicted in advance whether the rule of frontality or that of size will prevail. In 16 out of 17 such cases the rule of size appeared to be followed; the difference

TABLE 7.4

Correlation between sidedness and position of a sudden vertical dip (point A with respect to B) in an almost flat horizontal horizon-line (S.D.-colony)

Angle between sighting point and dip	Point A				Point B	
	Correspondence between side of dip and sidedness of wasp					
	Inconspicuous		Conspicuous		Conspicuous	
	Yes	No	Yes	No	Yes	No
0–20°	1	13	23	0	36	0
20–50°	5	21	3	8	1	3

The difference for the class of 0–20° between conspicuousness and inconspicuousness of point A is significant ($p < 0.0001$); class 20–50°: $p = 0.28$ (not significant).

The difference between class 0–20° and 20–50° of the conspicuous point A is significant at a $p < 0.0001$.

in angle, however, was at the most 20°, and in most cases not more than 5–10°. Which of the two factors will be dominant probably depends on some relation between the differences in angle and size. In 31 cases no clear difference existed between sides; no conspicuous object was present, nor a clear difference in indentation. 15 of the wasps were left-sided, 16 right-sided. Apparently much smaller details than those used in our classifications were important to the wasps.

Changes in the horizon-line

(a) It has been mentioned already that wasps may change their sidedness with time. In a number of cases it could be established that this change coincided with a change in their sighting direction. This may happen as a consequence of digging a new cell in the same nest after finishing one parental cycle, or as a consequence of repair to a heavily damaged nest-entrance. In all cases the new sidedness again correlated with the new horizon qualification.

(b) In the foregoing pages it was implicitly suggested that the horizon-line defined as the line of contrast between dark woodland and clear sky would be an especially important feature of the orientation pattern used by *Bembix*. In order to test this idea experimentally a vertical triangle (base 30 cm, height 33 cm) was presented to a number of wasps at right-angles to the nest-axis at a distance of 1 m from and in front of the nest. Viewed from ground-level, in no case did the top of such an object reach the horizon-line. In only one case did a wasp show a very short disorientation flight (1 s), whilst in 17 cases landing occurred without any sign of hesitation. In contrast, presentation of the same triangle on top of a thin rod (total height 1 m) resulted in short but clear disorientation flights in 16 out of 24 cases (difference: $p<0.001$) when the top of the triangle or a bigger part of it crossed the horizon-line.

(c) If, generally speaking, a greater richness in contrast at one side of the frontal horizon section determines the sidedness, then reduction of this richness by introducing a flat horizon-line over a great part of this section would be likely to change sidedness. We tried to achieve this (1) by placing a large black vertical sheet (length at least 50 cm, height at least 30 cm) at a distance of about 50 cm from the nest in such a position that almost the whole quadrant of the horizon on the preferred side was screened off; or (2) by placing a white vertical sheet (9 cm high) at 2 cm parallel to the nest-axis extending 15 cm to the front and to the back. After one or two nest visits the wasp appears to be accustomed to the new situation because she first makes a reorientation flight and subsequently lands without any hesitation. In 26 out of 30 cases (taking the situations together) the sidedness of the wasps did appear to change. Their sidedness was tested either with the new 'horizon' still in place (in the black sheet experiments, see above) or with the old situation restored (the white sheet was taken away during testing). Since these two methods yielded the same results we can conclude that a wasp 'qualified' its horizon according to its appearance as seen during its last return to the nest.

Conclusion

From the data given it is clear that on the one hand the sidedness of a wasp is not correlated with a number of factors (shadow, etc.) which could have been responsible for the asymmetry in disturbing effect of a vertical sheet, and on the other hand that sidedness is strongly correlated with the presence and position of elements in the frontal section of the horizon. Conspicuous objects projecting above the horizon-line, sudden vertical dips in the horizon, and the degree to which it is indented are all important features. The larger the objects are and the more frontal their position, the more they seem to determine the sidedness of the wasp. Too great an apparent size or 'massiveness' of the frontal object seems to induce an opposite sidedness. Further, some experimental evidence has been presented demonstrating the importance of the horizon-line.

7.5. The position of the sighting-point

The importance of conspicuous horizon elements

From Table 7.2 it can be seen that a fair proportion of wasps had their sighting-points in the direct vicinity of the 15 conspicuous horizon elements used for the analysis. Since we had the impression that sighting-points were definitely not randomly associated with these objects, we recorded their position with respect to as many nests as we could (Hu-colony, $n = 800$). To calculate an expectancy based on randomness the following procedure was followed. The Hu-colony consisted of about 5 subcolonies, i.e. dense concentrations of nests with less dense areas between them. From points situated approximately in the middle of the various subcolonies, measurements were taken of the sectors occupied by the selected horizon-objects. From the total of nests present in the subcolony the number of sightings per degree could be calculated and consequently the expected number of nests that should sight at the conspicuous objects on a random basis. For each subcolony the actual number significantly exceeded the value expected (for the whole colony the figures were: expected 28·2 per cent, observed 57·3 per cent, $n = 800$, $p < 0.001$). It is clear, therefore, that *Bembix* in many cases chooses its sighting-point with respect to the presence of a conspicuous object at the frontal horizon. An evaluation of the S.D.-colony is much more difficult, since (a) the horizon-line does not have an irregular appearance and (b) the direction of the nest-burrow is more or less forced upon the wasp by the steepness of the slope on which the colony is situated.

Considering now the remainder of the nests in the Hu-colony ($n = 341$) not sighting towards horizon-objects, calculation reveals that there is a deficiency of nests sighting in the sectors within 5° left and right of these objects (in all subcolonies, $p \leqslant 0.02$). (To calculate the expected value the number of remaining nests and the total number of degrees of the intervals between objects were used.) It could be, of course, that this result is due to a bias on our part. Locating the sighting-points might be too prejudiced towards the objects themselves. On the other hand it may mean that the wasps do indeed match their sighting-points very accurately to conspicuous horizon elements when their nests are being constructed. Whether this is the case or not, large horizon elements do attract significantly more sightings than expected on a random basis (62·6 per cent, including the 5° sectors on each side of the objects, $p < 0.0001$).

The precise position of the sighting-point

Within the group of wasps sighting at conspicuous objects, a number had their zero-points at the sides of these elements. To analyse this group of sighting points the objects, were divided in two groups, the first comprising objects of which at least one side showed a clear interruption of the horizon-line (see Fig. 7.4(a), for example, solitary birches), the second consisting of objects which reached their highest point gradually (see Fig. 7.4(b), for example a group of taller pines).

The proportions of side-sightings in the two groups appeared to differ significantly ($n = 469$, $p < 0.001$). Apparently, the type of objects as given

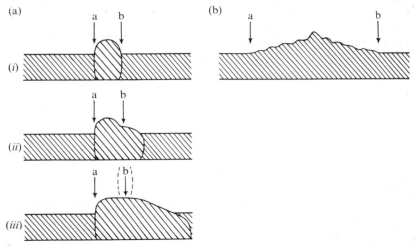

FIG. 7.4. Position of sighting-points (not falling on top of conspicuous objects) as related to conspicuousness of vertical interruptions of the horizon-line (see text).

in Fig. 7.4(a) 'attracts' much more side-sightings. It is the sudden interruption of the horizon-line which makes the sides of the group (a) objects attractive. This is demonstrated again when one compares the probabilities of sightings at side a and at side b (see Fig. 7.4). Of some objects (Fig. 7.4(a) and (b)) both sides were always visible, irrespective of the location of the nest; of other objects sometimes both, but in most cases only one side, could be clearly seen (Fig. 7.4(a) (ii)), of still others one side was conspicuous again irrespective of the location of the nest (Fig. 7.4(a) (iii)). Analysis shows that in the case of (b) and (a) (i), half of the side-sightings fall upon the left side (arrow a in Fig. 7.4(b): 50 per cent, $n = 57$; and Fig. 7.4(a) (i): 53·7 per cent, $n = 93$); and that this percentage in case (a) (ii) increases to 68·6 ($n = 35$), whereas in case (a) (iii) 94 per cent of all side-sightings coincides with the conspicuous side a. The latter increase is significant ($p < 0·01$) as is the difference between (a) (i) and (a) (iii).

Sighting-points at the early stages of nest-construction

Very occasionally one observes a change of sighting-direction by wasps that have already lived for days in a nest, caring for their offspring. This means that the choice of the sighting-direction has to occur during the phase of nest-digging. When a wasp starts to dig a new nest it walks or flies for short distances within a restricted area, scratching the surface here and there with its forelegs, sometimes digging a shallow burrow by biting through the hard upper layer of the soil, and at some places digging even deeper holes, only to abandon them again. It may take an hour or more before the wasp 'decides' to dig a really deep hole, which in the majority of

TABLE 7.5

Sightings of incomplete and abandoned nests as related to their depth; sighting towards a conspicuous horizon element+, not so sighting−(Hu-colony)

| | Depth of incomplete nest | |
(a)	(b)	(c)
0−<2 cm	2−<4 cm	4 cm upwards
+ −	+ −	+ −
23 26	27 7	32 8

Difference between (a) and (b): $p<0.01$; (a)−(c): $p<0.01$; (b)−(c): not significant.

Difference between completed nests and (a): $p<0.02$; and (b): $0.20>p>0.10$; and (c): $p=0.10$.

cases will become its definite nesting place. However, holes of 5 cm deep and more may be still abandoned rather easily. For 123 wasps the depth and the sighting direction of abandoned burrows were recorded. Only one burrow (the most shallow one) per wasp was used to collect the data of Table 7.5, in which the number of holes sighting towards one of the conspicuous horizon elements is given as related to the depth of the burrow. As can be seen from the table only very shallow holes (less than 2 cm) sight significantly less at the objects used in classifying the horizon; holes any deeper than 2 cm are already the same as completed nests in this respect. Though the percentage of holes sighted at conspicuous objects thus improves sharply with their depth, the sighting-direction of very shallow holes already exceeds the expected probability based on random sighting (28·2 per cent, see §7.5: *The importance of conspicuous horizon elements*). This would mean that even walking wasps do orientate themselves to horizon-objects at a distance. That walking and even sitting wasps are able to make use of (potential) landmarks at great distances is further suggested by two observations. First, wasps too cold to fly may walk up to their nests in a rather straight line from distances of at least half a metre. Secondly, wasps disoriented by a disturbing object show a particular type of 'nervous' walking from left to right and back over the closed nest-entrance. This 'dancing' in our opinion serves to locate the exact position of the entrance with the aid of far-distant cues. Dancing is often shown by wasps disturbed by vertical sheets, but hardly ever when horizontal disturbing objects are used. Baerends (1941, p. 262, 266) suggests that *Ammophila campestris* finds its closed nest-entrance by using tactile information over the last 3–4 cm when walking towards it. During this phase the wasp taps the ground with the antennae and scratches it with the forelegs, so that eventually the nest-entrance filled with loose sand is found. Although we observed similar behaviour in *Bembix*, we do not think dancing functions in gathering such tactile information in addition to (or instead of) the function suggested above. First, tapping the ground with the antenna is no part of the dancing movement; secondly, when nest-entrances are covered by a small glass plate (4×4 cm, with or without sand glued on it) undisturbed wasps dig furiously on the hard plate right after landing without any sign of dancing; thirdly, wasps unable to return to their nest for the night because of sudden heavy rainfall at the end of the day, show no dancing when alighting the next day but start immediately to bite their way through the sand-plug at the entrance, which has been hardened by the showers.

Although dancing apparently has no function in detecting spots of loose sand (i.e. the closed nest-entrance), *Bembix* seems to react to looseness of sand. A wasp disorientated by a disturbing object may dig furiously after alighting on a spot with loose sand. When alighting on hard surface it does not show digging. In such cases a wasp repeatedly returns to the wrong spot after being chased away, as if it 'feels' itself at home. In spite of an orientational mistake (induced by the disturbing object) it has apparently 'decided' that the spot chosen is the correct position of the nest-entrance. It is interesting that such a wasp can be released from its monomaniacal choosing of the wrong spot by chasing it away so drastically that it leaves the colony area. After return it shows real, good searching for the correct nest-spot as it did before the fixation to the wrong place. Since the wasp quickly returns to the same spot, the perception of looseness of sand at a landing place seems to give additional, though not necessary, information as to the correctness of the place chosen on the basis of optical orientation.

Summarizing the data of this section: it is quite possible that wasps starting to dig a nest select from the outset a spot to dig with reference to a sighting at a conspicuous landmark located far away, even if they are only walking to the site.

7.6. Concluding remarks

It seems likely that the phenomenon of sidedness, i.e. the asymmetrical disorientation of a wasp by a vertical sheet placed successively left and right of the nest, has to be explained by the asymmetrical position of conspicuous horizon elements situated in the section of the horizon the wasp is sighting at while landing. It is possible that the effective characteristics of these objects, like their size or their number, have to be described in terms of one crucial parameter: the length of the line of contrast between dark woodland and light sky. Likewise, the effect of indentation or the presence of a sudden vertical dip in the horizon-line can be understood by applying the same parameter. The positive effect of 'frontality' is at first sight more difficult to interpret. However, according to Zaenkert (1939) the eye of *Bembix rostrata* possesses a sharp vision field in the centre of its frontal section, the ommatidia of which have a small diameter and consequently small angles between their axes. Objects in front of a landing wasp, therefore, are seen in much more detail than objects of similar size positioned at the sides. The same can be said for wasps hovering above the nest-entrance; they nearly always are facing as if for landing. For this reason the effect of a frontal position in determining sidedness may also be understood in terms of 'length of line of contrast', more ommatidia being involved in the frontal section of the eye than in its lateral part.

On the other hand, it could well be that this way of thinking is far too simple. Real patterning might play an important role. Sudden changes in the direction of the horizon-line may add to conspicuousness of these parts as compared with portions of the line of contrast having equal total length but without sharp declines or rises. The relative clustering of sightings at sharp interruptions of the horizontal line of contrast is suggestive in this respect. Standardized experimentation with artificial horizons will be needed to disentangle the effective parameters.

Whatever may be the case, the interesting fact remains that the asymmetrically positioned conspicuousness in the frontal section of the horizon seems to determine the asymmetrical information value of all parts of the rest of the horizon. The non-frontal parts are not 'evaluated' by the wasps as such, neither is this part at the side of heavier disturbance evaluated less than the other side, because of the one-sided high information value present in the frontal section. All parts of one side have greater orientational value than all corresponding parts of the other, apparently by some kind of integration process.

A consequence of the fact that the wasp always lands from one direction (see Fig. 7.1(a)), though it may approach its nest from any direction without any difficulty, is that some horizon elements are more important than others because of their position. (The latter statement is based on observation and experimentation, watching the return flights of wasps caught at the nest and transported in many directions over distances up to 10 m.) The more often the wasp lands, the more often some specific parts of the horizon are seen in great detail. Moreover, the way the location of the nest is learned will increase the value of objects in the direction of the nest-axis. The orientation flights made during digging of the nest always start with hoverings; the wasp backs out from the nest, flies up, hovers on the spot for some seconds right above the nest-entrance, and then lands again. These hoverings, though they are performed again intermittently, are soon accompanied by sidewards movements over short distances at right-angles to the nest-axis. During these excursions the wasp is still facing the zero-point. Later on loops are performed in a number of directions, which lead the wasp away from the nest and back again towards the entrance and usually conclude with a quick turn just before landing in the sighting direction. Eventually circles (or spirals) are flown around the nest. From the order in which the different types of orientation flights are performed, it is conceivable that right from the beginning of the learning process the total picture of far-distant cues will have a conspicuous frontal part; hence, the importance of frontal objects for relocation later on can be expected.

From the foregoing, it does not follow that the view of frontal objects is indispensable for such relocations. When, for example, the frontal and both lateral views are screened off by vertical sheets high enough to prevent a flying *Bembix* from looking over them, a wasp is still able to locate its nest. After being successively 'adapted' (that is, not being disoriented any more) to such a complicated set-up, it may simply hover in a 180° reversed position sighting towards the caudal section, and then land in the correct position. Alternatively it may land in a more or less correct position on the nest-patch, walk a quick complete circle, and then start to open the entrance. It is important to note that in such cases no reorientation to the added objects is involved. Such behaviour is just another example of what van Iersel and van den Assem (1965) called 'shifting of attention': the wasp just looses interest in some parts of the orientation system and directs its attention to undisturbed parts of it.

It is clear that even the caudal part of this system has landmarks known in sufficient detail to relocate the nest without any difficulty. While hovering seems to serve especially for learning details in front of the nest, flying loops may serve for learning details in other parts of the orientation system, since the wasp making the loops directs its field of sharp vision towards these parts. The function of circling around the nest is presumably to locate roughly the area in which the nest lies by scanning successively the points of the far-distant orientation system.

It will be understood that during all types of learning flights, characteristics of the nest-surroundings are incorporated at the same time. We shall not deal here with the question of which elements of the total system engage the attention of *Bembix* most whilst digging its nest.

The ability of the wasp to 'adapt' rapidly to discrepancies between observed and expected orientation patterns, by shifting its attention to undisturbed parts of the orientation system, may also be responsible for the fact that repeated disturbance nearly always results in decreasing disorientation.

Similarly, this shifting of attention explains why it is difficult to apply the training method used with *Philanthus* in investigating which characteristics of landmarks are important for *Bembix* (see §7.1). This method is successful only when a wasp remains willing to follow the artificial landmarks to which it has been trained, when these are displaced over a short distance. The easier it does so, the better choice-experiments between two separate sets of landmarks can be carried out. *Bembix* wasps show some disorientation in such choice experiments. This suggests that they experience a discrepancy between an observed and an expected pattern of elements used by them for orientation. As said in the introduction *Bembix* soon stops choosing in such experiments. After some trials it alights on the nest-patch and pays no further attention to the displaced landmarks. They seem to shift their attention to far distant cues which do not change their position in relation to the nest spot. Blocking of the wasp's frontal view (by a small vertical sheet at right-angles to the sighting direction) induces immediately renewed choices of the sets of displaced landmarks.

The wasp appears to use a very extensive series of visual features to locate its nest, indeed about as extensive as its eyes can deal with. What could be the function of using such a complex orientation system? In the first place, its is likely that the precision of localization will be improved the more orientation points there are available. Secondly, the more points there are the better the chances for relocation of the nest when some kind of disturbance causes parts of the environment to be altered and landmarks eliminated.

Another question concerns whether it is adaptive to select a conspicuous far-distant object in the frontal section as a sighting point. This question is difficult to answer, because it is certain that wasps that have no such sightings do find their nest. I had the impression that undisturbed wasps in the S.D.-colony, with its rather monotonous horizon-line, had greater difficulty in relocating their nests than wasps in the Hu-colony. The former often showed some searching before finding, even if they had occupied their burrows for a long time. However, no strictly comparative measurements have been made. It might be adaptive to reduce the searching time as much as possible, since there is always the danger that parasitic flies (*Miltogramma* sp.), which are often abundant in the colony, will drop their eggs on the prey the mother wasp is carrying in for the larva. One observes *Miltogramma* trailing in numbers behind a flying wasp which is searching for its nest. The parasite's eggs are deposited during the period when the wasp is opening the nest-entrance. This might be a particularly relevant consideration when the wasp brings in her first prey on which she will deposit her own egg. At this stage wasps are often still rather disoriented and fly with great speed in a zig-zag fashion over the area around the nest.

This way of flying might in its turn be an adaptation to the particular danger of encountering parasites in this crucial phase of the parental cycle. A further adaptive value of having a good orientation mark in the frontal direction may arise still earlier when the nest is being built. During building the wasp may leave the nest to perform orientation flights, which may be extended for a metre or more, and also to feed at flowers. In the latter case digging is interrupted for rather long periods while the wasp leaves the colony area. When returning wasps often show quite a long period of searching. Further research is needed to evaluate these various possible functions of a strong frontal horizon marker; probably they all contribute to breeding success.

7.7. Summary

The Digger Wasp *Bembix rostrata* flying in the vicinity of its nest is strongly disoriented when the orientation system used by the wasp to relocate its nest is disturbed by erecting artificial vertical flat features in the neighbourhood of the nest. The wasp shows 'sidedness' in the sense that a feature placed successively right and left of the nest parallel to the nest-axis produces a difference in the degree of disorientation. Sidedness appears to be independent of (a) the position of the sheet with which the test is started; (b) the position of the sun and shadow at the moment of testing or during digging the nest; (c) a possible difference in characteristics (such as degree of contrast, etc.) of the surface immediately around the nest; (d) the extent to which parts of the horizon-line (the line of contrast between dark woodland surrounding the colony and clear sky) remain visible over the top-rim of the vertical feature; (e) the type of testing object (its colour, height, etc.); and (f) the position of the test object (lateral distance from the nest, caudal or frontal presentations).

Conversely sidedness appears to be highly correlated with the presence of large objects (for example trees) projecting above the horizon-line within an angle of about 30° at either side of the sighting-point of the wasp. The more frontal the position of the 'horizon-object', or the larger it is, the better the correlation. The degree of indentation of the frontal horizon-line (when conspicuous objects are absent) and the position of sudden vertical dips in an otherwise undetailed frontal horizon-line are also determinants of sidedness. The side of the horizon with more indentation or the side possessing the dip coincides with the side of heavier disturbance. When the apparent width (massiveness) of frontal objects increases too much (more than 20° the sidedness of the wasp tends to be at the opposite side from the object.

In one of the colonies studied—a colony surrounded by a horizon with conspicuous features—the sighting points of the nests were selectively clustered on these distant features. Sightings on horizon features were often directed at the sides of these objects when they abruptly disrupted the horizon-line. However, the majority of the sightings was directed at the middle parts of conspicuous features, when these reached their highest point above the horizon-line more gradually. Burrows abandoned at an early stage when only 2 centimetres deep were already sighted at con-horizon-points to as great an extent as were completed nests.

We concluded with a discussion of the possible functions of using such an extensive orientation system.

References

Baerends, G. P. (1941). Fortpflanzungsverhalten und Orientierung der Grab-wespe, *Ammophila campestris* Jur. *Tijdschr. Ent.* **84**, 68–275.

van Beusekom, G. (1946). Over de oriëntatie van de bijenwolf (*Philanthus triangulum*). Ph.D. Thesis, Leiden.

—— (1948). Some experiments on the optical orientation in *Philanthus triangulum* Fabr. *Behaviour* **1**, 195–225.

Iersel, J. J. A. van and J. van den Assem (1965). Aspects of orientation in the digger-wasp *Bembix rostrata*. *Anim. Behav., Suppl.* **1**, 145–62.

Siegel, S. (1956). *Nonparametric statistics for the behavioral sciences*. McGraw-Hill, New York.

Tinbergen, N. (1932). Ueber die Orientierung des Bienenwolfes (*Philanthus triangulum* Fabr.). *Z. vergl. Physiol.* **16**, 305–34.

—— Kruyt, W. (1938). Ueber die Orientierung des Bienenwolfes (*Philanthus triangulum* Fabr.)—III Die Bevorzugung bestimmter Wegmarken. *Z. vergl. Physiol.* **25**, 292–334.

Zaenkert, A. (1939). Vergleichende morphologische und physiologisch-funktionelle Untersuchungen an Augen beutefangender Insekten. *Sber. Ges. naturf. Freunde Berl.* 82–167.

8. Aggressive interactions in flocks of rooks *Corvus frugilegus* L.: a study in behaviour-ecology

I. J. PATTERSON

I FIND it difficult to define my rather hybrid field of study, suspended between behaviour and ecology; should it be called etho-ecology or eco-ethology? A rapidly expanding area, it is still not a clearly delimited one although several discrete subdivisions have emerged.

Perhaps the sub-division most closely related to ethology is that concerned with the adaptiveness or survival value of behaviour. Niko Tinbergen has been outstandingly successful in highlighting the question, 'how does this behaviour help the animal to survive?'—a question which had hitherto been overshadowed by questions of the causation, development, and evolution of behaviour. The recent book, which collects together some of his best-known work (Tinbergen 1972), contains several studies which show the fruitfulness of the ecological approach to behaviour. A good example is the demonstration (Tinbergen, Broekhuysen, Feekes, Houghton, Kruuk, and Szulc 1962) that the brief act of removing egg shells by the Black-headed Gull, *Larus ridibundus* L., functioned to reduce predation. This study of function has been applied by other members of the Tinbergen group to a complex of behaviours, for example, to the various anti-predator behaviours of the Black-headed Gull (Kruuk 1964).

In contrast to this single-species approach, other studies have concentrated on the comparison of several related species, arguing that differences in their behaviour must relate to differences in their ecology. The function of much of the breeding behaviour of the kittiwake, *Rissa tridactyla* L., studied by Cullen (1957), became clear when compared with ground-nesting gulls in the context of their contrasting environments.

Comparison of species has been applied, notably by Crook (1965), to entire social systems. By comparing the social organization of a range of related species of weaver-birds (*Ploceidae*) living in different habitat types, Crook was able to suggest the adaptiveness of each social structure to its own environment.

In the types of study considered so far there has been variation in the levels of behaviour studied, from the relatively simple sequence of egg-shell removal to the large 'functional blocks' of behaviour which make up a species's social structure. In studies of the latter there is generally concentration on relationships between individuals on agonistic behaviours and on the effects of these on dispersion. Further synthesis, by considering the patterning of agonistic behaviour between individuals and spatially, leads to the explanatory principles of dominance and territory.

Both of these are of particular importance in studies which consider the relevance of social behaviour to the limitation of the size of animal populations in the wild.

The remainder of this contribution is an example from this field of study; an investigation of a complex of agonistic behaviours which may function to limit the size of a population.

Many studies of vertebrate populations have shown or at least suggested that social behaviour can be important in limiting population density. Watson and Moss (1970) concluded: 'Although a general conclusion of this review is that behaviour is frequently a major factor limiting populations of vertebrates, there is still a great variety of interpretations about the relative importance of the various different limiting factors in the wild. One important reason contributing to this uncertainty is that too few studies include observations on enough different aspects to enable the author to decide conclusively the relative importance of each.' In the present study of the rook, *Corvus frugilegus* L., information is being collected on several aspects of population and social behaviour (Dunnet and Patterson 1968; Dunnet, Fordham, and Patterson 1970; Patterson, Dunnet, and Fordham 1971) and on seasonal variations in food supply (Feare, Dunnet, and Patterson 1974). This has enabled us to study relationships between food, behaviour, and population. This essay examines one aspect of social behaviour—aggressive interactions in feeding flocks—as a possible factor causing short-term changes in rook numbers in relation to food supply. The factors limiting rook populations over the long term are the main focus of other parts of the study.

'Fighting' in mixed feeding flocks of rooks, jackdaws (*Corvus monedula* L.), and carrion crows (*Corvus corone* L.) has been described by Lockie (1956). 'Fighting' is a convenient but perhaps misleading term since, as Lockie (1956) points out 'On most occasions aggressiveness was limited to threat...' In the present study one bird lunged, walked, ran, or flew at another which usually retreated immediately without retaliation, so that the result was a simple supplanting movement. Actual combat was very infrequent. The aggressor usually adopted a posture with back and belly feathers raised, wings slightly lifted, and the bill pointed strongly downward, and maintained this even while moving some distance toward the victim. This posture was the same as Lockie's (1956) 'bill-down', although the head was often held lower, level with the back, and the bill pointed more vertically downward than he shows in his Fig. 1 (Lockie 1956, p. 183). The victim usually had sleeked plumage and the bill pointing upward as it moved away. On the infrequent occasions when the victim retaliated, combat was confined to a brief exchange of short jabs either from the 'bill-down' position which then became Lockie's (1956) 'half-forward' or 'full-forward' by raising of the bill to the horizontal, or from a high standing 'bill-up' position sometimes with jumping or fluttering in the air (Lockie's 'take-off' posture).

Interactions were sometimes clearly over food items found by the victim or over a feeding site where the victim was probing particularly actively. In other cases the aggressor appeared to be defending a space (individual distance) immediately around itself or its mate, but in most interactions the context was not clearly identifiable.

Lockie (1956) measured the frequency of interaction in flocks of 15–40 birds near Oxford in the winters (October–February) of 1952–3 and 1953–4. He found more interactions in denser flocks and considered dispersed flocks (with more than 2 metres between individuals) separately from more compact ones. Frequency of interaction was inversely related to air temperature (mean of daily minimum and maximum over 2-week periods) with particularly high frequencies during periods when the ground was frozen and snow lay for some days. Lockie suggested that food might be scarcer in colder weather, that there might be a dominance hierarchy in the flock and that the frequent aggressive interactions might limit numbers in relation to food supply by selectively eliminating low-ranking individuals. This essay will extend these observations by considering interactions at other times of year in addition to winter and especially by testing for correlation between frequency of interaction and direct measures of food intake made by Feare, Dunnet, and Patterson (1973).

The main approach was similar to that of Lockie (1956)—measuring seasonal variation in the frequency of aggressive interactions. If these are important in limiting numbers in relation to food their frequency should be inversely related to food intake so that interactions should be most frequent in periods of food shortage (when the mean daily intake is below the daily requirement for maintenance). If this relationship exists, it will be necessary to show further that when interactions are frequent, individual rooks eat less, and for limitation to occur as Lockie (1956) suggested, certain individuals should eat less than others.

8.1. Seasonal variation in frequency of aggressive interaction in relation to seasonal variation in food intake

Seasonal variation in food intake

In our study the mean daily food intake by the birds was measured by C. J. Feare (Feare, Dunnet, and Patterson 1974), who watched rooks feeding throughout the year and estimated intake in kilocalories per day. A minimum value for daily food requirement was obtained by measuring the maintenance intake of a captive rook kept in a large outdoor aviary. Periods of food shortage were suggested when the estimated intake of the wild birds approached or fell below the requirement of the relatively sedentary captive.

Summer was shown to be the period of greatest food shortage, but breeding adults probably also had a low intake while rearing young. In one year there was a short period of low daily intake just before the breeding season. No food shortage as defined above was indicated in winter.

The data on intake were reflected in changes in the time spent searching for food at different seasons. The mean time spent feeding per day was much higher in summer, especially in dry periods, than in winter. Generally in summer rooks spent twice as long searching for about half as much food per day as they obtained in winter. These comparisons of course refer only to food quantity in terms of energy and do not take into account any requirements for particular nutrients.

Other evidence also suggested summer food shortage. Beshir (1971) showed that body condition, in terms of weight, total fat, and pectoral muscle weight, was lowest in summer and highest in late winter. This can be further related to the seasonal pattern of mortality. Most disappearance of individually marked breeding adult rooks occurred between March and October (Dunnet and Patterson 1968) and most recoveries of rooks ringed in the national ringing scheme occurred in the late breeding season and summer (Feare, Dunnet, and Patterson 1974).

It seems clear that since food intake in summer appeared to be sufficiently low to produce loss of condition and mortality, the hypothesis predicts that aggressive interactions in feeding flocks should be most frequent then.

Seasonal variation in frequency of aggressive interactions

Data on interactions were collected on standard transects of a study area near Newburgh, Aberdeenshire (for details of the transects see Patterson, Dunnet, and Fordham (1971)). Flocks of rooks were selected for study by their position relative to available roads. The birds had to be clearly visible from a car (used as a mobile hide) and at such a distance (200–400 metres) that interactions could be observed without alarming the birds. If rooks began to fly up or move away the observer drove on or waited for some time to allow the birds to settle.

Ten adjacent rooks were chosen arbitrarily from the central part of the flock and all interactions of the type described earlier between the 10 individuals were recorded on a tape recorder. The observation period was measured with a stop watch and varied between 2·0 minutes and 16·3 minutes (usually 5–10 minutes) depending mainly on how long the flock remained undisturbed. During this period some of the original 10 birds would move off among the rest of the flock and others took their place in the 'field of view', so that the sample was of 10 rooks with a varying individual membership. Since Lockie (1956) suggested an effect of flock density, this was measured by estimating the distance between each of the 10 birds and its nearest neighbour as described by Patterson, Dunnet, and Fordham (1971). Some observations were made on days with a trace (under 0·5 centimetres) of snow and some on days with over 5 centimetres of level snow fall.

The air temperature at 0900 hours, taken from routine meteorological records at Culterty Field Station, was used as an index to the temperature at the feeding sites which were within a few kilometres of the Station and at a similar altitude.

The frequency of aggressive interaction per group of 10 birds per minute (Fig. 8.1) was very variable at all times of year but was considerably higher in winter, particularly in February, than in the summer. The true summer values should be even lower since scattered flocks in long vegetation, where interactions were rare, were probably under-represented in the samples since they were difficult to see.

Thus in the summer months when food intake was sufficiently below

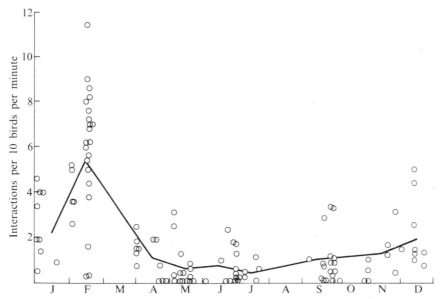

FIG. 8.1. Seasonal changes in frequency of aggressive interaction (per sample of 10 birds per minute) in feeding flocks of rooks. The points are individual samples; the line joins the monthly mean frequencies.

requirement to affect condition and mortality, interactions were relatively rare, and not at their highest frequency as would be expected if they were important in limiting numbers in relation to a restricted food supply. However, Fig. 8.1 shows only gross seasonal changes and a more detailed examination of some of the factors affecting frequency of interaction can help to explain variations in frequency within and between seasons.

Factors affecting frequency of interaction

The relationships between interaction frequency and air temperature, snow cover, and the spacing of birds in the flock were measured by examining interaction rates in the two commonest feeding sites, stubble or stacked grain and grass fields (including fields cut for silage and short-growing cereals). This restriction reduced variations associated with less commonly used field types yet retained a comparison between two contrasting feeding situations.

Air temperature.   In both grain crops and grass fields there was a significant increase in the rate of interaction with decreasing temperature (Fig.

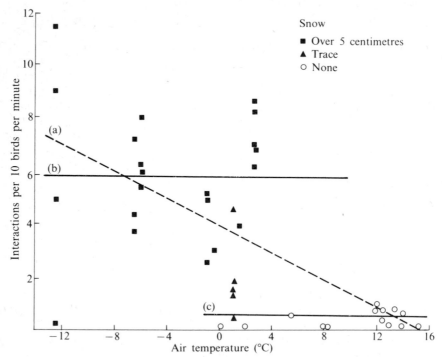

FIG. 8.2. Frequency of aggressive interaction in rook flocks on grain crops, in relation to temperature. The points are individual samples. The lines are calculated regressions; line (a) is for all points ($y = 3\cdot94 - 0\cdot08x$, $r = -0\cdot634$, $p < 0\cdot001$), line (b) for observations in snow over 5 centimetres deep ($y = 5\cdot82 - 0\cdot002x$, $r = -0\cdot063$, $p < 0\cdot1$).

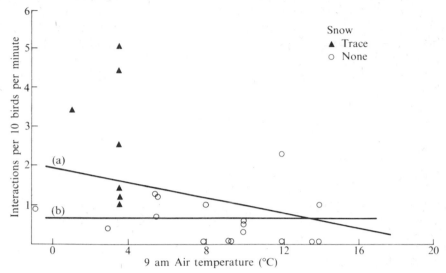

FIG. 8.3. Frequency of aggressive interaction in rook flocks on grass fields, in relation to temperature. The lines are calculated regressions; line (a) is for all points ($y = 1.90 - 0.031x$, $r = -0.418$, $p < 0.01$), and line (b) for observations on days with no snow only ($y = 0.68 - 0.001x$, $r = -0.031$, $p < 0.1$).

8.2, line a; Fig. 8.3, line a). The higher rates (over one interaction per 10 birds per minute) were mainly on days when air temperature was below 4 °C. However this relationship disappears when days with the same amount of snow are compared. In the absence of snow, rates of interaction showed no significant trend with temperature over a range from −1 °C to +16 °C in both types of crop (Fig. 8.2, line c; Fig. 8.3, line b). In snow more than 5 centimetres deep, interaction rates in flocks on grain crops again showed no systematic variation with temperature over a range from −12.5 °C to +3 °C (Fig. 8.2, line b).

Spacing. The more closely spaced groups of rooks had significantly higher frequencies of interaction both on grain crops (Fig. 8.4) and on grass fields (Fig. 8.5) than did more spaced-out groups. Rooks on grass were spaced more than those on grain crops, but had a similar frequency of interaction at the same spacing.

When days with different depths of snow were considered separately, interaction increased significantly with decreased spacing on grain crops in deep snow (Fig. 8.4, line a) and on grass fields without snow (Fig. 8.5, line b). On grain during a trace of snow there was an increase in encounters with closer spacing (Fig. 8.4, line b) but the correlation between these two factors was not statistically significant. On grain crops without snow (Fig. 8.4, line c) there was no change in the rate of interaction with changes in spacing.

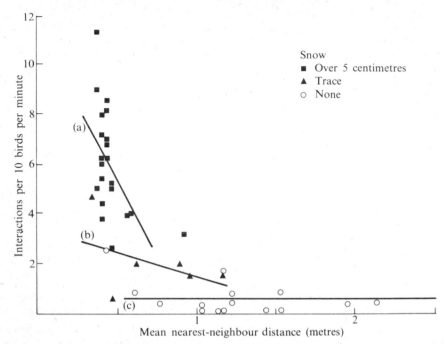

FIG. 8.4. Frequency of aggressive interaction in rook flocks on grain crops, in relation to mean nearest-neighbour distance. The lines are calculated regressions; line (a) is for observations in snow over 5 centimetres deep ($y=11{\cdot}07-1{\cdot}15x$, $r=-0{\cdot}632$, $p<0{\cdot}01$), line (b) for observations during traces of snow ($y=3{\cdot}38-0{\cdot}19x$, $r=-0{\cdot}440$, $p<0{\cdot}1$), and line (c) for observations on days without snow ($y=0{\cdot}56+0{\cdot}0002x$, $r=0{\cdot}003$, $p<0{\cdot}1$). The regression equation for all the observations of interaction on spacing is $y=5{\cdot}15-0{\cdot}20x$, $r=0{\cdot}523$, $p<0{\cdot}001$. This regression and line (c) take into account two extreme observations at spacings of 4 metres and 5 metres and interaction frequencies of 0·8 and 1·0 respectively.

Snow cover. It is clear from the preceding sections that frequency of interaction varied with the depth of snow. On days with temperatures within the range of $-1\,°C$ to $+2\,°C$, interaction rates in flocks on grain crops were significantly lower during traces of snow than during deep snow (Fig. 8.2, using Mann–Whitney U test, $p=0{\cdot}015$) and significantly higher than on days without snow ($p=0{\cdot}036$). On grass fields, interaction rates on days with traces of snow were also significantly higher ($p=0{\cdot}028$) than on days of similar temperatures ($-1\,°C$ to $+4\,°C$, Fig. 8.3) when there was no snow. Thus snow appeared to increase considerably the frequency of aggressive interaction.

Spacing in flocks was also related to snow. Neighbouring rooks on grain crops were significantly closer together in snow over 5 centimetres deep than in traces of snow (using Mann–Whitney U test, $p<0{\cdot}001$, data in Fig. 8.4) and closer with snow traces than with no snow ($p<0{\cdot}001$, Fig. 8.4). During periods of snow cover, Feare, Dunnet, and Patterson (1974)

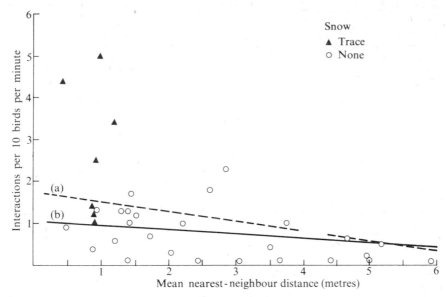

FIG. 8.5. Frequency of aggressive interaction in rook flocks on grass fields, in relation to mean nearest-neighbour distance. The lines are calculated regressions; line (a) refers to all observations ($y = 1\cdot74 - 0\cdot023x$, $r = -0\cdot486$, $p < 0\cdot01$) and line (b) to observations on days without snow ($y = 1\cdot03 - 0\cdot01x$, $r = -0\cdot415$, $p < 0\cdot02$). Both regressions take into account observations with mean spacings over 6 metres with the following pairs of spacing; interaction frequency $-7\cdot7:0$, $7\cdot8:0\cdot0$, $9\cdot5:0\cdot5$, $8\cdot5:0\cdot2$, $7\cdot7:0\cdot1$, $7\cdot3:1\cdot1$, $8\cdot0:0\cdot6$.

showed that rooks changed from feeding mainly on stubble and grass fields to feeding largely on much more localised sites (grain stacks and dung heaps) where the birds would be crowded together much more.

Since spacing was reduced during snow cover and since interactions were more frequent at closer spacings, the higher rate of interaction during snow can be explained through its relationship with spacing. However, frequency of interaction on fields with snow was higher than on fields without, even with no change in spacing. Flocks on days with a trace of snow had significantly higher rates of interaction than those with the same range of spacings ($0\cdot3$–$1\cdot3$ metres on grain crops; $0\cdot5$–$1\cdot3$ metres on grass) on days without snow (using Mann–Whitney U test, $p = 0\cdot03$ on grain, $p = 0\cdot012$ on grass, data in Figs 8.4 and 8.5).

These relationships between rate of interaction and environmental factors can explain the seasonal changes in frequency of interaction. High rates of interaction were recorded only in snow and so obviously could not occur except in winter. The low summer rates can also be partly attributed to the decrease in the frequency of interaction with increased spacing. Patterson, Dunnet, and Fordham (1971) showed that the mean nearest-neighbour distance in rook flocks increased from 2·2 metres on grain crops and 4·9 metres on grass fields in Junuary, to 9·4 metres on grass in July (when no grain was available). In addition to this increased spacing within flocks, over 40 per cent of rooks seen in August were feeding 'outside flocks' (defined as more than 50 metres from another rook). Since interactions became rare (fewer than 0·5 interactions per 10 birds per minute, the mean rate in July) when mean spacing exceeded 2 metres (grain crops) to 6 metres (grass fields) the low rate of interaction in summer can be explained by the much larger spacing between neighbours. Many birds were almost certainly unable to see when another had found a food item. In any case, approach would expend a large amount of energy relative to the value of the item (often small surface invertebrates in summer).

Aggressive interactions were thus very infrequent during the period of summer food shortage. However, it is important to establish whether such interaction affected the feeding rate of the birds.

8.2. Feeding rate in relation to frequency of interaction

The most direct way to assess the possible depression of feeding rate by aggressive interactions is to measure the rates of both at the same time in the same individual rooks. In most situations, however, except during snow cover which will be discussed separately, the rate of interaction was so low as to make this impracticable; any given individual fed for long periods without being involved in an interaction. However, the possible effect of the aggressive behaviour can be assessed by comparing its mean frequency of occurrence with the mean feeding rates measured by Feare, Dunnet, and Patterson (1974). In autumn and winter (October–February), excluding periods of snow the average rook was involved in an aggressive interaction once for every 102 items (grains) obtained in grain crops and once for 14 items (invertebrates) in grass fields (Table 8.1). Since only a proportion of these interactions resulted in loss or gain of a food item and since both aggressor and victim resumed apparently normal feeding rates within a few

TABLE 8.1

Feeding rate in relation to frequency of interaction

	Interactions per bird	Items per minute	Items per interaction
Grain Sep.–Feb	0·05	5·12	102
Grass Sep.–Feb.	0·09	1·32	14
Grass Jun.–Aug.	0·08	1·15	14

seconds of the start of the encounter, it seems very unlikely that feeding rate was affected unless the interactions were concentrated in a very small proportion of the population. Even if feeding rates were affected, at this time rooks were feeding for only 58 per cent of the available time (Feare, Dunnet, and Patterson 1974), so that it is likely that daily intake could be maintained by increasing the time spent feeding. Similarly, in summer each bird was again involved in only one interaction per 14 items obtained (or more than 14, if as suggested earlier the summer rates of interaction were overestimated), although at this season there was less opportunity to increase the already large amount of time spent feeding (Feare, Dunnet, and Patterson 1974).

Thus for most of the year aggressive interactions occurred at too low a frequency to depress the feeding rate of individuals. However, the much higher rates recorded during snow cover might be expected to have a greater effect. This was measured directly by C. J. Feare (Feare, Dunnet, and Patterson 1974), who found that on grain stacks birds involved in interactions during a 1-minute observation period had a significantly lower intake than those which fed undisturbed. Attackers had a higher intake rate than victims of an attack or threat although this difference was not statistically significant. However it leads to the need to determine whether particular individuals feeding rates are more affected by interaction.

8.3. Dominance

During the measurement of rates of interaction discussed earlier, the result of each encounter was recorded. In most cases the outcome was clear and could be classified as a win or loss for one of the birds when the other retreated. Some encounters had an indefinite ending with both birds moving away to resume feeding.

The results (Table 8.2) confirmed Lockie's (1956) finding that the attacking bird won most encounters; he interpreted this to mean that a hierarchy existed within the flock and that rooks attempted to supplant only their subordinates. However, a hierarchy, in its usual sense of stable dominance relationships between individuals known to one another, need

TABLE 8.2
Result of aggressive encounters

Result	Number of encounters	% (of those with known result)
Attacker wins	296	83
No clear winner	51	14
Attacked bird wins	10	3
Result unknown	33	

not be postulated to explain these results. It is possible that rooks were able to assess the relative size or aggressiveness of other, unknown birds and attempted to supplant only those unlikely to resist. In any case since most food items were small it may have been better for victims to retreat immediately rather than risk injury in actual combat, so that the same individuals may not have been consistently attackers or victims.

C. J. Feare (Feare, Dunnet, and Patterson 1974) found that rooks feeding at the top of grain stacks stayed longer without being supplanted and had higher feeding rates than those lower down. This can also be explained in terms of a hierarchy with the dominant birds holding the best (top) feeding sites, but as the authors point out, a number of other explanations are possible.

Observations on encounters between individually marked rooks, or between these and other members of the population, could of course establish whether stable dominance relationships exist. However, although very many rooks have been marked (Patterson, Dunnet, and Fordham 1971) the population was large and the proportion marked rarely exceeded 10 per cent. The feeding flocks were constantly mobile so that each marked individual was seen infrequently and for a short time, so that the usually low rate of interaction (Table 8.1) made it rare for a particular marked rook to be seen in more than a very few encounters.

These factors which made it difficult to study dominance also made it inherently unlikely that stable dominance relationships existed in the rook population. The birds formed a constantly mixing group of 700–1600 birds (Dunnet and Patterson 1968) without consistent sub-group formation. Since Guhl (1953) and others have found that the formation of stable hierarchies became impossible in groups of more than 200 domestic fowl, it is unlikely that each rook knew the relative status of each of the very large number of others in the population.

The question of dominance, and whether aggressive encounters differentially affect particular members of the population through this or any other mechanism, must be left open in the absence of direct evidence.

It is of course possible to explain changes in numbers without involving direct agonistic encounter between individuals. The measurement of seasonal changes in frequency of aggressive interaction in feeding flocks (Fig. 8.1) failed to confirm the original hypothesis that this frequency would increase with food shortage, which occurred principally in summer. The marked dispersal of rooks at that time (Dunnet and Patterson 1968) when they were seen in areas such as moorland not used in other seasons (Watson 1970) occurred when there was a minimum of aggressive interaction in the flocks. The very low density in summer (Patterson, Dunnet, and Fordham 1971) was probably achieved by each rook avoiding others without interaction while feeding, and the dispersal and death of particular birds may have been related to their individual success in finding food. During the summer, juveniles disappeared more rapidly than adults (Dunnet, Fordham, and Patterson 1970) and young birds of several species (for example, Sandwich Tern, *Sterna sandvicensis*, (Dunn 1972)) have been shown to be less efficient than adults in obtaining food. Thus there is little to suggest that aggressive interaction was involved in any changes in numbers during summer food shortage and it is much more likely that such changes were a direct response to the quantity and dispersion of food.

A closer examination of feeding in winter (Feare, Dunnet, and Patterson 1974) showed that there was likely to be some short-term food shortage during periods of snow cover, when there was also a high rate of aggressive interaction particularly in denser flocks.

This raises two questions; what was the cause of this change in behaviour, and was it likely to affect numbers by causing dispersal or mortality?

It is possible that interactions were caused differently at different spacings between neighbouring rooks. At the longer distances most approaches were probably elicited by the sight of another rook with a food item or with an obviously productive feeding site, though some may have resulted from the proximity of another bird to the attacker's mate. With decreasing inter-neighbour distances it is likely that more interactions were caused by infringement of individual distance, although the functional significance of this may still have been the defence of adequate feeding space. In periods of deep snow cover, feeding sites and possible feeding time at them were limited (Feare, Dunnet, and Patterson 1974) producing increased crowding and consequent increased interaction over space. On sites like grain stacks, where food was locally abundant and concentrated, it is unlikely that many interactions were over individual food items. Feare, Dunnet, and Patterson have shown that birds winning encounters had a feeding rate no higher than those which did not interact, so that there was probably no advantage, in terms of immediate rate of intake, in attacking another bird. It seems likely therefore that most of the increase in rate of interaction during snow was caused by the increase in aggressive encounters in snow without change in mean spacing. This increase could have been caused by a direct effect of a lowered daily food intake. Feare, Dunnet, and Patterson found that although food intake was well above requirement over the winter as a whole, daily intake during snow was lower than in snow-free periods. A hungrier rook might be more likely to attack another irrespective of the spacing between them.It is difficult to understand the possible function of this since, as discussed above, even when a bird won its encounter it did not seem to benefit by an increased food intake. It is possible that rooks successful in encounters were able to stay longer at the feeding site without being supplanted by others and so would have an increased feeding time even though feeding rate was not changed. On the other hand, hunger may well have been the factor causing the decreased spacing itself, making rooks approach feeding sites even when a high density already existed there and making them reluctant to flee when approached. The relationships between food shortage, spacing, and interaction rate during snow were thus complex and difficult to assess with the present data, but do emphasize the possible importance of aggressive behaviour during snow cover.

Considering now the second question, since attacks were rarely resisted (Table 8.2) and since rooks losing encounters had a reduced feeding rate, it is possible that the effects of interaction were concentrated on a proportion of the population. If so, this component of the population might have been more likely to suffer stress (Christian 1968) or to disperse to other feeding areas, as has been suggested in the wood-pigeon (*Columba palumbus*) by Murton (1968). In one period of snow, marked rooks were seen on stacks which were well outside their normal feeding range but no systematic observations were made on movement or changes in number during snow.

In the absence of more conclusive data on stress or dispersal during snowy weather and on the social status, food intake and movements of individually marked rooks, the question must be left open with aggressive interaction in feeding flocks remaining a potential regulatory mechanism during winter snow.

Acknowledgments

This paper reports part of a team study of the rook, directed by Professor G. M. Dunnet, and supported by a grant from the Agricultural Research Council. I am indebted to G. M. Dunnet, J. D. Lockie, and A. Watson for useful and constructive comment on the manuscript.

References

BESHIR, S. A. (1971). *Seasonal variations in some aspects of 'condition' in the rook* (Corvus frugilegus *L.*) *in North-East Scotland*. Unpublished M.Sc. thesis. Aberdeen University.

CHRISTIAN, J. J. (1968). The potential role of the adrenal cortex as affected by social rank and population density on experimental epidemics. *Am. J. Epidem.* **87**, 255–66.

CROOK, J. H. (1965). The adaptive significance of avian social organisations. *Symp. zool. Soc. Lond.* **14**, 181–218.

CULLEN, E. (1957). Adaptations in the kittiwake to cliff-nesting. *Ibis.* **99**, 275–302.

DUNN, E. K. (1972). Effect of age on the fishing ability of sandwich terns *Sterna sandvicensis* L. *Ibis* **114**, 360–6.

DUNNET, G. M., FORDHAM, R. A., and PATTERSON, I. J. (1970). Ecological studies of the rook (*Corvus frugilegus* L.) in North-east Scotland. Proportion and distribution of young in the population. *J. appl. Ecol.* **6**, 459–73.

—— PATTERSON, I. J. (1968). The rook problem in North-east Scotland. In *The problems of birds as pests* (eds R. K. Murton and E. N. Wright), Institute of Biology Symposium, Vol. 17, pp. 119–39.

GUHL, A. M. (1953). Social behaviour of domestic fowl. *Tech. Bull. Kans. agric. exp. Stn* **73.**

FEARE, C. J., DUNNET, G. M., and PATTERSON, I. J. (1974). Ecological studies of the rook (*Corvus frugilegus* L.) in North-east Scotland. Food intake and feeding behaviour. *J. appl. Ecol.* **11,** 867–96.

KRUUK, H. (1964). Predators and anti-predator behaviour of the black-headed gull *Larus ridibundus* L. *Behaviour, Suppl.* **11.**

LOCKIE, J. D. (1956). Winter fighting in feeding flocks of rooks, jackdaws and carrion crows. *Bird Study* **3,** 189–90.

MURTON, R. K. (1968). Some predator-prey relationships in bird damage and population control. In *The problems of birds as pests* (eds R. K. Murton and E. N. Wright), Institute of Biology Symposium, Vol. 17, pp. 157–80.

PATTERSON, I. J., DUNNET, G. M., and FORDHAM, R. A. (1971). Ecological studies of the rook (*Corvus frugilegus* L.) in North-east Scotland. Dispersion. *J. appl. Ecol.* **8,** 815–33.

TINBERGEN, N. (1972). *The animal in its world* Vol. 1: Field Studies. Allen and Unwin, London.

—— BROEKHUYSEN, G. J., FEEKES, F., HOUGHTON, J. C. W., KRUUK, H., and SZULC, E. (1962). Egg shell removal by the black-headed gull *Larus ridibundus*, L.; a behavioural component of camouflage. *Behaviour* **19,** 74–117.

WATSON, A. (1970). Comment in discussion. *Animal populations in relation to their food resources* (ed. A. Watson), British Ecological Society Symposium, Vol. 10, pp. 252.

—— MOSS, R. (1970). Dominance, spacing behaviour and aggression in relation to population limitation in vertebrates. *Animal populations in relation to their food resources* (ed. A. Watson), British Ecological Society Symposium, Vol. 10, pp. 167–220.

Part II
Comparison and evolution

9. An evaluation of the conflict hypothesis as an explanatory principle for the evolution of displays

G. P. BAERENDS

9.1. Introduction

I VIVIDLY remember when in the spring of 1948 Niko Tinbergen, shortly after I had joined him on the boat that would take us to the Dutch bird island Terschelling, enthusiastically told me that he believed he had traced, in co-operation with Jan van Iersel and some of his other pupils at Leiden University, the evolutionary origin of all five courtship activities of the male Three-spined Stickleback (zigzagging, pricking, leading, showing the nest-entrance, and quivering). This achievement struck me as a turning point in the study of animal behaviour for it meant that a period of observing, photographing, describing and enjoying, the peculiar behaviour patterns in the social life of animals from various groups, had now led to fruitful thinking about the processes through which behaviour patterns are shaped in the course of evolution.

The method followed was primarily a comparative one. First, this method had led to the discovery that an animal may perform morphologically the same behaviour pattern in different situations or functional contexts. One of these could usually, on logical grounds, be considered to correspond with the phylogenetically oldest function of the element in the behaviour repertoire, the others as secondary or derived. For instance, food-begging in gulls was considered to have been originally a juvenile behaviour pattern and thus to be a derived behaviour element when used by an adult female in courting a male; the manipulation of plant material by a gull was considered to be a nest-building activity in origin but a derived activity when serving threat in a boundary conflict or appeasement in a nest-relief ceremony. In ethology this appearance of an activity in a secondary context is known as the displacement phenomenon (Kortlandt 1940; Tinbergen 1940, 1952b). Secondly, it was recognized (Portielje 1928; Tinbergen 1940; Lorenz 1941; Daanje 1951) that incomplete behaviour patterns (intention movements, awakening movements) should also be considered as a source from which threat and courtship behaviour can be derived.

Intention movements of functionally opposed behaviour patterns (for example, approach and avoidance, or attack and flight) occur in combination in ceremonies by which a position is defended or a partner courted. The combination may take the form of a repeated alternation of incomplete versions of the opposite patterns or these may both, as far as physically possible, operate at the same time. The former type of combination (successive ambivalence) is common in boundary fights of territory-owning fish, birds, and mammals ('pendulum fights'); it can also be recognized in the 'zigzag-dance' of a male Three-spined Stickleback courting a female (Tinbergen 1952a). The classical example of the other type (simultaneous ambivalence) is the upright threat of the gulls (Tinbergen 1952a, b). This pattern can be understood as a mosaic-like compromise of conflicting, mutually inhibitive behaviour patterns. Unfortunately the name 'compromise activity'—which would nicely fit all cases of simultaneous ambivalence at least—has been claimed by Andrew (1956) for the rather extreme case of competition in which only elements common to both conflicting behaviour systems can be carried out.

Simultaneous ambivalence of species-specific behaviour components may concern the characteristic stereotyped shape of the pattern (fixed action pattern (FAP), according to Thorpe (1952), or the modal action pattern (MAP according to Barlow (1968)) as well as the orientation component (taxis). The latter possibility is a third source for the derivation of communication activities, recognized and named 'redirection' by Bastock, Morris, and Moynihan (1953). A substitute object present in the new 'compromise' direction may then evoke the completion of the response, as in the 'grass-pulling threat' of the Herring Gull (Tinbergen 1952b). But the mere turning away of a weapon (facing-away), as an appeasement signal (head-flagging in the Black-headed Gull, head-up postures in other birds; see Tinbergen and Moynihan (1952); Tinbergen (1959); also Beer, this volume), could also have originated from a conflict between opposite orientations.

Not only the phenomena of ambivalence and redirection, but also the occurrence of displacement have been explained (van Iersel and Bol 1958; Sevenster 1961) as consequences of a conflict between largely incompatible behaviour systems. Tinbergen (1952a, b) was of the opinion that interaction between the drive to attack and the drive to escape had led to the occurrence and shaping of threat displays, and interactions between one of these drives and the mating drive to sexual or courtship displays. Following these ideas, Morris (1956a) spoke of a three-point conflict between the incompatible tendencies to flee (F), to attack (A), and to mate (M)—the FAM-conflict—as a determinant of the patterns of courtship. He argued that the relative contribution of each of these tendencies differs between species (for instance, using the notation of a lower-case letter for a smaller contribution: fAM in the Three-spined Stickleback *Gasterosteus aculeatus*, and the bullhead, *Cottus scorpio*; FaM in certain passerine birds; faM in the swordtail, *Xiphophorus helleri*), but he was also aware that similar differences exist between different displays in one species. Differences in the absolute strengths of the tendencies to attack and to flee, as well as in their ratios, were extensively used by Moynihan (1955, 1958*a*, *b*, 1962) to interpret the evolutionary development and the function of different threat display within each of a number of gull species.

It should be stressed here that the conflict hypothesis was developed to explain the phylogeny of the various forms of displays. Arguments for its support were in the first place taken from similarities in their form to overt attack and overt fleeing behaviour found in comparative studies. However, this hypothesis necessarily involved speculations about causation. At least in the beginning of their phylogenetical development the displays must have been under the control of causal factors underlying the agonistic tendencies. Consequently several authors have used methods for studying causation to verify the conflict hypothesis. For instance Moynihan (1955), Manley (1960), Baerends and van der Cingel (1962), Stokes (1962*a*), Blurton Jones (1968), and Kortmulder (1972) have attempted to find temporal or sequential correlations between the occurrence of each type of display and of overt agonistic activities. This method, however, can serve this purpose only if the display patterns studied, after having obtained signal value, had not completely lost their relations with the systems originally controlling them. Nevertheless, there is a real possibility that important changes take place in the complex of factors leading to the occurrence of an activity when this pattern evolves into a communicative display.

For derived activities usually become ritualized, that is, their form changes in a way adaptive to the secondary signal function. They tend to become more conspicuous, while structures and colours underlining this conspicuousness may be added. The continuous variation in the form of displays which one would expect if the underlying conflicting tendencies interact in various ratios of strength was recognized by Lorenz in the threat behaviour of dogs. By contrast, Tinbergen (1959) found the freedom of variations to be very much restricted in the agonistic displays of the Black-headed Gull; an analysis of cine-film frames showed that the time spent in the oblique or in the forward display posture was 5–6 times as long as that of the transition postures between them. The hypothetical processes through which the causation of a display would become less dependent on the factors originally underlying it and more dependent on other (novel) factors, is called 'emancipation'.

In contrast to comparative anatomy, comparative ethology lacks the possibility of comparing recent forms with fossil ones in order to reconstruct phylogeny. However, the ethologists can compare, with the aim of finding phylogenetic relations, similar-looking behaviour patterns in related recent species, and—thanks to the evanescence of behaviour and to phenomena like displacement, redirection, and ambivalence—they can also compare similar-looking behaviour patterns in different behavioural contexts within one species. Such a procedure makes sense only when the patterns compared can in principle be considered homologous, that is, derived from the same behaviour pattern in an ancestor. Baerends (1958) and Wickler (1961a, b, 1967a) have discussed the criteria which can be used to distinguish homologous behaviour patterns.

It follows from the above that the heuristic and the explanatory validities of the 'conflict hypothesis' are considerable. According to this hypothesis the conflicts between functionally opposite tendencies—that always occur when potential rivals or mates meet—on the one hand create the causal basis upon which communication behaviour evolves (ambivalence, redirection, and displacement) and, on the other hand, provide the selective pressure (the survival value of mutual understanding) necessary for phylogenetic development. The conflict hypothesis seems to have led to a much 'better' understanding of the form of displays, particularly in fish (sticklebacks: van Iersel 1953; Morris 1958; Sevenster 1961; cichlids: Baerends and Baerends-van Roon 1950; guppy: Baerends, Brouwer, and Waterbolk 1955; *Barbus*: Kortmulder 1972) and birds (passerines: Andrew 1957; Hinde 1952, 1953, 1956a; Morris 1954, 1957; gulls: Moynihan 1955, 1958a, b, 1962; Tinbergen 1959). Its application for understanding the origin of threat displays in some mammalian species has already been mentioned; van Hooff (1962) used the conflict hypothesis in his explanation of the forms of facial expressions in primates.

However, during the last 10 years the conflict hypothesis and its applications have been repeatedly criticized and questioned, by, among others, some authors who formerly seemed to have used it to advantage. In particular Brown (1964a) became sceptical about the value of the conflict hypothesis when, in an ethological study of the social behaviour of the Steller's Jay (*Cyanocitta stelleri*), he attempted to interpret the form and occurrence of various visual and acoustic displays in terms of equilibria between tendencies to attack and to escape. He used morphology, temporal relations, and context of occurrence as criteria, but his results frequently led to different and contradictory conclusions when this method was separately applied to an entire display and to a particular part (for example, position of the crest) or an accompanying component (for example, vocalization). Andrew (1972) has also expressed his dissatisfaction with the conflict hypothesis. The criticisms mainly concern the following questions.

1. Is the application of the homology concept to behaviour legitimate?
2. Is it justified to attribute a dominating role to drives, motivations, or behaviour systems in the causation of behaviour?
3. Is the 'conflict hypothesis' in accordance with the available relevant neurophysiological evidence?
4. Are other possible sources for the derivation of communication behaviour sufficiently investigated?

In the following section I shall first attempt to answer these questions and second discuss some explanatory perspectives of the conflict hypothesis.

9.2. Consideration of the criticisms

The homology concept

Recently Atz (1970) and Gottlieb (1970) have critically discussed the application of the concept of homology, as borrowed from comparative anatomy, to behaviour. Although Atz does not seem to deny that the concept is in principle applicable to behaviour, he is of opinion that such an application would be 'operationally unsound and fraught with danger' until behaviour can 'critically be associated with structure' (Atz 1970, p. 69), that is, when the underlying machinery is morphologically and physiologically known. Gottlieb admits the value of homology as a tool in comparative morphology as long as use can be made of fossilized structures, but he seems to deny it when soft tissues like those of the nervous system, or evanescent behavioural acts are concerned. However, the fact that behaviour elements are evanescent and that our knowledge about the physiological machinery underlying them is still very incomplete may make the recognition of behavioural homologies more difficult and uncertain than that of structural ones, but does not invalidate the application of the concept of homology to behaviour itself. Actually, our knowledge about the physiological mechanisms controlling the ontogeny of structure is also incomplete and evidence exists that during phylogeny the set of causal factors underlying homologous structures may change considerably.

In our opinion the life-span of the phenomenon studied should not be used as a restrictive criterion for the application of the homology concept. Because a behaviour pattern is evanescent it can—particularly in the stereotyped behaviour we are concerned with here—also be recurrent in the same individual. This characteristic (which in itself facilitates the study of causation) opens up the possibility that an activity may appear in different contexts of the behaviour repertoire. The conflict hypothesis is a guide in the analysis of how this feature may have promoted the development of communication behaviour in the course of evolution. Our problem, how species-specific behaviour elements, particularly displays, have attained their shape, follows from the assumption that behaviour elements, like structures, will have had behavioural precursors in ancestors. This assumption is justified by the fact that in comparative studies of related recent species stereotyped behaviour elements can be considered in exactly the same way as structures and with similar results.

Atz and Gottlieb have concentrated their criticisms on the use of the homology concept for detecting, reconstructing, and establishing phylogenetic or taxonomic relationships. For this purpose behavioural homologues may indeed be a more hazardous and difficult tool than structural homologues, because the analysis necessary to establish them is often so difficult and time-consuming. But for detecting the mechanisms underlying the evolution of behaviour patterns we cannot avoid facing up to these difficulties when carefully developing a consolidating hypothesis of how the derivation and the shaping of the signal activity studied might have been brought about.

The role of behaviour systems

In the original formulations of the conflict hypothesis the conflict was thought to obtain between drives, that is, groups of common causal factors of complexes of activities (Thorpe 1952). It has frequently been stated that arguments for the postulation of drives should be obtained from causal analysis only. However, because of the effort, the time, and the methodological difficulties involved in such an analysis the rough impression that activities serving the same major function are largely controlled by common causal factors was often taken for granted. Without much further analytical investigation animals were assumed to have one attack drive, one fleeing drive, one sexual drive, one parental drive, one feeding drive, and so on. This 'short-cut' procedure, although erroneous, has not been altogether unfortunate because the models resulting from it offer fascinating prospects. The time has come, however, to correct this procedure. Although there is no doubt that common causal factors exist and that often they roughly correspond with common function, an exact and penetrating insight in the way signal behaviour has evolved will be possible only on the basis of a proper experimental analysis. For in several of such studies it has been shown that behaviour elements, which were originally considered by the investigator to serve one common function, were in fact not under the control of the same super-ordinated mechanism; part of them belonging to one and part to another causal behaviour system; for example, escape behaviour in the fowl (Hogan 1965), in the ruff, *Philomachus pugnax* (van Rhijn 1973), and in the cichlid fish, *Tilapia mariae* (Baldaccini 1973).

The causal analysis should take into account the criticism frequently expressed, particularly by Hinde (1956b, 1959a, 1970), of the use of overly simple models with unitary drive concepts. In cases where a causal analysis of the relationship between different activities of an animal has been undertaken (for example, reproductive behaviour of canaries: Hinde (1965); incubation behaviour of gulls: Baerends (1970, 1973); feeding behaviour of rats and mice: de Ruiter (1963)), groups of activities were usually found to share common causal factors, but in more complicated and more various ways than has often been concluded from the early simple hierarchy models for the functional organization of behaviour (see, for example, Tinbergen 1950). In particular it was found that elements of a lower integration level can be controlled by more than one of the super-ordinated, often mutually competitive, mechanisms, and it was realized that mechanisms of each level can be influenced by feedback stimulation resulting from ongoing activities (Kortlandt 1955; McFarland 1971; McFarland and Sibley 1972). Consequently it may be preferable to speak of a network structure instead of a hierarchy. Nevertheless, it is possible to distinguish on grounds of common causality different systems and sub-systems within this network, characterized by relatively close temporal and sequential relations between the activities and often also by a selective responsiveness to stimuli from the environment; as diagrammatically represented in Fig. 9.1. This common causality is connected with serving a common function. The various systems in the functional organization of an animal should not be considered as discretely separate; their morphological and physiological substrates cannot be expected to be uniform.

Apart from sharing common causal factors, each of the different activities of a system also has its own specific causal factors. As a result the correlation in the occurrence of various activities of the same system is usually not very high. This is understandable, for an activity is not likely to survive in the course of evolution if it does not serve a special detailed function with controlling factors of its own; it functions as part of a system, often in different systems, but it also exists in its own right.

When 'conflict between drives' is considered to be synonymous with a conflict between activated systems in the above sense it should be understood that no reference to one particular level of integration is meant. Thus the hypothesis leaves open the possibility of considering conflicts between very restricted systems, or even between single responses, as origins for signal behaviour—a possibility to which Andrew (1972) has drawn attention—besides conflicts in which more or less super-ordinated larger groups of elements are involved. Moreover, the possibility remains that a conflict between a relatively highly integrated group and a simple system, response or even stimulus may lead to communication behaviour.

The simplest possibility of a conflict is probably one between two opposite orientation components during locomotion; it may occur as an

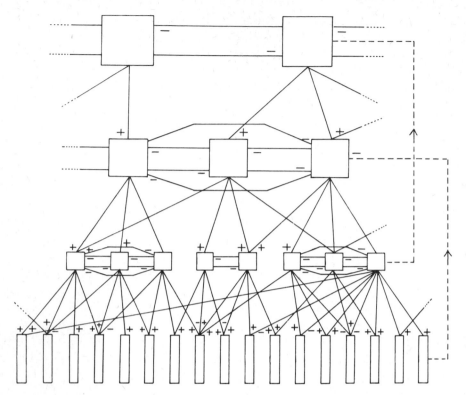

FIG. 9.1. Hypothetical schematic representation on the hierarchical organization (functional network) of behaviour. The rectangles on the bottom row represent fixed patterns; the squares represent systems of different order co-ordinating other systems or fixed patterns through activation (+) or inhibition (−). Systems of the same order have mutually inhibiting relations. The dotted lines represent two of the many feedback connections that are thought to be present in the network. The diagram can be extended on both sides.

unritualized exploratory approach–avoidance conflict in an animal confronted with a new kind of food or, sometimes highly ritualized, in a 'pendulum' boundary fight. However, when possible examples of such simple cases are observed with care, more elements than just locomotory components are often found to be involved, which suggests an interaction of systems. Among these are autonomic responses (respiratory, excretory, or thermo-regulatory responses, and in some animals colour-changes as well) not occurring in simple exploratory behaviour and which may be the result of stimuli received during approach or avoidance (Morris 1956*b*; Groen 1955). Morris suggests that some of them may be caused by frustration. However, several autonomic responses, seem, functionally speaking, to be anticipating the situations and stimuli that may arise as consequences of the approach or the retreat when actually carried out.

Therefore, the facilitation of such responses often precedes the perception of these stimuli and must then be part of the expressions of an activated system.

A good opportunity to study the linkage between somatic activities and autonomic responses is offered by the colour changes of fish. As an example we can take the mechanism of colour-pattern change in *Tilapia mariae* (Baldaccini 1973). In this species three major colour patterns can be distinguished (Fig. 9.2): (a) the barred pattern, which is typical for non-territorial individuals moving about in schools; (b) the spotted pattern, typical for fish aggressively defending a territory; and (c) the dark pattern, indicating a tendency to escape but to stay nevertheless in a certain area. Several other patterns are combinations of the three basic ones; they indicate the presence of an internal conflict. During a fight the dark pattern and/or the barred pattern may gradually develop and partly replace the spotted pattern, particularly in the fish which is likely to lose. The proportions of the barred pattern and the dark one are correlated with the probabilities that the fish will stay in or that it will leave the territory.

Furthermore the assumption that a signal activity is based on a conflict between systems, instead of simply on opposite orientations, can be supported by the fact that in such an activity non-locomotory somatic components are often incorporated in a consistent non-random way. A classic example is the upright threat of the Black-headed Gull in which there is an invariable combination of stretched neck, a certain position of the bill, and lifted wrist joints of the wing which are not necessary for the locomotory requirements of approach or avoidance. Other examples can be found in a quantitative study by Stokes (1962*a*) of the regularities with which various discrete behaviour elements occur in combination, as components in displays performed by Blue Tits, *Parus caeruleus*, during mutual encounters on a food table. Stokes considered 9 elements, concerning the orientation of the body, the posture of wings, tail, head, and beak, and the erection of the feathers of body and crest. Pairing them in the 36 ways possible, he found 18 combinations to be positively and 14 combinations to be negatively correlated. For instance, fluffing the body feathers and raising the crest were positively correlated and so were a horizontal body posture, raised wings, and a fanned tail, but each element of the former group was negatively correlated with each of the latter. Moreover, the occurrence of elements of the former group was found to increase the probability that the bird would retreat, and the occurrence of those of the latter group increased the probability that the bird was going to attack. Two elements, an open beak and an erect nape, were indicators of an increased probability of staying. The probability of correctly predicting the ensuing behaviour increased with the number of behaviour elements taken into account.

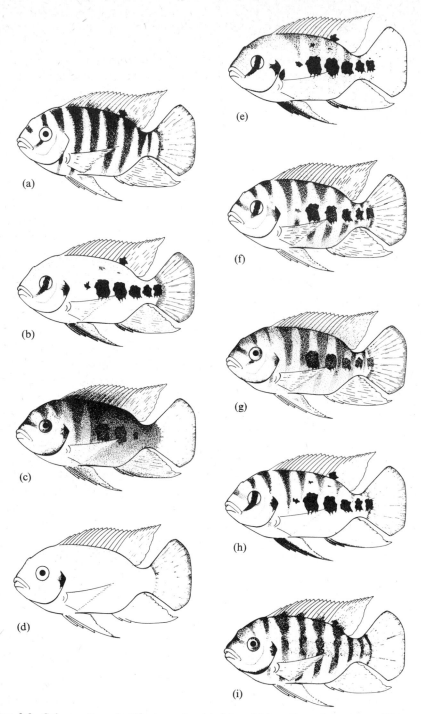

FIG. 9.2. Colour patterns in *Tilapia mariae*. (a), (b), and (c) are the major patterns; (d) is very rare and only occurs in transition periods when the fish is inactive. (e) is a combination of (b) and (c) in which the areas corresponding with the bars of type (a) are light coloured. (f) and (g) are transitions between (e) and (a). (h) and (i) are different combinations of (a) and (b) characteristic, respectively, for fish guarding young and fish beginning to defend a territory but still attached to the school. (After Baldaccini 1973.)

These facts indicate a certain amount of common causality between the complex of factors causing escape and that causing particular display patterns, and also of the complex of factors causing attack and that leading to other types of display. Some of these factors can be found in the external situation before and during the performance of the acts, but the considerable amount of independence of the external situation which appears to exist argues in favour of an at least equally important influence of internal common causal factors. Blurton Jones (1968) carried the analysis further by studying in a similar way the temporal association between entire displays in the Great Tit, *Parus major* (Fig. 9.3). These displays represent a higher level of integration than the smaller elements studied by Stokes—and Blurton Jones attempted to correlate each of them with overt attack, overt escape, or overt feeding behaviour (which implies a tendency to stay close to the food). He found three groups of co-varying actions: (1) attack, and the threat displays head-down, head-up, horizontal and wings-out; (2) hopping away, crest-raising, fluffing, looking around, turning; and (3) approaching food, feeding. Through manipulation of the agonistic tendencies by a special stimulus eliciting attack, and another stimulus releasing escape, and of the tendency to stay by feeding the birds or depriving them from food, Blurton Jones could induce experimentally each of the three co-varying groups. When the attack- and the escape-evoking stimuli were simultaneously presented the occurrence of the four threat displays studied was increased. A similar result had been obtained earlier by Blurton Jones (1960), when experimenting on the causation of threat postures in Canada geese.

Some of Blurton Jones's observations and experiments on tit displays make it possible to draw some conclusions on the way, or the level, at which the conflicting systems interact, which can be expected to vary from case to case. For instance, the occurrence of threat displays always depended on the presence of attack-evoking stimuli and obstructions of the attack movement. The latter could be of different kinds, for example, an interfering flee-evoking stimulus, a physical barrier, or an attractive stimulus (food) presented in competition opposite to the attack stimulus. However, only obstruction through a flee-evoking stimulus caused a change in the ratio between the head-up and the head-down displays (in favour of the former) and this result could not be attributed to interference of other activities evoked by the flee-evoking stimulus (like hopping away or turning). The occurrence of wings-out, but not of the other threat displays, was found to be controlled by the strength of the attack stimulus.

In his analysis of the agonistic and courtship displays of *Barbus* species Kortmulder (1972) found no arguments to postulate more than one major aggressive and one major fleeing system in these fish. But whereas for some activities, as the 'inhibited attack', it had to be assumed that the conflict

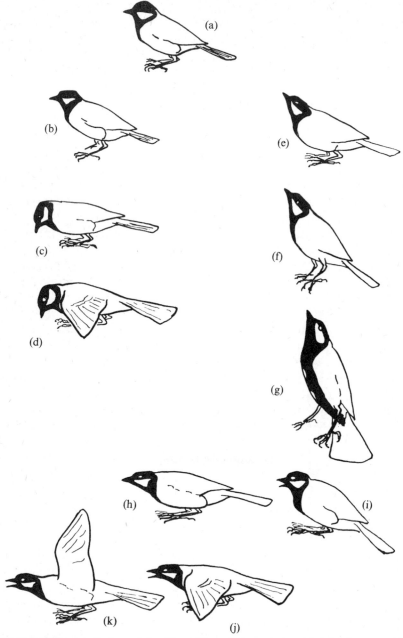

FIG. 9.3. Different displays of the Great Tit and the variation of their elements. (a) rest; (b)–(d) head-down; (e)–(g) head-up; (h)–(k) horizontal. (After Blurton-Jones 1968.)

took place at a relatively central level, for others it was more parsimonious to place it more peripherally, in the case of 'rolling', for instance, the conflict was considered more likely to involve an aggressively motivated fixation of the opponent combined with a tendency to flee.

The introduction of the localized food stimulus in Blurton Jones's experiments with Great Tits revealed the influence of a third factor, besides the tendency to attack or approach and the tendency to retreat, on the causation of displays, that is, the tendency to stay put. The horizontal display was found to depend on this tendency, and on the tendency to attack. The influence of this third dimension had also to be assumed in Baldaccini's explanation of the causation of the different colour patterns in *Tilapia mariae* mentioned above (p. 195). Baerends and van der Cingel (1962) found that the occurrence of the snap display of the common heron (*Ardea cinerea*) under three different circumstances was always characterized by the obvious presence of a tendency to stay put, in addition to both agonistic tendencies.

The results of the sophisticated measurements considered are essentially in agreement with Moynihan's (1955) original idea that each display would be characterized by a typical equilibrium, in an absolute and relative sense, of mutually interacting behaviour mechanisms. Further support can be found in Kruijt's (1964) study of the communication behaviour of Red Junglefowl. During fights cocks show a number of apparently irrelevant activities, some of which at least have signal value for the opponent. One of these activities, ground-pecking, is likely to have originated from displaced feeding and redirected attack (Feekes 1972); two others, preening and head-shaking, have probably entered the fighting ceremony through displacement; a fourth activity, head-zigzagging, was shown to result from simultaneous ambivalence between ground-pecking and head-shaking. Kruijt made an assessment of the ratios between the tendencies to attack and to escape typical for each irrelevant activity, by examining separately their frequency of occurrence in cocks winning and cocks losing their fight. His results, represented in Fig. 9.4, are highly indicative of a relation between the type of irrelevant activity or display and the ratio between the opposing agonistic tendencies.

The exact determination of the equilibria between tendencies is a particularly difficult matter in the case of displays which tend to be temporally linked with other displays and which are, therefore, rarely followed or preceded directly by overt agonistic activities. Correlations with colour patterns, like those described above, may be helpful in such cases, although such patterns are much more difficult to quantify than activities.

From a theoretical point of view, objections can be made against the assumption that temporally linked activities are likely to be similarly

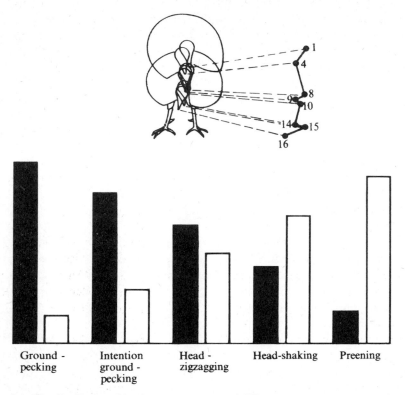

Fig. 9.4. The distribution of irrelevant movements of different types among winners (black) and losers (white) in a cock fight. A reconstruction of the course of the bill tip from cine film (24 frames per second, frame-numbers indicated) demonstrates the ambivalent character of head-zigzagging: downward-pecking and head-shaking are combined. (After Kruijt 1964.)

motivated. This need not always be true, for we know first that motivations fluctuate in strength with time and that the dominant role of one specific motivational state can rapidly and repeatedly be taken over by another, and secondly, that the phenomenon of rebound is not unlikely to occur at this integration level also (Nelson 1965).

Moynihan's (1955) original suggestion that the kind of display would be determined not only by the strength ratio between the conflicting tendencies but also by their absolute strengths is difficult to verify because of the operational problems involved in measuring separately absolute and relative strength. Nevertheless, some promising progress has been achieved here by a number of workers from the Tinbergen school. In the Three-spined Stickleback, Sevenster (1961) produced good evidence that the occurrence of 'displacement' fanning in the Three-spined Stickleback, an evolutionary precursor of the communication activities by which male sticklebacks show the entrance to their nests to females, depends on a special strength—relation between the tendencies to behave sexually and aggressively. Sevenster-Bol (1962) found the locomotory component of the spawning act, 'creeping through', of this species to depend on a particular ratio between the same tendencies, largely independent of their absolute strength. Blurton Jones (1968) in his experimental analysis of Great Tit display found indications for effects of both absolute strength and ratio: an increase of the strength of the escape stimulus applied caused a relative increase of the proportion of the 'head-up' display with respect to other displays, but a strengthening of the escape *and* the attack stimulus raised the proportion of 'wings-out'. In their study of preening in terns, *Sterna sandvichensis* and *S. hirundo*, van Iersel and Bol (1958) obtained data indicating that a particular ratio between the tendency to escape and the tendency to incubate is optimal for the occurrence of displacement preening, which is an evolutionary precursor of communicative activities in a great many birds. When this ratio resulted from relatively strong tendencies the displacement activities occurred more often than with weaker tendencies and a greater proportion of them were of a high-intensity type.

The latter study contains indications that beside these quantitative aspects, the quality of the conflicting tendencies can be a determinant for the kind of interruptive activity which is evoked and could consequently develop into a signal activity. In the terns a conflict between incubation and aggression was more likely to promote nest-building than one between incubation and escape, which more often facilitated preening (see also Baerends 1970).

Although the difficulties of measurement are considerable they should not be regarded as prohibitive or insurmountable. As shown by the above-mentioned examples it looks worthwhile to improve our methods for the quantitative study of motivation, for the conflict hypothesis seems to promise a better insight in the processes through which species-specific behaviour patterns have attained their present shape. A last example taken from a study by my associate S. T. Tjallingii (unpublished) of social behaviour in the sheld-duck, *Tadorna tadorna*, will confirm this impression and touch upon another phenomenon which still needs to be discussed here. Lorenz (1941, 1951) had drawn attention to the enticing display of the female sheld-duck, pointing out that it consists of two components: one directed towards a rival and the other towards the partner.

Tjallingii distinguished three different variations of the movement (Fig. 9.5), one (a) in which the head was mainly moved upwards with the bill lifted towards the partner, another (c) in which the neck was horizontally stretched with the bill pointing towards the partner, and an intermediate third variation (b) in which both components were equally present. Unpaired females predominantly show type (a), paired females type (c).

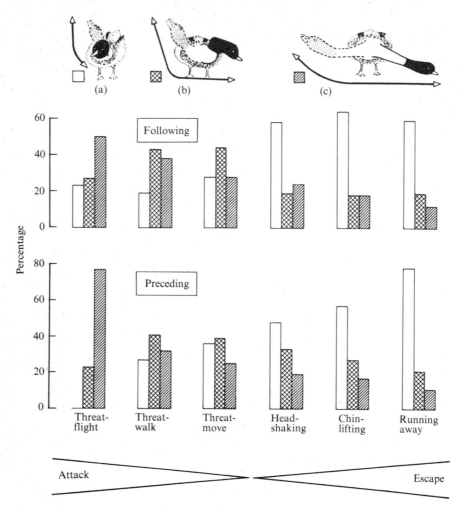

FIG. 9.5. The relation between the three forms of the enticing display in the sheld-duck and the tendencies to attack and to escape. The ordinates give in percentages the ratios between type (a), type (b), and type (c) when the enticing display follows or precedes each of three different intensities of threat and of three different intensities of escape. The probability of attack increases from threat-move to threat-flight, the tendency to escape from head-shaking to running away. (After S. T. Tjallingii, unpublished.)

Moreover, type (a) is more often preceded or followed by fleeing and chin-lifting (an incipient movement of taking to the wing), type (c) more often with an overt attack–flight or attack–run. Head-shaking is correlated equally often with attack as with escape behaviour. This display, which according to Tjallingii functions to establish and maintain the pair bond, can thus be considered as a mosaic of parts of the attack and of the flying-up or escape patterns. In contrast to what was said above about the upright and forward displays of the Black-headed Gull, transitions between the extremes (a) and (c) are common. This is in agreement with Lorenz's opinion that this display would be little ritualized, as opposed to the homologous enticing display in other ducks, for example, the mallard, where the orientation components are usually no longer directed with respect to other birds but only to the own body (Weidmann 1956). This attainment of independence from the external causal situation may be considered as emancipation; the enticing display has become a classical example of it.

Enticing in the sheld-duck is an example of a signal activity which has undergone some changes without losing its connections with the original systems from which it has been derived. Such a change is the introduction of a mechanism ensuring a regular alternation between the sidewards and upwards components. Another example is the mechanism controlling the regular alternations of approach and retreat with respect to the female in the zigzag-dance of the Three-spined Stickleback. The length of the approach components is proportional to the frequency with which the female is attacked during courting, that of the retreat components with the frequency of moving away towards the nest; the frequency of occurrence of the activity as a whole was used successfully as a measure for the tendency to fertilize a clutch (van Iersel 1953). I do not know of any instance for which it has been convincingly shown that emancipation is really complete, that is, that a derived activity has really become fully independent of the causal factors thought to have originally been underlying it. I shall even argue below that such a development would have a negative survival value.

Before proceeding further we should now realize that we have gradually yielded to the temptation to test a hypothesis on the phylogeny of an activity by studying its present causation, notwithstanding the fact that the ethological literature contains ample evidence that emancipation, at least to some extent, exists. This is legitimate as long as we draw conclusions in favour of the conflict hypothesis from positive findings of common causality between aspects of displays and the occurrence of agonistic behaviours. Negative findings, however, may not be used against this hypothesis, unless we can reject the possibility that emancipation has taken place. The conflict hypothesis was developed for explaining how some display activities obtained their form in the course of evolution and although it has often been extended for explaining causation, such attempts are likely to become less rewarding with increasing emancipation of the displays. As evolution is still proceeding, the causal mechanisms active in shaping display behaviour should still be present in unritualized behaviour. Absence or reduction of these mechanisms, however, may be expected when emancipation advances. This should be kept in mind in our discussion of the next category of criticisms to be considered.

Neurophysiological data

Brown and Hunsperger (1963), in particular, have drawn the attention of ethologists to physiological findings which in their opinion are not in agreement with the conflict hypothesis. Stimulating single loci in the brainstem of the cat electrically, Hunsperger and his associates (Hunsperger and Brown 1961; Brown and Hunsperger 1963; Hunsperger, Brown, and Rosvold 1964; Hunsperger 1965; Hunsperger and Bucher 1967) found areas from which they could elicit escape behaviour, but they never obtained pure attack in this way. However, behaviour patterns of threat, of the kind for which the conflict hypothesis was developed, did appear at single stimulation of certain sites. They reasoned that if the ethological conflict hypothesis as expanded to cover causation is correct, it must be possible to elicit pure attack and pure escape behaviour by electrical brain stimulation of separate areas of the brain, and threat behaviour should occur only when both of these areas are stimulated simultaneously.

Although Hunsperger (1965) correctly stated that the physiological facts found by him and his associates do not invalidate the conflict hypothesis when applied to the phylogeny of displays, I shall try to point out in the following that the neurophysiological objections made by Brown and Hunsperger are also untenable with regard to causation. My arguments are that: (1) the present neurophysiological evidence is in itself inconclusive; (2) most of this evidence is not accompanied by behavioural data sophisticated enough to be of much use in this context, (3) no essential disagreement between the ethological and neurophysiological evidence exists; (4) the behavioural programs and the neurophysiological mechanisms underlying them need not be isomorphic.

The effect of brain stimulation on aggression and flight in the cat has also been studied by other authors. It has further been studied in oppossums, in rats, and in birds. Sites from which escape and threat activities could be elicited were found in all cases. However, the data on attack found by different authors in the same or different species seem to contradict one another. In contrast to Hunsperger and associates, Wasman and Flynn (1962) and MacDonnell and Flynn (1964, 1968) report to have obtained attack in cats by electrical stimulation of one locus only. Woodworth (1971), Panksepp (1971), and Karli, Vergnes, Eclancher, Schmitt, and Chaurand (1972) elicited pure attack behaviour in rats from single sites in the lateral hypothalamus. Åkerman (1965) released attack (wing-blows) in pigeons through single-site stimulation, but Delius (1973) working with gulls obtained—as did Hunsperger *et al.* (1964)—escape and threat behaviour, but never pure attack, in this way.

Attempts to obtain displays by simultaneous stimulation of two areas failed in most cases. MacDonnell and Flynn (1968) alone were able to modify the attack pattern and to obtain hissing by a judicious balancing of simultaneous electric stimulation to the hypothalamus and thalamic areas of cats in the presence of a rat.

Interestingly, Hunsperger *et al.* (1961, 1964) obtained severe bouts of pure attack in addition to prolonged threat behaviour when the threat and escape areas in the cat were simultaneously stimulated. The same effect was produced when an external attack-releasing object was presented in combination with stimulation of the escape area. Delius (1973) performed a similar experiment with gulls and obtained the same result. He is inclined to conclude that fierce attack behaviour may be the result of a conflict between threat and fear; we shall come back to this interpretation later.

Other experiments also suggest that the external stimulus situation has an important influence on the effects of electrical stimulation. Roberts, Steinberg, and Means (1967), stimulating the hypothalamus and pre-optic zones of the opossum, obtained attack and other responses only when acceptable goal objects were present. In experiments by von Holst and von Saint Paul (1960) the presence of a stuffed polecat during continued stimulation of constant strength produced a phase of attack and threat which eventually merged into escape. Several explanations for the discrepancies between different neurophysiological studies can be suggested. First, the behaviour elicited has often been poorly described, thus making a reliable comparison between the results of different studies impossible. The names 'aggression' and 'attack' may refer to different behaviour patterns and only in a few studies do they refer to the intra-specific agonistic behaviour relevant for the conflict hypothesis. Many studies deal with inter-specific (predator–prey) attack and—as far as the studies on rats are concerned—with the mouse-killing of rats, which may be a laboratory artifact (Karli, Vergnes, Eclancher, Schmitt, and Chaurand 1972). One has to envisage the possibility that differences exist in the neurophysiological mechanisms underlying aggressive behaviour in these different functional contexts.

Second, the interfering influence of the external situation of past experience and of other activated behaviour systems on the effect of electrical brain stimulation has not been sufficiently taken into account. I shall illustrate this with two examples. Koolhaas (1973) released sequences of intra-specific agonistic behaviour in male rats through electrical stimulation of a definite area in the lateral hypothalamus and found the relative frequency of approach, avoidance, watching, and fighting in these sequences to be different when the opponent was a dominant male, a submissive male, or an oestrus female. The reaction also depended on experience with the individual male opponent in former fights (personal communication). Also in rats, Valenstein, Cox, and Kakolewski (1968) were able to change the nature of a response elicited by electrical stimulation of a particular hypothalamic locus by altering elements in the external situation to which the animal had previously been conditioned.

Although the interpretations of these new findings are still a matter of debate, they have shaken the original ideas on the functioning of the hypothalamus based on the early results of electrical stimulation of the brainstem. According to these the hypothalamus—beside playing a role in the control of vegetative functions—would be involved in particular in the organization of behaviour. Different areas or loci were thought to be connected with different spatial patterns of nervous activity, underlying different (functional) complexes of species-specific behaviour. Recent work, however, has produced a considerable amount of evidence for regulatory functions of parts of the brainstem of behaviour organized elsewhere; Panksepp (1971) is of the opinion that the hypothalamus mainly initiates and energizes the activity of other areas. Karli *et al.* (1972) believe that the lateral hypothalamus and the ventro-medial tegmentum are involved in the approach and, in contrast, the peri-ventricular structures in the avoidance of stimuli. The result of self-stimulation experiments connect the former areas with reward and the latter with punishment. Furthermore, indications suggest that the hypothalamus, together with other parts of the brainstem, acts as a link in the incorporation of external stimuli and previous experiences in the release of behaviour patterns.

Although these considerations at this moment seriously invalidate the neurophysiological basis on which Hunsperger and Brown built their criticism, it is nevertheless worthwhile to reflect on the consequences of their criticism on the conflict hypothesis if further research would prove their arguments to be true.

The possibility of eliciting threat behaviour from a single locus—on which most studies agree—does not necessarily argue against the idea of a dual motivation of such displays. As Hinde (1970) has already pointed out, in these cases the stimulus may still have reached and affected areas for attack and escape; also the electrode may have hit a locus which under natural conditions is elicited at a special strength relation of the tendencies to attack and to flee. Particularly in cases of emancipation one could expect such loci to exist. It is worth noting parenthetically here that the discovery of the modifiability of the response produced by a brain locus through conditioning opens the possibility of a mechanism for emancipation taking place by means of conditioning processes, an idea which will be further considered later.

The difficulties several physiologists have encountered in finding a neurophysiological substrate for pure attack should make the ethologist ask whether pure overt attack is actually ever observed in an animal. The evidence underlying the conflict hypothesis is concerned with elements of attack and escape behaviour; pure complete attack is the abstract extreme of a scale. Looking at escape and attack from the point of view of survival value, pure escape behaviour is likely to be of benefit to an animal which is in danger and free to move away. But the opposite, that is, the unrestrained running at another animal, is likely to be disadvantageous to the attacker. Consequently a safety mechanism, protecting the animal from the consequences of unrestrained attack, must be important. I can think of two possibilities for such a mechanism. First, during the approach of the object pursued various stimuli emanating from it (for instance the projection of its contour on the retina of the pursuer) become stronger and may gradually activate the mechanism for escape. Second, with increased activation of the attack system an internal mechanism may lower the threshold for escape, in a way comparable to the rebound phenomenon in alternating reflexes. Such a mechanism could also account for the above-mentioned observations that stimulation of the escape area in combination with an external attack-releasing object or with simultaneous electrical stimulation of the threat–attack area may produce prolonged threat behaviour and more severe attacks, just like those in a fleeing animal when it gets cornered. It seems fruitful to think of the possibility that the degree of mutual inhibition of functionally opposite tendencies might have some complicated non-linear relation with the strength of these tendencies, and it would be worthwhile to test this possibility using physiological and ethological methods.

The idea that at the height of attack the aggressive tendency weakens might explain observations made by H. Oosting (1970, unpublished) in our laboratory, on the appearance and fading of a dark cross-bar on the face of the cichlid fish, *Haplochromis burtoni*, during a ritualized fight between males. In this fight the males alternately take positions at right angles to each other, after which the male facing the other male's flank advances and butts its opponent. Before the male starts his approach his nose-bar is maximally dark. Oosting found darkness to be correlated with a maximal tendency to attack and Heiligenberg, Kramer, and Schulz (1972) demonstrated in experiments with models that the value of this bar for signalling high attack tendency is understood by the reactor. As soon as the actual attack is started, however, the band begins to disappear.

Finally, it seems relevant to mention here that during the development of attack behaviour in kittens, between the fourth and seventh week of age a gradual inhibition of strong attack by a tendency to retreat takes place as a result of experience with the effects of fierce attacks (Baerends-van Roon, in preparation). One is tempted to relate this finding to the rarity of pure attack responses with electrical brain stimulation and the influence of conditioning on the kind of response from a brain locus.

It seems to me that the neurophysiological criticism of the conflict hypothesis was due to naïve interpretations both of this hypothesis and the physiological facts. The present physiological evidence points to the existence of a system in the brainstem which, in close interaction with the external situation and with previous experience, determines, within a range in which attack and flight are extreme positions, the type of an agonistic response shown. The hypothalamus plays a part in the system, but nobody knows exactly how. There is no disagreement between the system postulated and the conflict hypothesis.

Brown and Hunsperger (1963) and Brown (1964b) expressed themselves with great confidence about the superiority of neurophysiological data over ethological ones. But the above brief survey has shown that in spite of—or rather due to—the fascinating development of brain research, the difficulties and uncertainties in this field are no less than in ethology. Brown and Hunsperger are of opinion that 'whatever is known about neural mechanisms in behaviour should be utilized by ethologists in the improvement of research hypotheses and theories' (ibid., p. 447). This point of view has been further debated upon by Brown (1964b) and by Rowell (1964), who stated that the mechanism by which the central nervous system transfers the ethological description of the stimulus–response relationship 'is not necessarily of interest to the ethologist at all' (ibid., p. 535).

I am of opinion that in the building of representational models for complicated behaviour events the ethologist should only make use of well-established morphological and physiological facts (like the presence of sensory organs and muscles), but that he should definitely not incorporate the array of hypotheses, many of them undoubtedly short-lived, which at any one moment may be current in neurophysiology. If the logical analysis of the organization of complicated behaviour sequences were to be too heavily based on current ideas in neurophysiology, ethology *and* physiology would be in danger; the physiological approach will only benefit from ethological evidence when this results from methodologically independent research. Brown's (1964, p. 538) statement that in the study of motivation it is legitimate only 'in the preliminary analysis to ignore physiology' contains for me the same kind of truth as a statement that it would be ligitimate only in the preliminary phases of physiology to ignore molecular biology. An approach by which the laws of complicated behaviour are deduced only from behavioural data is in my opinion (Baerends 1970) absolutely essential; Lehrman (1953) and others have warned against assuming isomorphism between the behavioural rules and the phylogenetical mechanisms effecting them. And that is precisely what has been done by neurophysiological criticisms of the conflict hypothesis. The ethological and physiological levels of integration should be studied quite independently in their own right. However, a considerable advance can be expected from teams in which physiologists and ethologists are working in close co-operation and with physiological and ethological methods which are equally sophisticated. Disagreements between ethological and physiological hypotheses should not be solved by favouring the one over the other but by attempts to bridge the discrepancies.

Alternative hypotheses

It is probably true that the conflict hypothesis has often been applied too readily to insufficiently investigated cases. This will always happen with an attractive explanatory principle, but it is unfortunate, particularly because it easily inhibits more penetrating research and consequently the discovery of new alternative hypotheses.

Andrew's (1972) plea that we should also look for principles simpler than conflict between incompatible tendencies for the understanding of the phylogeny of displays, is therefore justified. He considers simple responses serving the maintenance of the body (like protective, thermo-regulatory, locomotory, or cleaning responses) as possible sources for derived behaviour and he himself found this approach useful for the understanding of the derivation of facial expressions in primates (Andrew 1963a, b). One certainly follows the principle of parsimony by first exploring such simpler possibilities and resorting to more complicated solutions, like the participation and interaction of systems, only when the data require this. Such would be the case, for instance, when one or more of these maintenance responses appear in an anticipatory fashion, that is, before they could have been released by the adequate stimuli.

Miller and Hall (1968), in their analysis of the reproductive behaviour of *Trichogaster leeri*, considered the conflict hypothesis invalid as a causal explanation for the 'lateral spread display' a prominent constituent of agonistic and courtship behaviour in this and other anabantoid fish. Kortmulder (1972) rejected the conflict hypothesis for the comparable lateral threat display of *Barbus* species. He argues that this display—in which the body of the fish is assuming an S- or C-curve, while all fins are maximally extended—results from spinal reflexes to tactile stimulation of the lateral body surface. Although Kortmulder considers it possible that in agonistic encounters these reflexes would be facilitated by an attack–escape conflict, he considers a direct influence of agonistic tendencies unlikely; for instance, because the threat posture is maintained continuously also when the fish is alternately approaching and avoiding his opponent, and because the lateral position does not result from ambivalent orientation but is obviously actively sought by the threatening fish. Kortmulder was able to evoke the threat posture through tactile stimulation with a ball-shaped model moved alongside the fish, even when the fish had been blinded.

Andrew (1962, 1963a, b, 1964, 1972) states that displays, particularly vocal ones, are sometimes 'spontaneously' shown when stimuli that could evoke attack or escape are fully absent. Particularly on the basis of his study on the causation of chick vocalizations Andrew is strongly inclined to attribute the occurrence of such displays to what he calls 'stimulus contrast', a relatively strong effectiveness of a stimulus, either because of its striking physical properties, or because striking experiences have been associated with it. However, this concept leaves open the problem how such an unspecific quantitative effect would lead to the performance of a particular display pattern in each member of a species; the possibility that an approach–avoidance conflict is intercalated between the perception of the stimulus contrast and the performance of the display is not necessarily excluded with this explanation. Andrew apparently prefers to entertain another possibility, namely, a linkage between a certain value of stimulus contrast and a behaviour pattern acquired through conditioning. This alternative is worthy of much further investigation. Kruijt (1964) and Feekes (1972) have shown that apparently irrelevant activities can become operants for obtaining social rewards. The latter author has produced evidence that ground-pecking during an agonistic encounter of cocks leads to a reduction of alarm and a rise in the tendency to attack in the performing bird, and can thus restore the original balance after a motivational shift. Such conditioning processes could bring about the phenomenon of emancipation, which is usually thought of as the result of a purely genetically fixed reorganization of the neural network.

These considerations lead to the interesting question how a receiving animal 'extracts' the information content (Marler 1961) from the communicative activities of another. One can easily imagine that intention movements can be recognized as incomplete versions of overt activities to which the receiver is responsive, whether via a conditioning process or not. The more the communicative activity deviates from the functionally relevant overt activity with a direct (instrumental) function—as, for instance, in case of signal activities caused through displacement—the more likely is it that the receiver must have acquired its knowledge about the information content of the signal through conditioning. One may expect that through conditioning a discrepancy can arise between the information content of an activity and the motivation originally underlying it.

Finally, mention should here be made of an interesting attempt by Nelson (1965) to explain the temporal pattern of occurrence of zigzagging, nest-building, creeping-through, and fanning in the Three-spined Stickleback with a model consisting of two variables, excitation and threshold, without postulating interactions between aggression and sex tendencies. However, this kind of model is only an attempt to explain the causes for the occurrence, not the form of the behaviour.

9.3. The explanatory power of the conflict hypothesis

The origin of divergence and adaptive radiation of displays

Because the conflict hypothesis involves the interaction between the systems for attack and escape it enables us to postulate an important cause of the radiation of signals serving for intra-specific communication and in particular for intra-specific recognition. Hinde (1956*a*, 1959*b*) for instance, could associate some differences between the forms of courtship displays of various species of finches with differences between these species in their aggressiveness towards conspecifics, particularly towards the mate. Stokes (1962*b*) correlated differences between homologous displays of four species of titmice with relative timidity, whilst Tinbergen and Broekhuysen (1954) found similar correlations when comparing different species of gulls. Baerends and Blokzijl (1963) attributed differences between the forms of homologous courtship displays of the cichlid fishes, *Tilapia mossambica* and *Tilapia nilotica*, to the extent to which the median fins were erected during these displays: *T. nilotica* often erected them considerably in situation where *T. mossambica* kept them laid down. As the grade of erection influences the degree to which sideways body movements are slowed down by the resistance of the water, the fin position is an important way of modifying these movements. This fin position was found to be under control of the balance between the tendencies to attack and to flee; *T. nilotica* showed a higher average level of attack behaviour than *T. mossambica* under comparable circumstances.

Such observations suggest that species may differ in the average threshold at which attack or fleeing behaviour is elicited. As some species are generally called shy and others aggressive, the existence of such specific differences in the tendencies to attack or to flee is actually common knowledge. Recently Kortmulder (1972) made a sophisticated attempt to measure relative differences in aggressiveness and shyness, in which various possible parameters for these tendencies were used, in a comparative study of the behaviour of seven *Barbus* species. Because of qualitatively different ways in which the aggressiveness was expressed the author did not find it possible to compare the degree of aggressiveness of all species on a simple scale, but comparisons could be made between some pairs of species (for example, the males of *B. nigrofasciatus* were found to be more aggressive than those of *B. conchonius*, and those of *B. tetrazona* more than males of *B. stoliczkanus*). Fleeing motivation was measured independently of aggressive motivation and not found to be different in dominant males of the four species mentioned. This type of work is extremely demanding, but the results achieved indicate that further studies along these lines are likely to give important results.

The ultimate causes of specific differences in the average readiness to attack or to flee must be sought in ecological factors; these tendencies must primarily be adapted to factors in the environment, like the presence of predators and the amount of cover; perhaps also to the kind of food which is taken. For the cichlid fish inhabiting the great African lakes evidence is available (Lowe-McConnell 1969; Fryer and Iles 1972) that speciation was promoted by the repeated appearance and disappearance of geographical barriers within their biotopes caused by fluctuating water-level as a result of long-term meteorological cycles. Every time part of a population had become isolated (for example, in a bay separated from the main lake), selection must have led to an optimum threshold for aggression and flight adapted to escape from predatory fish (Worthington 1937) and to staking out territories in the new environment for breeding and possibly feeding. A causal linkage of the specific thresholds for aggression and for flight, with the forms of signal activities, must be of considerable importance in promoting speciation. For as soon as selection for certain thresholds has started in the new environment, such a causal linkage will automatically change the form of displays determined by the balance of aggression and flight and thus erect an ethological barrier. This barrier is likely to endure and to remain effective, even if the geographical barrier later disappears. However, in cases of a genetically completely fixed neurophysiological emancipation of a ritualized display, the development of an ethological barrier would no longer possess the automatic linkage to ecologically determined changes in the attack and fleeing threshold. This rigidity would not be favourable for speciation and consequently perfect emancipation is likely to be a disadvantage for further evolution. As a matter of fact I question whether full emancipation has ever been satisfactorily demonstrated. If complete neural emancipation were common it is difficult to explain why correlation of certain displays with a typical aggression–flight equilibrium proved to be possible in almost all cases where it has seriously been attempted. Moynihan (1970) has argued that in the course of evolution displays, just as morphological structures, do not only develop but also disappear. If I am right that complete emancipation has a negative survival-value component, perfection of the emancipation process may be one of the causes of the disappearance of displays.

Another example of the linkage of ecological adaptations with the threshold for agonistic behaviour and the form and occurrence of displays can be found in the behaviour-genetical work of van Oortmerssen (1971) on two laboratory strains of mice. This author made a good case for the contention that the strain CPBs stems from a house-mouse population adapted to living above the ground surface, and the strain C_{57} Black from a population living in burrows dug in the ground. As a consequence CPBs is territorial and has a more sensitive aggression–flight balance than C_{57} Black, and in correspondence with the relatively high tendency to perform agonistic behaviour it has a more elaborated courtship repertoire, consisting of a greater number of signal activities. Sufficient evidence is available to say that, in general, species with a relatively low threshold for attack and/or flight must possess a relatively rich stock of communication activities, to reduce the amount of attack and fleeing in social situations and to make peaceful approaches of conspecifics possible.

The origin of sexual and age differences in displays

The evolution of differences between the display repertoires of the sexes of one species, particularly in courtship, can be similarly understood as based on derivation from conflicts between the causal behavioural mechanisms for attack and for escape. Sexual dimorphism in structures and in behaviour usually go hand in hand and both are related to the degree of division of labour between male and female of the species. In a great many species the average level of aggressiveness in intra-specific encounters is higher in the males than in the females and the average tendency to flee lower. This corresponds with the fact that the males usually take the greater share in defending a territory or a social position whereas the females take care of eggs or young. In such cases the male displays contain more attack, the female displays more escape elements, and the number of display activities in the male repertoire tends to be greater. In those rare cases where the female defends the territory and the male cares for the offspring—for example, the Red-necked Phalarope (*Phalaropus lobatus*; see Tinbergen (1935))—the balance between the agonistic motivations seems correspondingly reversed.

If both partners take care of the young (for example, gulls, substrate-spawning cichlid fish) the sexual dimorphism in behaviour (and external morphology) is often small, but when division of labour is present the dimorphism increases with the differences in tasks between the mates. In anatid birds and oral-incubating cichlid fish, for instance, it has become very marked. This dimorphism in the agonistic balance is one way in which species with parental care satisfy the conflicting demands that, whereas on the one hand, these animals must be able to defend their young effectively against predators, on the other hand, in handling the young the tendency to attack should be sufficiently low to prevent it to be released by or directed against them. Furthermore, the tendency to escape should be high enough to enable the parent to flee with the young and to take cover, but not so high that it would impede the parent from approaching them. Other ways in which the occurrence of disastrous agonistic behaviour is prevented are the display of appeasing social releasers by the young and the development of personal acquaintance.

Juvenile animals often have, like females, a relatively high escape tendency which can be expressed in the form of their displays (for example, hunched in gulls, food-begging in dogs). The similarity in motivational balance between juveniles and adult females often corresponds with strong resemblances between juvenile and female displays. I find it more parsimonious to think that this correspondence is caused by identical parallel causation than by a process of displacement in which one of the displays would have to be considered the original and the other the derived version of the signal activity; I shall return to this problem later (in *The occurrence of similar displays in different functional contexts*).

A relatively low tendency to attack in a female during social encounters does not exclude the possibility of the release of strong aggressive responses towards other individuals of the same species, or towards members of other species, particularly when they are potential predators. In such cases an otherwise shy female may perform threat displays (for example, the frontal threat of a rooster or of a mouth-breeding cichlid fish when defending young) with a fierce frontally-directed attack component. It is likely that strong releasing elements of the external situation and the general context play an important role in evoking such an attack. The reaction recalls some results of the experiments by Hunsperger *et al.* (1964) in which electrical stimulation of the flight area in a cat with simultaneous presentation of an opponent led to strong attack responses (§ 9.2, *Neurophysiological data*).

The performance of display behaviour typical for the other sex

It has frequently been reported in the literature—even for species with a marked behavioural sexual dimorphism—that an animal may occasionally perform a behaviour pattern belonging to the repertoire of the opposite sex. Hinde (1956*a*) has reported this for various species of finches and I have myself observed it in several species of fish, birds, and mammals. By means of the conflict hypothesis it is possible to understand how this phenomenon can occur, usually under peculiar circumstances.

The snapping display of the common heron is an example which has been analyzed to some extent. Normally this display occurs only in the male; it expresses his willingness to accept a soliciting female on the nest. Baerends and van der Cingel (1962) have presented arguments that this display is caused by interaction of three different tendencies: those to attack, to escape, and to stay put. The snap display is performed by females in rare cases (Baerends and Baerends-van Roon 1960), it can be argued that a conflict between these tendencies has a low probability of occurrence in this sex.

Morris (1952, 1955) has shown (mainly in the Ten-spined Stickleback and in the zebra-finch) how such pseudo-male or pseudo-female behaviour can occur when a female becomes extremely aggressive or a male extremely submissive. My associate G. J. Blokzijl and I have observed a very striking example in groups of hybrids between *Tilapia mossambica* and *T. tanganicae* of which the offspring was found to be exclusively female. When such a group was kept in a tank under crowded conditions, much chasing took place and one of the females developed the male nuptial dress and started to show male sexual behaviour. These examples indicate that at least in many cases, behavioural sexual dimorphism occurs not so much because an animal is not potentially able to carry out the behaviour of the other sex, but because it hardly ever reaches the motivational situation (the aggression–flight balance) necessary to produce it.

The occurrence of similar displays in different functional contexts

I suggested earlier that full emancipation of displays derived from conflicts between agonistic motivations has never satisfactorily been demonstrated and that total detachment from the originally underlying motivations might even be selectively disadvantageous. Nevertheless, it is common usage to speak of sexual and parental display activities and to assume that these are primarily motivated by a relatively autonomous sexual or parental system respectively. The fact that separate hormones inducing a sexual or a parental phase may control such activities provides some justification for this conclusion. A compromise within this apparent contradiction seems possible, for the different behaviour systems may influence each other mutually. Kruijt (1964) in his study of the development of these systems during the ontogeny of the social behaviour of Red Junglefowl, found that the sytsems for escape, aggression, and sexual behaviour became operative in succession. When the influence of the sexual system became apparent, this system was found to influence quantitatively the balance of the systems for attack and for escape. On the basis of these observations Kruijt postulated the hypothesis that the state of activation and mutual inhibition of attack and escape is influenced and stabilized through the action of the sexual system. He believes that with this hypothesis the periodical changes in the dominance relation between male and female in several species of finches (Hinde 1956a) can be interpreted as the result of seasonal shifts in the state of the sexual system. This hypothesis also helps to explain why the zigzag display of the male Three-spined Stickleback, in spite of being a ritualized ambivalent activity (see §9.2), can be used to measure the tendency to behave sexually (van Iersel 1953; Sevenster 1961; Sevenster-Bol 1962).

In the stabilization hypothesis the sexual system can easily be replaced by other systems or tendencies, like the parental system of a tendency to stay put (see §9.2). One can imagine that through this third factor quantitative values of the aggression–flight balance are brought about which would not commonly occur otherwise. Thus the resulting behaviour patterns can become sexual or parental displays, instead of serving a primarily agonistic function. In this way the pattern—as long as its causation is not drastically changed by emancipation—would be directly caused by the conflict between the systems for aggression and escape, but the parameters of this conflict would be manipulated by other systems. The activation of the agonistic systems at almost every encounter between conspecifics (also when they are mates, or parents and young) is indicated by the overt occurrence—often only incipient—of elements of attack and fleeing.

With this hypothesis it also becomes possible to understand why it is often so difficult to distinguish sharply between sexual, parental, or juvenile displays and certain types of dominant or submissive behaviour. For instance, Kortmulder (1972) mentions this in his study of *Barbus* species and we ourselves experienced it in a preliminary study of the courtship of Phasianidae. Furthermore, Wickler (1966, 1967a, 1967b) and Eibl-Eibesfeldt (1970) have directed attention to the strong resemblances in primates between some threat and copulatory behaviour patterns of males and between submissive and presentation behaviour of females. By means of the stabilization hypothesis it becomes possible to see these sexual behaviour patterns as originated from different interactions between attack and escape modulated by the system for sex. I prefer this view over that of Wickler and Eibl-Eibesfeldt who, on the contrary, interpret the threat and appeasement patterns as derivates from sexual behaviour, chiefly because the genitalia are demonstrated. I am inclined to consider this demonstration as a ritualization superimposed on displays which had originally an agonistic causation and function.

In cats the activities used by juvenile males in catching a mouse, attacking a littermate, and attempting to copulate are largely identical; the appeasement gestures of the littermate are similar to those of the sexually motivated female. The aggressive behaviour of a tomcat is similarly transformed when it perceives moderately strong flight-raising stimuli, a well-known playmate or sex-specific stimuli from a female (Baerends-Van Roon, in preparation).

To my knowledge the best experimental support obtained hitherto for the idea that for the performance of signal activities the sexual system operates through the manipulation of a conflict between aggression and escape comes form the analysis by Sevenster-Bol (1962) of the form of the activity by which the Three-spined Stickleback fertilizes the eggs deposited in its nest by a female. This activity consists of two components. One of them is creeping-through, a nest-building activity for making the nest tunnel. The probability of its occurrence was found to be highest when the sexual and aggressive tendencies—as measured independently—were matched. As the measurements of aggression probably produced a parameter for the combined effect of the aggressive and escape systems, creeping-through can be interpreted as brought about by disinhibition, due to the interaction of the three systems. The other component is ejaculation; it proved to be dependent on the sexual tendency only. Thus, in this case a displacement activity acts as a vehicle for the transfer of sperm.

Several observations can be found in the literature which make it likely that the tendencies to attack and to escape influence in turn the sexual and parental tendencies. For the Three-spined Stickleback it has been experimentally shown (van Iersel 1953; Sevenster 1961; Sevenster-Bol 1962) that spawning is facilitated by a certain range of the aggression–flight balance. Further experimental evidence for the dependence of sexual behaviour on the attack–escape equilibrium was obtained in our laboratory in cichlid fishes by W. J. Rowland working with *Hemichromis bimaculatus* and N. Vodegel working with *Pseudotropheus zebra*. In both studies a series of models was offered to isolated males. The models were of three sizes, each in two colours; the reproductive colour and the neutral colour of the species. When a male during confrontation with one of these models had reacted in a predominantly aggressive way, his behaviour could often be shifted towards sexual display by offering a more intimidating model (for example a model larger, more reproductively coloured, or both) which also evoked a certain amount of overt escape behaviour.

In the view presented above the behaviour systems for aggression and escape would contribute to the occurrence of communication behaviour by creating both need and opportunity and by influencing its form—every time animals meet in social, sexual, or parental situations. Systems other than those for aggression and escape would mainly contribute by bringing the animals together and thus raising the conflict from which the communication behaviour emerges. We are inclined to think that the number of activities directly controlled by the sexual and parental systems is very small indeed and may be restricted to behaviour patterns like intromission and actual spawning. Food-begging in young mammals and birds is certainly very much linked with aggression and escape and so is the behaviour of the adults before yielding to the begging of young.

This implies that the occurrence of a similar display in, for example, a sexual, parental, or agonistic context, can be explained as parallel expressions of the same process: the manipulation of the agonistic systems by another system activated in the animal. This explanation seems simpler than the derivation of a display activity from a similar activity in a different functional context through displacement, as was, for instance, originally suggested by Tinbergen (1940) for begging and feeding behaviour in gulls during caring for young and during courtship.

Against this idea one may quote the finding that homologous activities, carried out in an agonistic context, may differ in several aspects from such activities in a sexual or a parental context (see data of Manley (1960) in Tinbergen (1965), and also Beer's contribution in this volume). However, such modifications can be attributed to minor differences in the aggression–flight balance, due to differences in external stimulation, to context, and to differences in the degree to which elements other than agonistic are involved.

However, the systems for aggression and flight are not the only systems influencing the form of communication behaviour. Adaptations within other behaviour systems (for example, for taking special kind of food, for nest-building, hiding, etc.) can also be expected to influence the form of communication behaviour. For instance, as D. A. Kreulen has pointed out to me, the fact that goats are able to balance on their hind legs must be primarily adapted to browsing from higher branches, but enables fighting behaviour between males in upright position. The 'necking' display of giraffes must have become available only after the adaptation of the long neck to feeding high up in the trees.

Crook (1964) has shown in his comparative studies on weaver-birds how ecological changes during the Pleistocene, due to alternation of pluvial and arid periods, have led to radiation in diet, population dispersion, and social behaviour. Several behaviour systems must have been involved in this process.

The evolution of maintenance activities

Far more research has been done on the evolution of communication behaviour than on the evolution of activities with a primarily non-communicative function—instrumental activities. The reason is that the signal activities are more conspicuous and often of a more or less bizarre form. One therefore questions their function and evolutionary origin more readily than those of the maintenance activities the appearance and function of which seems so self-evident. However, it is of equal interest to know how different types of locomotion, of feeding behaviour, of nest-building, of feeding young, of cleaning the body, etc. have evolved. At the present state of our knowledge we can only speculate on the possible evolutionary origin of the instrumental patterns and this brief discussion has the aim of stimulating research in this field.

Primitive vertebrates moved through the water by means of oscillatory body movements, were filter-feeders, and ejected the gonadal products freely in the water. Orientation was possible by means of taxes or tropisms. A simple pattern derived from locomotion was digging in the sand. This probably belongs, with jumping-away, to the primitive escape patterns. Combinations of these patterns with a specific sensitivity for certain kinds of sensory information (movements of objects, sudden changes of light intensity, vibrations, etc.) may have led to the developement of one or more escape systems. A change from filter-feeding to more selective feeding on particulate food objects implies the acquisition of a quick approach followed by snapping. The former originated from the locomotory patterns; the latter perhaps from respiratory movements, which were probably also involved in filter-feeding. Active searching for food and taking it—animals as well as plants—probably evolved hand in hand with the development of agonistic behaviour to defend food sources and territories. As a consequence of the latter social behaviour became necessary and possible and—as we have seen above—it evolved on the basis of the already elaborate mechanisms for attack and for escape. Baerends (1966) has given arguments that in some groups of birds and fish nest-building might have originated from agonistic conflicts. For instance, we have seen that Herring Gulls in a boundary conflict, through redirection, pull off grass and manipulate it. The way this is done looks very similar to primitive nest-building activities in related forms, such as the skuas and the terns. Most nest-building occurs in the beginning of the season when the partners are together for short periods and often not yet familiar with each other. This suggests that nest-building might have been derived from redirected aggression. In the Three-spined Stickleback redirected aggression leads to digging (Tinbergen and van Iersel 1948) and digging is the first phase of building a nest in this species. Many cichlid fish dig in agonistic conflicts and tear off plants. Males of the oral incubating *Tilapia* species court females on arenas in which each male has its own small territory where it has dug a pit. It can be experimentally shown that the near presence of a neighbour has a stimulating effect on digging.

There are strong indications that frustration in the completion of an activity stimulates aggression. This may have led to the evolution of certain kinds of specially adapted feeding behaviour, such as crushing or opening shells, removing insects from wood, etc. Blurton Jones (1968) has pointed to the great similarity between some forms of attack and food-manipulating activities with the bill in the Great Tit. Albrecht (1966) has derived certain fighting patterns in cichlid fish from feeding patterns.

Groen (1955) has argued that feeding young through regurgitation (as in dogs, for example) is also the result of an agonistic conflict, which in this case mainly finds expression through the autonomous nervous system. His explanation could also be applied to birds like gulls, herons, cormorants, etc. feeding through regurgitation. Regurgitation and defaecation can be parts of the escape pattern (for example, monkeys, gulls, eider-ducks), where they probably serve to reduce the body weight during fleeing, but at the same time may be used as a means to deter the pursuer.

In the Herring Gull we know that the visceral pattern of regurgitation occurs in combination with a typical call, the mew call, and a somatic pattern, bowing, in the feeding ceremony, which is used by male and female in attracting and feeding the young, and by the male in courting the female. Bowing as well as a modification of the mew call also occur in mild threat (Stout, Wilcox, and Creitz 1969; Beer, this volume). As already explained I now favour the idea that the same complex (or parts of it) appear in these different contexts because of the similarity in the underlying causal factors, rather than ascribing it to displacement or sparking-over as originally suggested (Tinbergen 1940, 1952b).

In our considerations on behaviour systems we have implicitly assumed that maintenance behaviour is highly emancipated. But then—as J. P. Kruijt has pointed out to me—the idea that several maintenance activities have arisen from conflicts between other systems might seem to contradict my view that extreme emancipation would be of negative survival value. However, I would object to extending this idea from communication to maintenance behaviour, because of the important difference that the shape of the latter is to a much greater extent dependent on special (instrumental) functions than is that of the former. When, for instance, an ecologically advantageous change of the aggression–flight balance is able to change an important nest-building activity it may no longer allow the construction of the nest. A similar change which modifies part of a signal may also have some harmful effect but this may be outweighed by its contribution to an ethological barrier of the new mutant.

The evolution of behavioural complexes

In several related gull species there is a feeding ceremony involved in the care of the young as well as in courting. We must question whether this similarity in the occurrence of feeding ceremonies has developed in parallel in these species, or whether in at least a number of them they are of common origin. Why should not functionally and, particularly, causally related behaviour patterns evolve as composite units, just as do the skull, the hand, or the blood circulatory system? The morphological complexes mentioned here are common to many taxonomically rather distant groups. Can complexes of behaviour patterns be found that are similarly basic to several taxons? This certainly holds for types of locomotion. It probably also holds for head-scratching with the hind leg which occurs in reptiles, mammals, and birds (see discussion in Wickler (1966)). And with regard to communication ceremonies, for instance, one may ask whether the order frequently occurring in pre-sexual behaviour: ambivalent locomotory activities → (redirected) nest-building → mutual feeding, may be homologous in many groups of birds and not merely parallel. This leads to the question whether complex communication ceremonies, already evolved in fish or reptiles, might still have their homologues in birds and mammals. Hithertho this possibility has not, apparently, been considered, but it is certainly worthwhile to do so.

9.4. Summary and conclusion

The hypothesis developed by Niko Tinbergen and his associates that internal conflicts between different activated behaviour systems must play a role in the occurrence of communicative behaviour and, in the course of evolution, in the shaping of signal activities has been critically discussed.

First, four categories of objections have been considered: (1) the validity of the homology concept in case of behaviour, (2) the validity of the concept of behaviour systems, (3) the relevance of neurophysiological evidence, and (4) the possibilities for alternative explanations. Second, a number of examples have been given of the heuristic value of the hypothesis and of its explanatory potential for several problems.

I am of the opinion that the idea of considering the interaction between behaviour systems as an important cause for various developments in communicative and also in instrumental behaviour remains extremely valuable. It has not yet been equalled or superseded by the few other alternative suggestions made.

Criticism mainly concerns the interpretation of the concept 'system' both from the behavioural and the physiological points of view. The results of the few penetrating studies carried out in this field make it clear that the mechanism underlying a system within the network of the functional organization of the entire behaviour of a species is likely to vary from case to case, both with regard to the type of components as well as its level of complexity. The hypothesis, therefore, should be considered as a guide to further analysis, a method of thinking to reach an exact description of the mechanism at work in each single case. Used in that way it will lead to understanding; used as a blanket explanation it is likely to serve only superficiality.

The essence of the hypothesis is in the attainment of various states through the interaction of roughly opposite tendencies, a principle often applied in nature as well as in technology. In a considerable number of cases these tendencies are obviously related to overt attack and escape. The question of whether this justifies our thinking in terms of attack and escape systems or not can be expected to become really answerable only after our knowledge about the causal mechanism behind these functions has been considerably improved. This, however, should not discourage those who try to apply the hypothesis in evolutionary thinking about behaviour, for the value of such attempts is no longer open to doubt. It should only make them sufficiently cautious.

Acknowledgements

My interest in the problems discussed in this chapter originates from the period I had the privilege to be Niko Tinbergen's pupil and assistant at the University of Leiden. Since then these matters have for more than 25 years played an important part in the research programme of the Zoological Laboratory of the University at Groningen. Consequently the ideas expressed in this paper have resulted from intensive contact with many colleagues and associates; I feel greatly obliged for their stimulating contributions and co-operation. I particularly wish to thank Dr. R. H. Drent and Dr. J. P. Kruijt for reading and criticizing my manuscript; their assistance has led to many improvements. I am indebted to Mr. J. M. Koolhaas and Dr. P. R. Wiepkema for valuable advise in neurophysiological problems and to those of my co-workers, mentioned in the text, who kindly allowed me to use their unpublished data.

References

ÅKERMAN, B. (1965). Behavioural effects of electrical stimulation in the forebrain of the pigeon. II. Protective behaviour. *Behaviour* **26**, 339–50.

ALBRECHT, H. (1966). Zur Stammesgeschichte einiger Bewegungsweisen bei Fischen, untersucht am Verhalten von Haplochromis (Pisces, Cichlidae). *Z. Tierpsychol.* **23**, 270–302.

ANDREW, R. J. (1956). Some remarks on behaviour in conflict situations, with special reference to Emberiza sp. *Br. J. Anim. Behav.* **4**, 41–5.

—— (1957). The aggressive and courting behaviour of certain Emberizines. *Behaviour* **10**, 255–308.

—— (1962). The situations that evoke vocalization in Primates. *Ann. N.Y. Acad. Sci.* **102**, 296–315.

—— (1963a). Trends apparent in the evolution of vocalization in the old world monkeys and apes. *Symp. zool. Soc. Lond.* **10**, 89–101.

—— (1963b). The origin and evolution of the calls and facial expressions of the Primates. *Behaviour* **20**, 1–109.

—— (1964). Vocalization in chicks, and the concept of "stimulus contrast". *Anim. Behav.* **12**, 64–76.

—— (1972). The information potentially available in mammal displays. In *Non verbal communication* (ed. R. A. Hinde), pp. 179–206. Cambridge University Press.

ATZ, J. W. (1970). The application of the idea of homology of behaviour. In *Development and evolution of behaviour* (ed. T. C. Schneirla), pp. 53–74. Freeman, San Francisco.

BAERENDS, G. P. (1958). Comparative methods and the concept of homology in the study of behaviour. *Arch. neerl. Zool.* **13**, 401–17.

—— (1966). Ueber einen möglichen Einfluss von Triebkonflikten auf die Evolution von Verhaltensweisen ohne Mitteilungsfunktion. *Z. Tierpsychol.* **23**, 385–94.

—— (1970). A model of the functional organization of incubation behaviour. In 'The Herring Gull and its egg'. *Behavior, Suppl.* **17**, 263–312.

—— (1973). Moderne Methoden und Ergebnisse der Verhaltensforschung bei Tieren. *Rheinisch–Westfälische Akademie der Wissenschaften*, Vorträge N. 210, pp. 7–27. Westdeutscher Verlag, Opladen

—— BAERENDS-VAN ROON, J. M. (1950). An introduction to the study of the ethology of Cichlid fishes. *Behaviour, Suppl.* **1.**

—— —— (1960). Ueber die Schnappbewegung des Fischreihers Ardea cinerea L. *Ardea* **48**, 136–50.

—— BLOKZIJL, G. J. (1963). Gedanken über das Entstehen von Formdivergenzen zwischen homologen Signalhandlungen verwandter Arten. *Z. Tierpsychol.* **20**, 517–28.

—— BROUWER, R., and WATERBOLK, H. T. (1955). Ethological studies on *Lebistes reticulatus* (Peters). I. An analysis of the male courting pattern. *Behaviour* **8**, 249–334.

——van der CINGEL, N. A. (1962). On the phylogenetic origin of the snap display in the common heron (*Ardea cinerea* L.). *Symp. zool. Soc. Lond.* **8**, 7–24.

BALDACCINI, N. E. (1973). An ethological study of reproductive behaviour, including the colour patterns of the Cichlid fish *Tilapia mariae* (Boulanger). *Monitore zool. ital.* (*N.S.*) **7**, 247–90.

BARLOW, G. W. (1968). Ethological units of behaviour. In *Central nervous systems and fish behavior* (ed. D. Ingle), pp. 217–32. University of Chicago Press.

BASTOCK, M., MORRIS, D., and MOYNIHAN, M. (1953). Some comments on conflict and thwarting in animals. *Behaviour* **6**, 66–84.

BLURTON JONES, N. G. (1960). Experiments on the causation of the threat postures of Canada Geese. *Rep. Wildfowl Trust* **11**, 46–52.

—— (1968). Observations and experiments on causation of threat displays of the Great Tit (*Parus major*). *Anim. Behav. Monogr.* **1**, 2, 75–158.

BROWN, J. L. (1964a). The integration of agonistic behavior in the Steller's Jay *Cyanocitta stelleri* (Gmelin). *Univ. Calif. Publs. Zool.* **60**, 223–328.

—— (1964b). Goals and terminology in ethological motivation research. *Anim. Behav.* **12**, 538–41.

—— HUNSPERGER, R. W. (1963). Neurethology and the motivation of agonistic behaviour. *Anim. Behav.* **11**, 439–48.

CROOK, J. H. (1964). The evolution of social organization and visual communication in the weaver birds (Ploceinae). *Behaviour, Suppl.* **10**.

DAANJE, A. (1951). On locomotory movements in birds and the intention movements derived from them. *Behaviour* **3**, 48–98.

DELIUS, J. D. (1973). Agonistic behaviour of juvenile gulls, a neuroethological study. *Anim. Behav.* **21**, 236–46.

EIBL-EIBESFELDT, I. (1970). Männliche und weibliche Schutzamulette im modernen Japan. *Homo* **21**, 175–88.

FEEKES, F. (1972). "Irrelevant" ground pecking in agonistic situations in Burmese Red Junglefowl (*Gallus gallus spadiceus*). *Behaviour* **43**, 186–326.

FRYER, G. and ILES, T. D. (1972). *The Cichlid fishes of the Great Lakes of Africa.* Oliver and Boyd, Edinburgh.

GROEN, J. (1955). De mens als psycho-somatische totaliteit van reactiepatronen. *Ned. Tijdschr. Psychol.* **10**, 187–223.

GOTTLIEB, G. (1970). A stranger in the land of homologies. Draft of oral presentation for Philosophy of Biology Symposium, University of North Carolina, Chapel Hill, U.S.A.

HEILIGENBERG, W., KRAMER, U., and SCHULZ, V. (1972). The angular orientation of the black eye-bar in *Haplochromis burtoni* (Cichlidae, Pisces) and its relevance to aggressivity. *Z. vergl. Physiol.* **76**, 168–76.

HINDE, R. A. (1952). The behaviour of the Great Tit (*Parus major*) and some other related species. *Behaviour, Suppl.* **2.**

—— (1953). The conflict between drives in the courtship and copulation of the chaffinch. *Behaviour* **5**, 1–31.

—— (1956a). A comparative study of the courtship of certain Finches (Fringillidae). *Ibis* **98**, 1–23.

—— (1956b). Ethological models and the concept of 'drive'. *Br. J. Phil. Sci.* **6**, 321–31.

—— (1959a). Unitary drives. *Anim. Behav.* **7**, 130–41.

HINDE, R. A. (1959b). Behaviour and speciation in birds and lower vertebrates. *Biol. Rev.* **34**, 85–128.

—— (1965). Interaction of internal and external factors in integration of canary reproduction. In *Sex and behaviour* (ed. F. A. Beach), pp. 381–415. Wiley, New York.

—— (1970). *Animal behaviour: a synthesis of ethology and comparative psychology* (2nd. edn.). McGraw-Hill, London.

HOGAN, J. A. (1965). An experimental study of conflict and fear: An analysis of behavior of young chicks towards a mealworm. Part I. The behavior of chicks which do not eat a mealworm. *Behaviour* **25**, 45–97.

HOGAN-WARBURG, A. J. (1966). Social behaviour of the Ruff *Philomachus pugnax* (L). *Ardea* **54**, 109–229.

von HOLST, E and von SAINT PAUL, U. (1960). Vom Wirkungsgefüge der Triebe. *Naturwissenschaften* **18**, 409–22.

HOOFF, J. A. R. A. M. van (1962). Facial expressions in higher primates. *Symp. zool. Soc. Lond.* **8**, 97–125.

HUNSPERGER, W. (1965). Neurophysiologische Grundlagen des affektiven Verhaltens. *Bull. schweiz. Akad. med. Wiss.* **21**, 8–22.

—— BROWN, J. L. (1961). Verfahren zur gleichzeitigen elektrischen Reizung verschiedener subkortikaler Areale an der wachen Katze. Abwehr- und Fluchtreaktion. *Pflüg. Arch. ges. Physiol.* **274**, 94–5.

—— —— ROSVOLD, H. E. (1964). Combined stimulation in areas governing threat and flight behaviour in the brain stem of the cat. *Progress in Brain Research* **6**, 191–7.

—— BUCHER, V. M. (1967). Affective behaviour produced by electrical stimulation in the forebrain and brain stem of the cat. *Progress in Brain Research* **27**, 103–27.

van IERSEL, J. J. A. (1953). An analysis of the parental behaviour of the male Three-spined Stickleback (*Gasterosteus aculeatus* L.). *Behaviour, Suppl.* **3**.

—— BOL, A. C. A. (1958). Preening of two tern species, a study on displacement activities. *Behaviour* **13**, 1–88.

KARLI, P., VERGNES, M., ECLANCHER, F., SCHMITT, P., and CHAURAND, J. P. (1972). Role of the amygdala in the control of "mouse-killing" behavior in the rat. *Adv. behav. Biol.* **2**, 553–80.

KOOLHAAS, J. M. (1974). Intraspecific aggressive behaviour elicited by electrical stimulation of the lateral hypothalamus in the rat. *Brain Res., Osaka* **66**, 364.

KORTLANDT, A. (1940). Wechselwirkungen zwischen Instinkten. *Arch. neerl. Zool.* **4**, 442–520.

—— (1955). Aspects and prospects of the concept of instinct (vicisitudes of the hierarchy theory). *Arch. néerl. Zool.* **11**, 155–284.

KORTMULDER, K. (1972). A comparative study in colour patterns and behaviour in seven Asiatic *Barbus* species. *Behaviour, Suppl.* **9**.

KRUIJT, J. P. (1964). Ontogeny of social behaviour in Burmese Red Junglefowl (*Gallus g. spadiceus*). *Behaviour, Suppl.* **12**.

LEHRMAN, D. S. (1953). A critique of Konrad Lorenz's theory of instinctive behavior. *Q. Rev. Biol.* **28**, 337–63.

LORENZ, K. (1941). Vergleichende Bewegungsstudien an Anatinen. *J. Orn. Lpz.* **89**, Suppl., 194–294.

—— (1951). Ueber die Entstehung auslösender "Zeremonien". *Vogelwarte* **16**, 9–13.

LOWE-MCCONNELL, R. H. (1969). Speciation in tropical freshwater fishes. *Biol. J. Linnean. Soc. Lond.* **1**, 51–75.

MacDonnell, M. F. and Flynn, J. P. (1964). Attack elicited by stimulation of the thalamus of cats. *Science, N.Y.* **144**, 1249–50.

—— —— (1968). Attack elicited by stimulation of the thalamus and adjacent structures of cats. *Behaviour* **31**, 185–202.

Manley, G. (1960). Unpublished thesis. Oxford University. Referred to in Tinbergen (1965).

Marler, P. (1961). The logical analysis of animal communication. *J. theor. Biol.* **1**, 295–317.

McFarland, D. J. M. (1971). *Feedback mechanisms in animal behaviour.* Academic Press, New York.

—— Sibley, R. (1972). "Unitary drives" revisited. *Anim. Behav.* **20**, 548–63.

Miller, R. J. and Hall, D. D. (1968). A quantitative description and analysis of courtship and reproductive behavior in the Anabantoid fish *Trichogaster leeri* (Bleeker). *Behaviour* **32**, 83–149.

Morris, D. (1952). Homosexuality in the ten-spined stickleback (*Pygosteus pungitius* L.). *Behaviour* **4**, 233–61.

—— (1954). The reproductive behaviour of the zebra finch (*Poephila guttata*) with special reference to pseudofemale behaviour and displacement activities. *Behaviour* **4**, 271–322.

—— (1955). The causation of pseudofemale and pseudomale behaviour: a further comment. *Behaviour* **8**, 46–56.

—— (1956a). The function and causation of courtship ceremonies. In *Symposium Fondation Singer-Polignae: L'instinct dans le comportement des animaux et de l'homme*, pp. 261–86. Masson, Paris.

—— (1956b). The feather postures of birds and the problem of the origin of social signals. *Behaviour* **9**, 75–113.

—— (1957). The reproductive behaviour of the Bronze Mannikin (*Lonchura cucullata*). *Behaviour* **11**, 156–201.

—— (1958). The reproductive behaviour of the Ten-spined Stickleback. *Behaviour, Suppl.* **6**.

Moynihan, M. (1955). Some aspects of reproductive behavior in the Black-headed Gull (*Larus r. ridibundus* L.) and related species. *Behaviour, Suppl.* **4**.

—— (1958a). Notes on the behavior of some North American Gulls. II: Non-aerial hostile behavior of adults. *Behaviour* **12**, 95–182.

—— (1958b). Notes on the behaviour of some North American Gulls. III. Pairing behavior. *Behaviour* **13**, 112–30.

—— (1962). Hostile and sexual behavior patterns of South American and Pacific Laridae. *Behaviour, Suppl.* **8**.

—— (1970). Control, suppression, decay, disappearance and replacement of displays. *J. theor. Biol.* **29**, 85–112.

Nelson, K. (1965). After-effects of courtship in the male three-spined stickleback. *Z. vergl. Physiol.* **50**, 569–97.

Oortmerssen, G. A. van (1971). Biological significance, genetics and evolutionary origin of variability in behaviour within and between inbred strains of mice. *Behaviour* **38**, 1–92.

Panksepp, J. (1971). Aggression elicited by electrical stimulation of the hypothalamus in albino rats. *Physiol. Behav.* **6**, 321–9.

Portielje, A. F. J. (1928). Zur Ethologie bezw. Psychologie der Silbermöwe Larus a. argentatus Pont. *Ardea* **17**, 112–49.

van Rhijn, J. G. (1973). Behavioural dimorphism in male Ruff (*Philomachus pugnax*). *Behaviour* **47**, 153–229.

ROBERTS, W. W., STEINBERG, M. L., and MEANS, L. W. (1967). Hypothalamic mechanisms for sexual, aggressive and other motivational behaviors in the opossum, *Didelphis virginiana*. *J. comp. physiol. Psychol.* **64**, 1–15.

ROWELL, C. H. F. (1964). Comments on a recent discussion of some ethological terms. *Anim. Behav.* **12**, 535–7.

RUITER, L. de (1963). The physiology of vertebrate feeding behaviour; towards a synthesis of the ethological and physiological approaches to problems of behaviour. *Z. Tierpsychol.* **20**, 498–516.

SEVENSTER, P. (1961). A causal analysis of a displacement activity (Fanning in *Gasterosteus aculeatus* L.). *Behaviour, Suppl.* **9.**

SEVENSTER-BOL, A. C. A. (1962). On the causation of drive reduction after a consummatory act (in *Gasterosteus aculeatus* L.). *Arch. néerl. Zool.* **15**, 175–236.

STOKES, A. W. (1962A). Agonistic behaviour among Blue Tits at a winter feeding station. *Behaviour* **19**, 118–38.

—— (1962b). The comparative ethology of Great Blue, Marsh and Coal Tits at a winter feeding station. *Behaviour* **19**, 208–18.

STOUT, J. F., WILCOX, C. R., and CREITZ, L. E. (1969). Aggressive communication by *Larus glaucescens*. Part I. Sound communication. *Behaviour* **34**, 29–41.

THORPE, W. H. (1952). The definition of some terms used in animal behaviour studies. *Bull. Anim. Behav.* **9**, 33–40.

TINBERGEN, N. (1935). Field observations of East Greenland Birds. I. The behaviour of the Red-necked Phalarope *Phalaropus lobatus* (L) in spring. *Ardea* **24**, 1–42.

—— (1940). Die Uebersprungbewegung. *Z. Tierpsychol.* **4**, 1–40.

—— (1950). The hierarchical organization of nervous mechanisms underlying instinctive behaviour. *Symp. Soc. expl. Biol.* **4**, 305–12.

—— (1952a). A note on the origin and evolution of threat display. *Ibis* **94**, 160–2.

—— (1952b). Derived activities: their causation, biological significance, origin and emancipation during evolution. *Q. Rev. Biol.* **27**, 1–32.

—— (1959). Comparative studies of the behaviour of Gulls (Laridae): A progress report. *Behaviour* **15**, 1–70.

—— (1965). Some recent studies of the evolution of sexual behaviour. In *Sex and behaviour* (ed. F. A. Beach), pp. 1–34. Wiley, New York.

—— BROEKHUYSEN, G. J. (1954). On the threat and courtship behaviour of Hartlaub's Gull (*Hydrocoloeus novae-holandiae hartlaubii* (Bruch)). *Ostrich* **25**, 50–61.

—— IERSEL, J. J. A. van (1948). "Displacement reactions" in the three-spined stickleback. *Behaviour* **1**, 56–64.

—— MOYNIHAN, M. (1952). Head flagging in the Black-headed Gull: its function and origin. *Br. Birds* **45**, 19–22.

VALENSTEIN, E. S., COX, V. C., and KAKOLEWSKI, J. W. (1968). Modifications of motivated behaviour elicited by electrical stimulation of the hypothalamus. *Science, N.Y.* **159**, 1119–20.

WASMAN, M., and FLYNN, J. P. (1962). Directed attack elicited from hypothalamus. *Arch. Neurol.* **6**, 220–7.

WEIDMANN, U. (1956). Verhaltensstudien an der Stockente (*Anas platyrhynchos* L.). I. Das Aktionsystem. *Z. Tierpsychol.* **13**, 208–271.

WICKLER, W. (1961a). Oekologie und Stammesgeschichte von Verhaltensweisen. *Fortschr. Zool.* **13**, 303–65.

—— (1961b). Ueber die Stammesgeschichte und den taxonomischen Wert einiger Verhaltensweisen der Vögel. *Z. Tierpsychol.* **18**, 320–342.

WICKLER, W. (1966). Ursprung und biologische Deutung des Genitalpräsentierens männlicher Primaten. Z. Tierpsychol. **23,** 422–37.

—— (1967a). Vergleichende Verhaltensforschung und Phylogenetik. In *Die Evolution der Organismen* (3rd edn, ed. G. Heberer), Bd. 1.

—— (1967b). Socio-sexual signals and their intra-specific imitation among primates. In *Primate ethology* (ed. D. Morris), pp. 69–147. Weidenfeld and Nicolson, London.

WOODWORTH, C. H. (1971). Attack elicited in rats by electrical stimulation of the lateral hypothalamus. *Physiol. Behav.* **6,** 345–53.

WORTHINGTON, E. B. (1937). On the evolution of fish in the great lakes of Africa. *Int. Revue ges. Hydrobiol. Hydrogr.* **35,** 304–17.

10. Evolutionary aspects of orientation and learning

M. LINDAUER

10.1. Introduction

IT IS an obligation for every biologist to consider the evolutionary context of the knowledge which he acquires by description and analysis.

There are two ways in which the behaviour involved in orientation and associated learning processes can affect evolution:

1. During the process of evolution, orientation and learning achieve increasing specificity and can thus function as mechanisms for reproductive isolation allowing speciation to occur; at the same time orientation and learning become adapted to the particular requirements of the species' habitat.

2. Orientation and learning actively shape evolution in so far as they provide the mechanisms allowing pioneering individuals from geographically isolated populations to occupy new habitats and thereby provide the basis for allopatric speciation. Appropriately, Ernst Mayr (1970) emphasizes the major importance of food and habitat selection in speciation rather than stressing just reproductive behaviour, which has been so much the preoccupation of ethologists. In many species exploratory behaviour, individual experience, and learning by trial and error are highly significant for habitat and food selection.

If we are to consider orientation in an evolutionary context, then we ought to think of two basic aspects. First the degree of structural organization and complexity of the orientation mechanisms, and secondly the web of causal relationships between orientation and specific situations in the habitat, that is, its adaptiveness. Some examples will serve to illustrate these two aspects.

The orientation performed in the escape reaction of a fly towards light is on a primitive level, as is the avoidance reaction of a planarian or an earthworm elicited by a light stimulus. These are simple phototaxic reflexes, elicited by a single-sense modality. Since there is no defined goal in the environment, the success of such orientation is largely left open to chance. In contrast, predation by *Ammophila*, with its subsequent storage of prey in the nest chamber, reveals a highly goal-directed orientation structured by innate components and complex learning processes. Here not simply one modality but at least three are involved: the visual, olfactory, and tactile senses. The wasp's searching in habitats suitable for prey, its recognition and capture of prey, its re-orientation to the home nest and correct introduction of the caterpillar into the nest chamber—all these require a centrally-steered feedback programme of action which demands information at the right time and location from correspondingly differentiated sensory inputs. Innate releasing mechanisms in co-operation with situation-specific learning processes make orientation less uncertain. These examples illustrate very different levels of organization; however, both are in their biological function of equally high adaptive value.

I must make clear that it is not my intention to sketch a phylogeny of orientation—I do not believe that this can be done. Orientation and learning do often show phylogenetic characters which, however, can only be understood in the context of the evolution of the whole organism. The following examples, taken from the life-cycle of parasites, illustrate the problems inherent when considering the evolution of orientation mechanisms taking only the levels of organization into account (see Fig. 10.1).

Filaria bancrofti, the nematode which causes elephantiasis, has to adapt precisely to the activity rhythm of its intermediate host—the mosquito, *Aedes*—for the transfer to a new host to occur. By day the microfilarias live in the lung capillaries; at dusk they swarm to the skin capillaries where they can be ingested by nocturnal mosquitos feeding on the host. How does the internal parasite's activity become synchronized with photo-period? The movement to the skin capillaries occurs when the changing partial pressure of oxygen in the human lungs, which decreases in the evening, releases the diurnal block to swarming in the parasite.

The selection pressure inherent in the need to adapt to the specific activity pattern of the intermediate host is particularly well demonstrated in the closely related species *Loa loa*, whose intermediate host is the diurnal *Chrysops*. The parasite again uses the change in partial pressure of oxygen as a time cue. However, it is not the *absolute* difference between the partial pressure during the day and night which causes migration; the parasite's sensitivity to the partial pressure changes with temperature. At lower temperatures the microfilarias are more sensitive to the partial pressure of oxygen and remain in the lung tissues; as the body temperature rises during the day, the sensitivity threshold decreases and the microfilaries migrate into the subdermal connective tissue.

How can the evolutionary stage of this adaptation be assessed? Do we have to look at it as a regression to a more primitive level of organization, as is the case in many parasitic species? I think that the term 'organizational level' should be used only in a restricted and relative way; the sensory physiologist and neurophysiologist will regard the described adjustments of the parasite to the measured parameter as being a highly

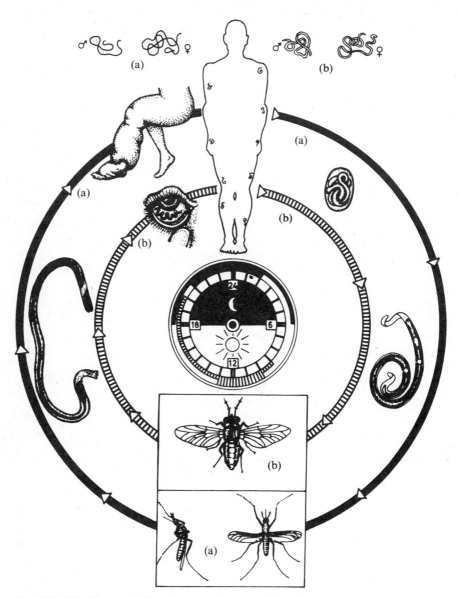

FIG. 10.1. Life history of *Filaria bancrofti* (solid line (a)) and *Loa loa* (barred line (b)). The central figure shows the times when the intermediate mosquito host of both species are active. The daily migration of the micro-filarias is adapted to the particular activity rhythm of each species of intermediate host.

specialized and high-ranking orientation performance, as well as representing a fine adaptation to an ecological niche which is of high survival value.

In the remainder of this essay I want to illustrate these more general considerations by looking at two sensory modalities, the chemical and the visual, in order to examine the relationships of orientation and learning to evolution.

10.2. The chemical sense
Genetically fixed learning dispositions

Orientation using chemical signals often involves innate knowledge of species-specific and food-specific odours, but this must be supplemented by learning and memory so that formerly neutral odours which come to have significance for survival can be assigned to appropriate responses. Thus one consideration out of the manifold relationships between learning and evolution appears to me to be particularly important. Does storage of potential signal parameters result from an 'unrestricted learning ability', or is there a genetically-fixed learning disposition selecting potential parameters? If a honey-bee is offered odour, colour, and form as food signs, we find that, whereas the odour is learned after only 1 recorded approach (on a 95 per cent confidence level), colour signs require 3 rewarded approaches, while form requires 20 or more. There is a hierarchy in the bee's response to the sign parameters, which is not only present at the level of the individual but is also extremely consistent within one colony and even within a race of honey-bees. Furthermore, a hierarchy was found within a single sense modality: odours which are related to flowering plants are learned more readily than those from fatty acids (Fig. 10.2). These are basic rules in the bee's coding of potential cues for conditioning, which still do not fully answer questions concerning restraints on cues it can use for learning. Yet our most recent experiments have shown that his learning disposition is strongly race-specific, and thus probably has a genetic component in its development.

In species *Apis mellifera Carnica*, *A.m. ligustica*, and *A.m. fasciata*, as well as the species *Apis cerana* were trained with 17 different odours (Koltermann 1973). Naïve bees of the different races assess and learn these odours with varying degrees of ease; each race learns with the greatest facility scents which are characteristic of its own native flora. Some examples may illustrate this. Orange, fennel, jasmine, and oil of rosewood are learned by *A.m. carnica* and also by *A.m. ligustica* and *A.m. fasciata* much less readily than lavender and rosemary; the Indian species *A. cerana*, however, learns the scent from oil of rosewood most readily, followed by orange, fennel, and jasmine. Thyme and anise are also learned better by *A. cerana* than by the honey-bees of the West. These preferences

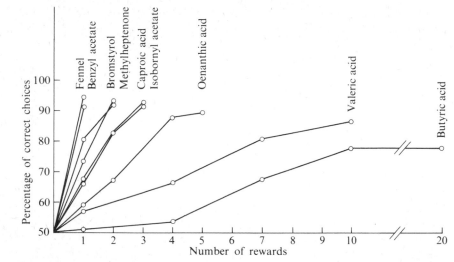

Fig. 10.2. Under standard environmental conditions, honey-bees do not learn with equal facility to associate different odours with food. After each reward (abscissa) learning success was tested by recording the number of approaches to each conditioned odour (ordinate). Flowerlike odours are more readily learned than those such as valeric and butyric acids.

correspond to the relevant native food plants, with the exception of the odour of orange. Oranges originated in south-east Asia and have been cultivated in Mediterranean countries for only 300 years. Evidently this time period has not been sufficient to allow the learning disposition of western honey-bees to become adapted to this addition to the available food plants.

There is a second point of importance, for the genetically determined preferences can be modified by learning to some extent. If bees of the *A.m. carnica* race are rewarded with 0·5-molar sucrose solution at their preferred odour (rosemary), but at the same time offered a rich 2-molar sucrose solution at the least preferred odour (caproic acid), the caproic acid is approached as frequently as rosemary after only 2 learning trials. With an increasing number of trials it becomes approached even more often than rosemary odour. Again, the races and the species vary with regard to the strength of the reward required to achieve such an alteration of preference through training. *Apis cerana* requires the greatest reward before any alteration of its natural preferences becomes measurable. It is followed by the Egyptian race *A.m. fasciata*, and the Italian *A.m. ligustica*, with the mid-European *A.m. carnica* bees having the most easily modified preferences. Doubtless the various degrees of flexibility shown by the diverse populations are adaptations to the nectar supply of their native habitats. In the tropical regions of east Asia nectar is available in surplus throughout the year, while in more seasonal climate of mid-Europe it is advantageous to be able to adjust to seasonally varying food plant compositions and to accept food of low concentration.

Chemical signals as reproductive isolating mechanisms

The evolution of signal systems used in intra-specific communication has to overcome a basic dilemma. On the one hand, in order to avoid intra-specific misunderstanding a signal should be maintained as constant as possible, while on the other hand, during speciation evolution signals must be modified to achieve distinctiveness. This paradox is particularly evident in communication by invertebrate sex attractants. Priesner (1973), using both measurements by behavioural assay and by antennograms, recorded the responses of more than 700 species of Lepidoptera from 27 families to inter-and intra-specific sex odours. Although intra-specific responsiveness is far the most marked, closely related species do show some degree of inter-specific responsitivity. Priesner grouped these species showing inter-specific responsitivity into 'reactive groups', which he interpreted as new taxonomic units. These specific and generic groups correspond neatly to those derived from a morphological classification (Fig. 10.3). Reproductive isolation between the different species has therefore to be guaranteed by additional isolating mechanisms. Those mechanisms are different for allopatric as opposed to sympatric species. The genus *Platysamia*, for example, is restricted to four North-American species. They all produce the same sex attractant but they are geographically dispersed, however hybridization has been recorded in the zones of overlap.

Sympatric species have evolved various types of isolating mechanisms. They may show seasonal or diurnal separation of their reproductive behaviour. Thus species of the genus *Rebelia* communicate with the same sex attractant, but some species swarm in the evening while others swarm in the morning. Females experimentally released at the wrong time of day are approached by males of the other species. The same situation holds for the species *Arctia villica* and *Parasemia plantaginis*; *P. plantaginis* is active during afternoons while *A. villica* is nocturnal. Normally, of course, sex attractants are released by virgin females only during the flight activity periods of the males. Species of Noctuidae and Tortricidae using similar sex attractants are reproductively isolated because they are active at different seasons of the year.

Courtship behaviour which is shown only on a species-specific host plant can also serve as an isolating mechanism. Females of *Telea polyphemus* release their sex attractant only when their antennal perceive *trans-2-hexenal*; this is the odour of oak leaves, their host plant. A similar phenomenon was found in *Choristoneura* species, which live on various

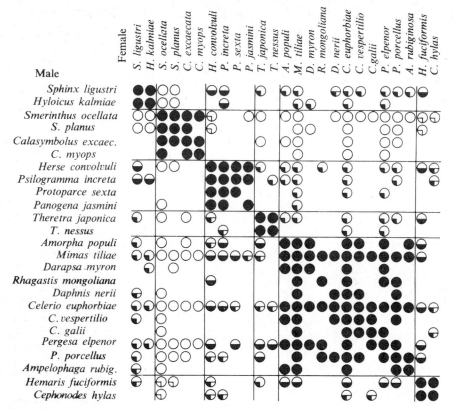

FIG. 10.3. 'Reactive groups' which were established by testing the effectiveness of inter- and intra specific sex attractants of 25 species of Sphingidae. An empty circle represents no attraction, and fully black circles represent complete intra- or inter-specific attraction; intermediate conditions are partly shaded.

conifers. When experimentally separated from their natural host plant for long periods these species come to interbreed since, apart from their different host preferences, there are only ineffective ethological barriers to hybridization.

One possible way in which the evolution of sex attractants may have occurred is outlined by Roelofs and Comeau (1968, 1970, 1971, 1972). They suggest that the main component of the sex attractant first becomes modified by side-components. Two species of Tortricidae, *Atoxophyes fasciata* and *A. orana*, both produce *cis*-9-tetradecenyl acetate as the major component of their sex attractant. Again both species have *cis*-11-tetradecenyl acetate as a side-component, but the proportion of the major component is in *A. fasciata* 1·8 : 1 and in *A. orana* 3·4 : 1. The single components are ineffective as attractants, only the right species-specific composition elicits the full response.

Allopatric populations do not require any fundamental alteration of the sex attractant for speciation to occur. The normal modification and recombination of genetic material is sufficient for species evolving in geographical isolation. Roelofs and Comeau (1972) have shown, however, that isomers of one single carbon atom in the sex-attractant molecule of *Bryotopha* (family Gelechiidae) gave rise to reproductive isolation between two closely related species, now separated by their response to the two isomers. A combination of both isomers still has an inhibiting effect on the males.

10.3. Visual sense
Orientation mechanisms

Both the visual sense organs themselves and the orientation responses to visual stimuli are much more varied than those of the chemical senses. One can detect not just the evolution of adaptations to a particular ecological niche but also an evolution of visual function. The dermal photo-receptors of an earthworm can only perceive light of different intensities, whilst the cup-shaped ocelli of *Planaria* or *Amphioxus* already allow an additional perception of direction. There are further major advances with the evolution of the image-forming eye, as in the eye of *Nautilus*, and finally the development of the lens and the compound eye. Various mechanisms of the dioptric apparatus achieve high visual acuity at different light intensities and independent of distance. In this context, an amazing development is found in the neural superposition compound eye of the fly in which homologous photo-receptor cells from a group of individual ommatida are linked together in a single optic cartridge and thus achieve a lower threshold for poor light conditions (Kirschfeld 1973).

There are two points of major interest with regard to the evolution of visual orientation: the evolution of those mechanisms which concern the *localization* of the goal, and those which involve the *identification* of the goal.

Localization of the goal. Jander (1965) sees the evolution of phototropotaxis, first defined by Kühn, as occuring in three steps:
  (1) archyphototaxis: sensory inputs of two receptors are compared with each other (for example, *Nauplius*);
  (2) prophototaxis: intensity and direction are decoded by each receptor (for example, *Myriapoda, Apterygota, Polyneoptera*);
  (3) metaphototaxis: the direction of the most intense stimulation determines the degree of the turning reaction (for example, *Rhynchota, Holometabola*).

While in case (1) input differences are compensated absolutely, in cases (2) and (3) a sinusoidal evaluation of the light direction occurs, and for this form of decoding central 'sinusoidal modulators' are suggested.

According to Jander's 'compensation theory' menotaxis developed from phototaxis in such a way that a central turning command turns the animal away from the basic direction by an angle proportional to the turning excitation. Intensive research on *Apis, Bombus,* and *Calliphora* has shown that these so-called original types are rather variable and non-uniform in themselves, neither are they characteristic of one species or another (Horn 1974). In lower light intensities some insects (for example, *Apis*) orient prophototactically but with increasing light intensities metaphototactically. This applies as well to those experiments in which light and gravity are offered in competition. When *Panorpa, Sialis,* and *Chrysopa* are presented with small angles on an inclined plane (30°–90°) and concurrently offered light, progeotactic orientation is shown; but with increasing angles of inclination, metageotactic orientation occurs.

Astrotaxis consists of compass menotactic orientation using the sun, moon, or stars. In addition to the simpler menotaxis this involves the true bearing of the angular velocity of the azimuth of the celestial body which is being responded to. Astrotaxis has been found in Crustacea, Arachnida, Insecta, Mollusca, Pisces, Reptilia, and Aves. Doubtless convergent evolution of this particularly demanding orientation mechanism has occurred in the various groups. As one would expect in view of its wide distribution, astrotaxis has come to mediate a variety of adaptive responses: for example, orientation to the sea or to banks in *Talitrus* and *Velia,* respectively; orientation to the food source and back to the nest in ants and bees; long-distance migration and orientation to spawning grounds in fishes; orientation to spawning grounds in tortoises; orientation during bird migration.

Sun-compass orientation with all of its components can be partly innate, as in fishes, *Talitrus,* and reptiles, or partly learned, as in ants, bees, and birds. Undoubtedly, the adaptive value of this type of orientation becomes improved by individual experience. Only individual experience can provide the exact calculation of the seasonal and geographical characteristics of the sun's movement.

Trans-modality transposing of menotactically maintained angles of orientation. During their dances in the hive honey-bees transpose exactly the menotactically maintained angle between the sun and the direction of flight into the same angle with respect of the gravitational field on the vertical comb taking vertically upwards as the 'equivalent' of directly towards the sun. Karl von Frisch, the discoverer of this ability of honey-bees, could hardly believe that this was confined to this one insect (von Frisch 1965). Later work showed that such transposition was an elementary orientation mechanism occuring frequently in the Insecta. However, as Fig. 10.4 illustrates, the transposition is sometimes ambiguous; in ants,

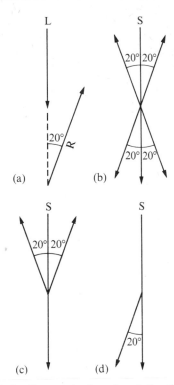

FIG. 10.4. Transposing of visually perceived angles of direction into the gravity field: (a) animal walks or flies menotaxically 20° to the right of the light source; (b) ants transpose true to the angle, but at any of four alternative directions; (c) ladybirds transpose true to the angle, but at either of two directions; (d) dung-beetles transpose like the honey-bee in its waggle dance, true to the angle and in one direction only. However, the direction maintained towards the light source becomes transposed into the same angle directly downwards with respect to gravity not directly upwards as in honey-bees.

for example, it is four-directional. That is, the angle run with respect to a light source, say α, is transposed into anyone of four angles with respect to gravity. Taking vertically up as 0°, the ant tends to run on a course of α, 360° $-\alpha$, 180°$-\alpha$, or 180°$+\alpha$. Clearly any such ambiguity could not be permitted once the evolution of a communication system like that of the honey-bee had begun.

Trans-modality transposing has developed within the genus *Apis* in relation to the different types of social organization shown. *Apis florea* does not build its combs into hives and needs to be able to see the sun or the blue sky if it is to communicate direction, that is, this species requires a horizontal dance plane where the goal can be communicated by direct indication (Lindauer 1971). An intermediate position is seen in *Apis dorsata*; again this species nests in the open, but it has no horizontal plane on the ridge of its comb since this is attached at the ridge to overhanging rocks or gutters. *A. dorsata* do dance on the vertical plane of the comb but require a constant view of the sky. This simplifies the task for recruits, as they do not have to remember the communicated angle with respect to gravity as do *Apis mellifera* in their dark hives. These bees can transpose the angle back into the visual modality only after having reached the hive entrance.

The orientation mechanisms I have mentioned so far serve as the localization of the goal; the identification of the goal requires further capacities.

Identification of an orientational goal-recognition pattern.   Pattern recognition is a basic requirement for the identification of a goal. This presupposes that the five so-called 'coherent factors' are recognized and integrated into a form: spatial position, uniformity, similarity in details, symmetry, and connection of single points to contours.

Basic processes in pattern recognition are initiated when partial decoding occurs in the receptive fields of the retina. The retina of the vertebrate eye or the optic ganglia of the compound eye can, in general, perform the following operations on the visual input:

(1) increase contrasts by lateral inhibition;

(2) code the size of the visual angle as the size and extension of excitatory and inhibitory processes in the receptive fields;

(3) recognize 'movements' of a light–dark zone by specific movement-sensitive neurons;

(4) localize a light–dark zone by means of the oval arrangements of direction-sensitive neurons;

(5) recognize the direction of movement by unilateral inhibition in direction-sensitive neurons.

The degree of information decoding in the retina however, differs between species and thus becomes important from the evolutionary viewpoint. Amphibia, for example, are capable of exhibiting operations (1), (2), and (3); pigeons, rabbits, and ground-squirrels exhibit (1), (2), (3), (4), and (5); but cats and monkeys exhibit only (1) and (2). These differences supposedly correspond to the degree of neocortical development. In Anura and Aves there is no visual cortex, and thus a great deal of input decoding occurs in the retina itself. Ground-squirrels and rabbits represent an intermediate level; they do possess a cortex but it is not as highly developed functionally as in higher mammals. Because of their high cortical development, the retina of cats and monkeys is correspondingly simply structured.

In mammals, the subcortical visual centres for pattern analysis are of less importance. Investigations by Humphrey (1970) on monkeys and by Schneider (1970) on Golden Hamsters indicate that lesions in the visual cortex cause loss of the ability to recognize forms, although differentiation between light intensities and stimulus localization are still possible. On the other hand, in toads form-identification itself takes place in the pretectal thalamic region and in the optic tectum (Ewert 1973). Ewert proposes that there are, in addition, 'master units' in defined regions of the visual pathways which respond to particular key stimuli such as prey or predators and trigger the appropriate behavioural responses.

Comparative anatomy illustrates the evolution of visual-orientation mechanisms in one other way. The localization and identification of pattern by vertebrates occurs in two different visual systems. In lower vertebrates decoding and recognition of a visual pattern is effected by the interconnection of those neuronal tracts in the retina and tectum which are specified for localization with a third tract of the pretectal thalamic region. Localization and identification systems are here strongly aligned anatomically. Accordingly in toads no pattern discrimination is detectable after lesioning the optic tectum. In mammals, pattern analysis takes place in phylogenetically 'younger' cortical neurons, which are largely independent of the collicular localization system. Thus, after lesioning the superior colliculus, the Golden Hamster cannot localize patterns in space but can still distinguish between them.

For distinguishing patterns it is not only visual acuity and image formation on the retina which are of importance; centrally controlled factors are also required. In stick-insects Jander found a preferential approach to 'bush-like' dummies, and six specific detectors ensure the recognition of pattern contours. It would be important to analyse to what extent individual experience can alter these preferences. This question as to how far individual experience can alter genetically controlled predispositions can be discussed using some of our own more recent findings.

Genetically-fixed learning dispositions in the recognition of pattern

At the centre of a circular table a 2-molar sucrose solution is offered to honey-bees. The feeding jar is positioned on a black star, and the bees are trained to recognize the star form as a food sign. However, the star is only a part of a greater complex of patterns, for example, 'circular table with centrally placed black star'. Which visual feature is of greater importance, the star form or its central position on the circular table? This can be tested by positioning the star at the edge of the table. If there have been only a few learning trials prior to such a test, the form proves superior to the relative position of the star. However, with an increasing number of rewarded approaches to the star in the central position, tests show that the relative position gains importance and is chosen more often than the star form itself, when this latter is presented off-centre. Such a relative assessment of spatial position versus form is race-specific, that is, it has a

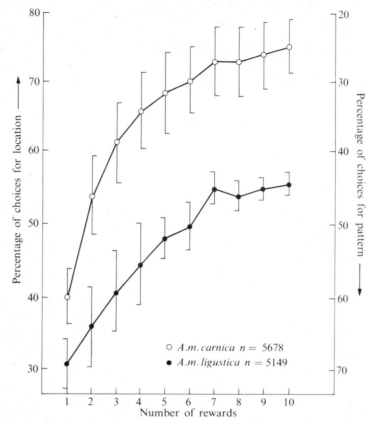

FIG. 10.5. Different learning curves are obtained for *Apis mellifera carnica* and *A.m. ligustica* when a star form and its central position on a table are offered in competition as orientation marks for a food source. Means and standard errors of a large number of trials for each race.

genetic component. In *A.m. carnica* the percentage of approaches to the off-centre star rises rapidly to some 73 per cent after 10 trials. *A.m. ligustica* adapts much less well to the changed position and improves more slowly and to a lower level (Fig. 10.5). This result applies only to those experiments in which the nearby surroundings of the training area are visually very diverse, that is, offer obvious landmarks such as trees, bushes, and walkways which means that the star form used in training is set amongst a complex pattern offering many visual parameters. The learning curves of both races are identical when the star is positioned in a uniform clear area.

The learning of form as a food sign thus depends on the visual structure both of close and more distant surroundings. *A.m. ligustica* preferentially uses the form of the goal itself for orientation while *A.m. carnica* learns more readily to make use of auxiliary features in the surroundings which become associated with the goal. For this reason, under normal circumstances *A.m. carnica* proves superior to *A.m. ligustica* in its ability to locate the goal. However, if there are no clearly defined landmarks in the area around the goal both races have to make use of sun-compass orientation from a distance and, when close enough, react to the pattern of the goal itself. In this situation both races show the same learning capacities (Lauer and Lindauer 1971).

These inter-racial differences appear to be adapted to the normal environment of the bees. *A.m. ligustica* originating in the Mediterranean area, are able to fly in sunny weather throughout the year. Sun-compass orientation can thus be relied upon for guidance directly to the goal. *A.m. ligustica's* learning capacities themselves are focused on the recognition of patterns. *A.m. carnica* bees, natives of mid-Europe, have to orientate themselves more often under a partially or fully overcast sky and cannot rely on the sun compass alone; thus it is adaptive for them to use landmarks which are associated with the food goal.

10.4. Conclusion

Orientation and learning have a significant impact on evolution. Each orientation mechanism and learning capacity, of course, involves genetically based information. However, reactions based on a genetic program are not simple passive results of selection operating in one particular situation. The organism has the capacity to select actively new ecological niches, feeding areas, or social partners. Exploratory behaviour, and in particular learning and memory which are the basis for avoidance as well as positive responses to individual experience, lead to what Baestrup (1971) has called 'modificatory steering' towards the environment. Selection operates on learning dispositions also, which may become genetically fixed if they are adaptive to a particular niche. However, the greater flexibility the individual has in learning the more effective is its orientation to new habitats and exploration of new niches. This will provide the possibility of isolation leading to new speciation.

References

BAESTRUP, F. U. (1971). The evolutionary significance of learning. *Vidensk. Medd. dansk naturh. Foren.* **134**, 89–102.

EWERT, J.-P. (1973). Lokalisation und Identifikation im visuellen System der Wirbeltiere. *Fortsch. Zool.* **21**, 307–33.

VON FRISCH, K. (1967). *The dance language and orientation of bees.* Harvard University Press, Cambridge, Mass.

HORN, E. (1974). On gravity orientation in insects. In *Mechanisms of space orientation and space perception related to gravity. Fortsch. Zool.* **23**, 1–20.

HUMPHREY, N. K. (1970). What the frog's eye tells the monkey's brain. *Brain Behav. Evol.* **3**, 324–37.

JANDER, R. (1965). Die Phylogenie von Orientierungmechanismen der Arthropoden. *Verh. dt. Zool. Ges.* **59,** 267–305.

KIRSCHFELD, K. (1973). Das neurale Superpositionsauge. *Fortsch. Zool.* **21,** 229–57.

KOLTERMANN, R. (1973). Rassen-bzw. artspezifische Duftbewertung bei der Honigbiene und ökologische Adaptation. *J. comp. Physiol.* **85,** 327–60.

KRISTON, I. (1971). Zum Problem des Lernverhaltens von Apis mellifica L. gegenüber verschiedenen Duftstoffen *Z. vergl. Physiol.* **74,** 169–89.

LAUER, J. and LINDAUER, M. (1971). Genetisch fixierte Lerndispositionen bei der Honigbiene. *Inf. Org.* **1,** 1–87.

LINDAUER, M. (1971). *Communication among social bees.* Harvard University Press, Cambridge, Mass.

MAYR, E. (1970). Evolution und Verhalten. *Verh. dt. zool. Ges.* **64,** 322–6.

PRIESNER, E. (1973). Artspezifität und Funktion einiger Insektenpheromone. *Fortsch. Zool.* **22,** 49–135.

ROELOFS, W. L. and COMEAU, A. (1969). Sex pheromone specificity: taxonomic and evolutionary aspects in Lepidoptera. *Science, N.Y.* **165,** 398–400.

—— —— (1971). Sex attractants in Lepidoptera. In *Pesticide chemistry,* Vol. 8: Chemical releassers in insects (ed. S. Tahoried), pp. 61–114. Gordon and Breach, London.

SCHNEIDER, G. E. (1970). Mechanisms of functional recovery following lesions of visual cortex or superior colliculus in neonate and adult hamsters. *Brain Behav. Evol.* **3,** 295–323.

11. The evolutionary significance of early experience

K. IMMELMANN

11.1. Introduction

IN RECENT years, the study of imprinting has found increased interest among students of animal behaviour. Apart from its ethological importance, however, there is accumulating evidence that imprinting, especially sexual imprinting, can also act as a powerful speciating mechanism and may thus be of direct evolutionary significance.

The aim of this essay is to summarize the available literature and to discuss the different ways in which imprinting exerts an influence upon speciation. As there are several other early learning mechanisms that share some of the characters of 'classical' imprinting, they will also be included into the discussion.

11.2. Imprinting and other early learning processes

In a great variety of species, it has been found that experience early in life has a stronger and more permanent effect upon adult behaviour than the same or even a greater amount of experience at a later age. This applies, above all, to sexual imprinting. Numerous cross-fostering experiments have shown that, in several groups of birds and possibly also in some other animals, sexual preferences are established by means of social contact with the (foster) parents or (foster) siblings during a comparatively brief period during juvenile life. Such preferences tend to be very stable and cannot very easily be changed through subsequent experience with different social objects (for review, see Immelmann 1972b). In some cases, even an absolute resistance to any alteration has been observed. In the Australian Zebra Finch, *Taeniopygia guttata*, for example, a definite irreversibility of mating preferences acquired during adolescence has been proved for a period of time that markedly exceeds the life expectancy of the species under natural conditions (Immelmann 1970).

Sexual imprinting, however, is not the only example of juvenile experience crucially affecting later adult behaviour. It has been found that experience during the first days or weeks of life may also have long-lasting effects on other object preferences. Most data are available for the early establishment of food preferences (Hess 1962; Burghardt 1967), habitat preferences (Löhrl 1959; Hildén 1965; Klopfer 1965; Braestrup 1968; Catchpole 1972) and, in parasitic animals, host preferences (Thorpe and Jones 1937; Nicolai 1964). Many of these early acquisition processes have some striking similarity to sexual imprinting. They are reported to be more or less restricted to periods very early in the individual's life, and in many cases, they have also been found to be highly resistant to alteration through subsequent experience. Apart from these object preferences, early experience has likewise proved to exert a strong influence on a variety of other phenomena of very different and sometimes rather comprehensive nature, such as the development of normal copulatory behaviour, the development of contact behaviour, the organization of maternal behaviour, the level of aggressiveness, wildness, fearfulness, and tameness, or the degree of socialization and aggregative behaviour (for reviews see Beach and Jaynes 1954; Collias 1950; Harlow and Harlow 1962; Denenberg 1963; Salzen 1966). Finally, even song learning in some species of passerine birds may occur very early and may be restricted to a sensitive period (Thorpe 1959; Mulligan 1966).

Due to the similarities to the main characters of sexual imprinting—occurrence of well-circumscribed sensitive periods and subsequent stability of preferences—such early learning has likewise been called imprinting or 'imprinting-like processes'. Although the experimental data for many of these learning processes are still too few to permit definite conclusions about the degree of conformity with sexual imprinting—especially as far as the underlying physiological mechanisms are concerned—the available evidence clearly points to the fact that experience very early in life does have a fundamental influence not only on the choice of a mate but also on a whole spectrum of other developmental processes.

11.3. Selection pressures favouring early learning

From the evolutionary point of view, this fact leads to the question of which selection pressures may have caused the development of learning processes of this type. This question comprises three different sub-problems: (1) Why does the envionment exert such an essential influence? (2) Why is this influence restricted to an early developmental stage? (3) Why does it cause such a stable result? These questions can be discussed most easily with reference to sexual imprinting and other object preferences, for which more data are at present available than for the more comprehensive developmental processes mentioned above.

1. That environment exerts a great amount of influence may be due to two different reasons. They are concerned with the mechanisms of information storage and with the degree of adaptability to changing environments.

(a) A number of experimental analyses have shown that naïve perceptual preferences tend to be of a rather general nature and usually refer only to very few and conspicuous characters of the object. Obviously the amount of information that can be stored in the genome is rather small as compared with the possibility of information storage in the memory (cf. Lorenz 1943; Baerends, Bril, and Bult 1965; Gottlieb 1965; Liley 1966). Therefore, object preferences based on learning will, as a rule, include more details and will thus be more precise than those based on genetic factors.

An example, among others, along this line is provided by the sexual differences in imprintability that have been found in several groups of birds, mainly in ducks, pigeons, and estrildid finches. In these species, the sexual preferences of the female are much less affected by early experience than those of the male. This probably is a consequence of the sexual dimorphism occurring in the species concerned. Due to its courtship song or courtship calls, its bright plumage markings, and its more elaborate courtship behaviour, the male offers more distinct signals for species recognition. The female, therefore, may rely to a certain extent on its unlearned preferences. The female has fewer and less conspicuous distinguishing characters because she has been selected for crypticity. Consequently the male needs a more precise knowledge of the opposite sex that, obviously, can be obtained only through individual learning (Schutz 1964; Immelmann 1972a).

Finally, another possible positive effect of learning the species-specific characters may be correlated with the fact that in the case of learned preferences, the most effective stimuli are always provided by those objects that bear the closest similarity to the object the individual had originally been exposed to. As a consequence, no 'supernormal' stimuli can be created and no exaggerated, 'luxuriant' characters are required for species-recognition.

(b) Any changes in the environment of a species, including the appearance of the species itself, are followed automatically by a corresponding change in the relevant object preferences, if the latter are acquired through personal experience. Object preferences that are transferred from one generation to the next by means of tradition are thus more quickly adaptable than those coded in the gene-pool of the species. This is of special importance in rapidly evolving groups of animals as well as in unstable environments where a fair degree of opportunism will be favoured as compared with absolutely rigid preferences for particular conditions. Similar thoughts have been expressed with regard to the determination of song in passerine birds by learning rather than by inheritance (Lemon and Herzog 1969).

2. Apart from the mere existence of learned object preferences there may be an additional advantage if their acquisition is restricted to early

developmental stages of the organism. There may be two aspects to this advantage.

(a) While the young animal is still a member of the family group its opportunities to learn species-specific characteristics as well as certain features of the species-specific environment are greater than when it later has to live on its own or in much looser groups. Selection pressure, therefore, will certainly favour rapid learning at an early age.

(b) An early sensitive period like this ensures the availability of relevant information well before its first application and in this resembles genetically coded preferences. This is of special importance in the case of sexual imprinting and 'host imprinting' that has to guarantee recognition of a mate or of a host already at the very beginning of the first reproductive period in the individual's life.

3. Finally, the very stable result of early experience probably serves to terminate some of the essential learning processes before dispersal from the natal area and thus to prohibit a subsequent influence of environmental stimuli at a time when—owing, for example, to deteriorating environmental conditions after the end of the reproductive season—the animal lives in mixed species flocks or in sub-optimal habitats. As, in general, the habitat of a species, especially if it is distributed over vast geographic areas, is far from being entirely uniform but tends to consist of a cline or mosaic of sub-habitats with slightly different environmental conditions, the return of an individual to its natal area offers at least a two-fold advantage:

(a) Owing to its early experience with the relevant sort of habitat, any one individual will be specially adapted to this particular kind of environment and will thus be able to make optimal use of hides, food resources, or nesting-sites;

(b) The population as a whole will be divided into a number of sub-populations each of which is adapted to a particular kind of sub-habitat. This in turn results in optimal utilization of the habitat as a whole as well as in a restriction of intra-specific competition to members of the same sub-population.

An additional advantage of early restriction to specific environmental conditions refers to the colonization of new habitats. If, for certain reasons (for example, owing to increased population pressure), an adult pair has settled in a marginal area of the species's distribution or if it has invaded a new type of habitat, the offspring will become imprinted on some of its characteristics and, for reproduction, will try to return to a similar kind of environment. If the habitat permits successful reproduction this may give rise to a new sub-population that ecologically is more or less separated from the original stock.

An impressive example of a process like this is provided by the European Mistle-thrush. Originally being a forest inhabitant, this species has recently invaded open parkland areas in large parts of north-western Europe. Peitzmeier (1951) has collected good evidence that the parkland form did not originate from local stock but from a parkland population in northern France that spread in north-easterly direction with a speed of about 5–9·5 kilometres per year until it reached the North Sea shore in northern Germany. As a consequence, two separate sub-populations with a mosaic pattern of distribution are at present to be found. In a detailed discussion of the relevant data, Peitzmeier comes to the conclusion that the respective preferences of the sub-populations have to be regarded as a result of habitat imprinting. The occurrence of ecologically distinct sub-populations has likewise been described for a number of other species (cf. Stresemann 1943; Peitzmeier 1949; Löhrl 1965). In these cases, similar mechanisms may have allowed for fast and stable adaptations to newly colonized habitats. The same probably applies to the rather quick colonization of urban habitats that occurred in several species of birds (cf. Erz 1966; Braestrup 1968).

11.4. The role of reproductive isolation

In order to maintain specific adaptations to certain types of environmental conditions, selection will favour a restriction of gene-exchange between the sub-populations. In establishing and supporting genetic isolation, early experience may once more exert an important influence. As a first step, imprinting to the same sort of habitat may promote pair formation between individuals of similar origin. If, in addition, the sub-populations also differ in some of the species-specific patterns for mutual recognition (for example, vocalizations and appearance), sexual imprinting on these characters may further contribute to achieving a sub-division of gene-pools by means of non-random mating.

In passerine birds, a widespread mechanism for separating sub-populations and reducing intra-specific gene flow is provided by the occurrence of song dialects. Marler and Tamura (1962) have suggested that a system of song dialects may act in two possible ways: through assortative mating due to a stronger reaction of females to the dialect of their own sub-population, as well as by means of an *ortstreue* effect due to the fact that the young bird will be attracted to breed in areas where the familiar dialect is to be heard. Dialect groups have repeatedly been found to differ also in a number of other characteristics, and thus to represent true sub-populations with slightly different ecological adaptations (King 1972; Nottebohm and Selander 1972). As the characteristics of the song, as a rule, are learned during youth and adolescence, early experience again participates in achieving sexual isolation between such sub-populations. A detailed discussion of the possible isolating function of dialects and sub-dialects is provided by Nottebohm (1969) and Baptista (1975).

Apart from local separation of sub-populations, a mating system partly based on early experience may also lead to restriction of gene-exchange within a population or sub-population and may thus contribute to maintaining adaptive polymorphism or polyethism within any one area. (The occurrence of polymorphism in birds has been discussed in detail by Selander (1971).) A well-known example is provided by the North-American Snow Goose, *Anser caerulescens*, which in its smaller sub-species is polymorphic and has two colour phases, a blue form and a white form. In mixed populations, birds of the same morph mate with one another more frequently than with birds of the opposite morph (Cooch and Beardmore 1959; Cooke and Cooch 1968). This system of positive assortative mating is based on sexual imprinting, as the males select mates which have plumage patterns similar to those of one of their parents (Cooke, Mirsky, and Seiger 1972). In the case of the Snow Goose, the two forms do not differ only in plumage colour; there are also slight differences in breeding biology, the white morph being better adapted for a rigorous environment. Under the present amelioration in climate, the blue phase is favoured and rapidly increases in numbers; but altogether the situation presents a system of balanced polymorphism, which—according to the climate—in some seasons or decades favours the blue morph and in others the white one (Cooch 1961).

Although no relevant data about the origin and extent of non-random mating are at present available a similar situation seems to exist in other polymorphic species. Apart from the Lesser Snow Goose, assortative mating has been described in the dark and light colour forms of the Arctic Skua, *Stercorarius parasiticus* (Bengtson and Owen 1973). As there is a well-defined cline in the frequency of the two morphs, the light form being more common in the north and the dark morph in the south, similar differences in ecological adaptations as in the white and blue Snow Geese are to be expected. Further examples for non-random mating have been described for a number of birds and mammals. Its significance for increasing the genetic variance of a population has been discussed by Bengtson and Owen (1973).

11.5. Influence on speciation

Altogether, it can be concluded that early experience has two possible evolutionary functions: initially it leads to the formation of habitat and other ecological preferences in any one individual through environmental imprinting; as soon as natural selection has led to the evolution of slightly different gene-pools adapting groups of individuals to local conditions, it also serves to preserve such gene-pools by means of sexual imprinting. This will finally lead to the division of populations into a continuous or mosaic system of sub-populations with habitat-linked differences in various characters: clutch size (Klomp 1970), body size (King 1972), agonistic, exploratory, and nest-building behaviour (Oortmerssen 1970), and many other characteristics.

The importance of differential habitat preferences as a cause of speciation has been stressed by many authors. According to Miller (1947) a group of small, partly isolated sub-populations, as created by such preferences, 'is the most favourable situation for developing new combinations of genes with, occasionally, new and perhaps high selective values'. If genetical isolation between sub-populations becomes more and more complete, the ecological preferences that initially were based on early experience, together with sexual imprinting, will finally be able to initiate speciation. An example of a borderline case in this direction is provided by the parasitic African Combassous, *Hypochera*. The taxonomy of this genus has always been a puzzle to systematists, because there are several forms that exist sympatrically without interbreeding but none the less have been shown to intergrade through intermediate 'races'. The sympatric forms have been found to be separated by ethological barriers on the basis of adaptations to different hosts which, as Nicolai (1964) has proved, are the result of vocal mimicry and partial imprinting on the host species. Obviously, therefore, early experience leads to a separation of different 'host populations', which in some cases have reached (almost) complete isolation. This in turn has promoted the development of slight morphological differences. The fact that there is no general agreement as to the number of recognizable species clearly indicates that several of these 'forms' are just at the point of reaching species level (for reviews, see Traylor 1966; Nicolai 1967).

The speed with which complete isolation will be achieved depends on whether imprinting is absolute or only partial. Mathematical models of the influence of imprinting on population structure, and of its possible evolutionary consequences, have been developed by O'Donald (1960), Kalmus and Smith (1966), Mainardi, Scudo, and Barbieri (1965), Scudo (1967), and Seiger (1967). The question of whether imprinting involves one or both sexes may also be of importance. If, as mentioned for the Zebra Finch (§11.2), only the male is rigidly imprinted on the characters of the mother whereas the female partly relies on innate preferences, a mutation in the appearance of the female sex can be expected to assert itself quicker within a population than a similar alteration in male characters. This difference may be partly concealed, however, if pair formation is initiated exclusively by the male. (A more detailed discussion of the role of sexual selection is given in Immelmann (1970).)

Apart from promoting adaptive polymorphism and polyethism within a species and from reducing gene-flow between different sub-populations, sexual imprinting, owing to its precise determination of subsequent mating preferences, does of course continue to act on the species level and contributes to maintaining barriers to hybridization in sympatric species (Immelmann 1970).

It has to be stressed, finally, that many of the above considerations still have more or less speculative character as the majority of experimental investigations on imprinting have been carried out in the laboratory, and long-range studies under natural conditions are still very few. Many conclusions, therefore, are drawn only from the comparison of laboratory findings with those field data that point in the same direction. The exact degree to which environmental imprinting and mating preferences based on sexual imprinting are important in speciation remains, in most cases, still to be shown.

One interesting correlation, however, has to be mentioned. From the evidence presented above it appears that wherever rapid, stable learning is of great adaptive value, natural selection will favour the development of imprinting and similar processes. Such value can be expected to be greatest in any rapidly evolving group as well as where several closely related species of similar appearance occur in the same region. Interestingly enough, both statements do seem to apply at least to all groups of birds in which imprinting has been found to be a widespread phenomenon (ducks, gallinaceous birds, pigeons and doves, and estrildid finches). Perhaps the occurrence of imprinting represents some sort of a pre-condition for rapid and extensive adaptive radiation, at least in vertebrates. The same opinion has been expressed by Thielcke (1970) with regard to the role of song learning in birds. It is to be expected that closer analysis of the great faunal diversity of tropical regions will reveal many examples of similar kind.

11.6. 'Accidents'

The great ecological and evolutionary significance of early learning processes is a consequence of their main distinguishing character, their strong resistance to subsequent alteration. Apart from its ecological advantages and its 'positive' influence on sexual isolation described so far, such stability may occasionally also work in the opposite direction. This happens if the young animal is exposed to a 'wrong' environment during its sensitive period.

Most examples of an 'accident' like this are of the sexual-imprinting type. Under laboratory conditions, sexual attachments of numerous species have been observed to be directed towards another species or—in domesticated forms—towards another colour variety of the same species as a consequence of early experience with that particular species or colour variety (for review, see Immelmann 1972b). Similar mating preferences, however, have also been produced under completely natural conditions, if clutches of two closely related species are exchanged. In the Herring and

Lesser Black-backed Gulls, *Larus argentatus* and *L. fuscus*, this led to the formation of mixed pairs (Harris 1970), and in the House and Tree Sparrows, *Passer domesticus* and *P. montanus*, several cases of actual hybridization have been described (Cheke 1969).

Although interspecific mating preferences of gulls and sparrows occurred under natural conditions, the previous cross-fostering of the individuals concerned was still induced artificially by the exchange of eggs. Similar 'exchange experiments', however, do also occur without human interference. In the arctic Snow and Ross Geese, *Anser caerulescens* and *A. rossii*, competition for nesting-sites in large mixed colonies has been observed to result in mixed clutches. Young hatched in these nests obviously become sexually imprinted to the wrong species, and this has led to frequent interbreeding between the two species; not less than 1400 hybrids are estimated to have been born annually in recent years. They are known to be fully fertile with each other and the parent species (Trauger, Dzubin, and Ryder 1971).

It can be concluded that under certain circumstances—due, for example, to recent changes in the distribution and abundance of a species that provide increased opportunities for interbreeding, or due to the rarity of conspecific mates in the marginal areas of distribution—sexual imprinting, as a consequence of its very rigid fixation of mate preferences, may occasionally also contribute to facilitate the breakdown of isolating mechanisms in closely related species and may thus have a clearly 'negative' effect upon speciation.

References

BAERENDS, G. P., BRIL, K. A., and BULT, P. (1965). Versuche zur Analyse einer erlernten Reizsituation bei einem Schweinsaffen. *Z. Tierpsychol.* **22**, 394–411.

BAPTISTA, L. (1975). Demes, dispersion and song dialects in sedentary populations of white-crowned Sparrows (*Zonotrichia leucophrys nuttalli*). *Univ. Calif. Publs Zool.* **105**. (In press)

BEACH, F. A. and JAYNES, J. (1954). Effects of early experience upon the behavior of animals. *Psychol. Bull.* **51**, 239–63.

BENGTSON, S. A. and OWEN, D. F. (1973). Polymorphism in the Arctic Skua *Stercorarius parasiticus* in Iceland. *Ibis* **115**, 87–92.

BRAESTRUP, F. W. (1968). Evolution der Wirbeltiere. Ökologische und ethologische Gesichtspunkte. *Zool. Anz.* **181**, 1–22.

BURGHARDT, G. M. (1967): The primacy effect of the first feeding experience in the Snapping Turtle. *Psychon. Sci.* **7**, 383–4.

CATCHPOLE, C. K. (1972). A comparative study of territory in the Reed warbler (*Acrocephalus scirpaceus*) and Sedge warbler (*A. schoenobaenus*). *J. Zool.* **116**, 213–31.

CHEKE, A. S. (1969): Mechanism and consequences of hybridization in sparrows *Passer. Nature, Lond.* **222**, 179–80.

COLLIAS, N. E. (1950): Social life and the individual among vertebrate animals. *Ann. N.Y. Acad. Sci.* **51**, 1074–92.

COOCH, G. (1961): Ecological aspects of the Blue Snow Goose complex. *Auk* **78**, 72–89.

—— BEARDMORE, J. A. (1959): Assortative mating and reciprocal difference in the Blue-Snow-Goose complex. *Nature, Lond.* **183**, 1833–4.

COOKE, F. and COOCH, F. G. (1968). The genetics of polymorphism in the goose *Anser caerulescens. Evolution* **22**, 289–300.

—— MIRSKY, P. J., and SEIGER, M. B. (1972). Color preferences in the lesser snow goose and their possible role in mate selection. *Can. J. Zool.* **50**, 529–36.

DENENBERG, V. H. (1963): Early experience and emotional development. *Scient. Am.* **208**, 138–42.

ERZ, W. (1966). Ecological principles in the urbanization of birds. *Ostrich, Suppl.*, No. **6**, 357–63.

GOTTLIEB, G. (1965). Imprinting in relation to parental and species identification by avian neonates. *J. comp. Physiol. Psychol.* **59**, 345–56.

HARLOW, H. F. and HARLOW, M. K. (1962). Social deprivation in monkeys. *Scient. Am.* **207**, 137–46.

HARRIS, M. P. (1970). Abnormal migration and hybridization of *Larus argentatus* and *L. fuscus* after interspecific fostering experiments. *Ibis* **112**, 488–98.

HESS, E. H. (1962). Imprinting and the 'critical period' concept. In *Roots of behavior* (ed. E. L. Bliss). Harper, New York.

HILDÉN, O. (1965). Habitat selection in birds. *Ann. Zool. Fenn.* **2**, 53–75.

IMMELMANN, K. (1970). Zur ökologischen Bedeutung prägungsbedingter Isolationsmechanismen. *Verh. zool. Ges., Köln 1970*, pp. 304–14.

—— (1972a). The influence of early experience upon the development of social behaviour in estrildine finches. *Proc. int. orn. Congr.* **15**, 316–38.

—— (1972b). Sexual and other long-term aspects of imprinting in birds and other species. *Adv. Study Behav.* **4**, 147–74.

KALMUS, H. and SMITH, S. M. (1966). Some evolutionary consequences of peg-matypic mating systems (imprinting). *Am. Nat.* **100**, 619–35.

KING, J. R. (1972). Variation in the song of the Rufous-collared Sparrow, *Zonotrichia capensis*, in northwestern Argentina. *Z. Tierpsychol.* **30**, 344–73.

KLOMP, H. (1970). The determination of clutch-size in birds. *Ardea* **58**, 1–124.

KLOPFER, P. (1965). Habitat selection in birds. *Adv. Study Behav.* **1**, 279–303.

LEMON, R. E. and HERZOG, A. (1969). The vocal behavior of Cardinals and Pyrrhuloxias in Texas. *Condor* **71**, 1–15.

LILEY, N. R. (1966). Ethological isolating mechanisms in four sympatric species of Poeciliid fishes. *Behaviour, Suppl.* **13**, 1–197.

LÖHRL, H. (1959). Zur Frage des Zeitpunktes einer Prägung auf die Heimatregion beim Halsbandschnäpper (*Ficedula albicollis*). *J. Orn. Lpz.* **100**, 132–40.

—— (1965). Zwei regional und ökologisch getrennte Formen des Trauerschnäppers (*Ficedula hypoleuca*) in Südwestdeutschland. *Bonn. Zool. Beitr.* **16**, 268–83.

LORENZ, K. (1943). Die angeborenen Formen möglicher Erfahrung. *Z. Tierpsychol.* **5**, 235–409.

MAINARDI, D., SCUDO, F. M., and BARBIERI, D. (1965). Assortative mating based on early learning: population genetics. *Aten. Parm* **36**, 583–605.

MARLER, P. and TAMURA, M. (1962). Song 'dialects' in three populations of White-crowned Sparrows. *Condor* **64**, 368–77.

MILLER, A. H. (1947). Panmixia and population size with reference to birds. *Evolution* **1**, 186–90.

MULLIGAN, J. A. (1966). Singing behavior and its development in the song sparrow *Melospiza melodia*. *Univ. Calif. Publs. Zool.* **81**, 1–76.

NICOLAI, J. (1964): Der Brutparasitismus der Viduinae als ethologisches Problem. *Z. Tierpsychol.* **21**, 129–204.

—— (1967): Rassen- und Artbildung in der Viduinengattung *Hypochera*. *J. Orn. Lpz.* **108**, 309–19.

NOTTEBOHM, F. (1969). The song of the Chingolo *Zonotrichia capensis* in Argentina: description and evaluation of a system of dialects. *Condor* **71**, 299–315.

—— SELANDER, R. K. (1972). Vocal dialects and gene frequencies in the Chingolo Sparrow (*Zonotrichia capenis*). *Condor* **74**, 137–43.

O'DONALD, P. (1960). Inbreeding as a result of imprinting. *Heredity* **15**, 79–85.

VAN OORTMERSSEN, G. A. (1970). Biological significance, genetics and evolutionary origin of variability in behaviour within and between inbred strains of mice (*Mus musculus*). *Behaviour* **38**, 1–92.

PEITZMEIER, J. (1949). Über nichterbliche Verhaltensweisen bei Vögeln. In *Ornithologie als biologische Wissenschaft* (eds E. Mayr and E. Schüz), Carl Winter Universitätsverlag, Heidelberg.

—— (1951). Zum ökologischen Verhalten der Misteldrossel (*Turdus v. viscivorus* L.) in Nordwesteuropa. *Bonn. zool. Beitr.* **2**, 217–24.

SALZEN, E. A. (1966). Imprinting in birds and primates. *Behaviour* **28**, 232–54.

SCHUTZ, F. (1964). Über geschlechtlich unterschiedliche Objektfixierung sexueller Reaktionen bei Enten im Zusammenhang mit dem Prachtkleid des Männchens. *Verh. dt. zool. Ges. München* 1963, pp. 282–7.

SCUDO, F. M. (1967). L'accoppiamento assortativo basato sul fenotipo di parenti: Alcune conseguenze in popolazioni. *Rc. Ist. lomb. Sci. Lett.* **B101**, 435–55.

SEIGER, M. B. (1967). A computer simulation study of the influence of imprinting on population structure. *Am. Nat.* **101**, 47–57.

SELANDER, R. K. (1971). Systematics and speciation in birds. In *Avian biology* (eds D. S. Farner and J. R. King), Vol. 1, pp. 57–147. Academic Press New York.

STRESEMANN, E. (1943). Oekologische Sippen-, Rassen- und Artunterschiede bei Vögeln. *J. Orn. Lpz.* **91**, 305–24.

THIELCKE, G. (1970). Lernen von Gesang als möglicher Schrittmacher der Evolution. *Z. zool. Syst. Evolutionsforschung* **8**, 309–20.

THORPE, W. H. (1959): Learning. *Ibis* **101**, 337–53.

—— JONES, F. G. W. (1937). Olfactory conditioning in a parasitic insect and its relation to the problem of host selection. *Proc. R. Soc.* **B124**, 56–81.

TRAUGER, D. L., DZUBIN, A., and RYDER, J. P. (1971). White geese, intermediate between Ross' Geese and Lesser Snow Geese. *Auk* **88**, 856–75.

TRAYLOR, M. A. (1966). Relationships in the combassous (sub-gen *Hypochera*). *Ostrich, Suppl.*, No. **6**, 57–74.

12. On strategies of behavioural development

P. MARLER

OF THE problems confronting the student of behaviour that demand attention in the next decade or so, none is more pressing than the understanding of behavioural development. Many investigators feel that the proper philosophical balance has been struck between genetic determinants of behaviour on the one hand, and environmental determinants on the other. But we have hardly begun to appreciate how, in fact, these two sets of influences interact with one another to achieve the extraordinary diversity of the natural behaviour of animals.

Perhaps most urgent is our need to comprehend the interaction that underlies learning. Learning is widespread in behavioural ontogeny, perhaps universal if one's definition is loose enough to include all environmental influences. But it is also clear that no organism can properly be thought of as a *tabula rasa*, approaching learning tasks as a totally free agent, without constraints. As a first consideration, learning cannot take place without a certain kind of machinery, and that machinery cannot develop without genetic guidance. I believe that genetic constraints on learning will prove to be more pervasive than just the provision of the basic machinery. Despite reminders from Tinbergen (1951, 1963, 1968), Lorenz 1969) and others that 'there seem to be more or less strictly localized dispositions to learn' the subject is still largely neglected by students of animal learning, with a few notable exceptions (Garcia, McGowan, and Green 1972; Seligman and Hager 1972).

If any species relies on a complex set of learned traits as the fundaments upon which its ecology and behaviour are constructed, traits around which its society is designed, it would be surprising is the form which learning takes, its timing, and the direction in which it most readily occurs in generating these patterns of behaviour should be left only to chance.

I shall try to illustrate this point by some studies of bird-song development which I believe indicate a fruitful approach to this problem. But I shall start not with laboratory experiments on ontogeny but rather with studies in the field. This is not an accidental choice. With any evolutionary problem the best source of hypotheses is an appreciation of the context in which the behaviour has evolved, both environmental and social, as has so often been demonstrated by the work of Niko Tinbergen. Perhaps the most serious indictment of psychological studies of animal learning is their divorce from an understanding of the problems that confront their subjects in nature. The fact that their subject is typically the Norway Rat and its domesticated descendants need impose no restriction here, as Garcia has so nicely demonstrated (Garcia, McGowan, and Green 1972). Another criticism, to my mind, is an unduly myopic focus on phylogenetic series, a point I hope to make by starting with a few remarks about primates.

12.1. Monkey vocalizations and their development

Plate 12.1 (p. 256) and Fig. 12.1 show sound spectrograms of four sounds I recorded a few years ago in Africa (Marler 1973). They were given by two species of monkey that live together in close proximity in the rain forest of East Africa. I use them to illustrate what are I believe selection pressures for different rates of evolutionary change in social signals with different functions. The first two are given only by adult males, the 'pyow' of the Blue Monkey and the 'hak' of the Red-tailed Monkey. The second two are 'chip' calls of females and sub-adults, first of the Blue then of the Red-tailed Monkey. The adult male calls, the 'pyow' and the 'hak', are easy to distinguish, while the 'chip' calls are difficult. Although measurements establish significant differences between the 'chips', they are less striking than those that separate the male loud calls of the two species.

A consideration of the different functions served by these two pairs of calls suggests that the members of one pair differ so much more from each other than those of the other pair do because the two species have interacted with one another in the past (and indeed they do still live in very close proximity). Standing at one place in the forest, one could readily hear all four sounds within a period of a few moments. Going through my field notes I find, for example, that of 129 initial sightings of Blue Monkeys in the forest, Red-tailed Monkeys were present in the same tree or group of trees on 61 occasions—that is, almost half the time. They spend much of the time within earshot of one another, and are active at the same time of day. They are vulnerable to the same predators, the most serious probably being man, leopards, and eagles.

The 'chip' is an alarm call. On many occasions I noted that monkeys of one species were responding to alarm calls of the other, becoming alert, starting to call themselves and moving into cover. If we imagine the two species of monkey as arising from a common ancestor it is a reasonable hypothesis that the slow rate of divergence of 'chip' calls during evolution, implied by these data, correlates with the incidence of inter-specific communication of danger.

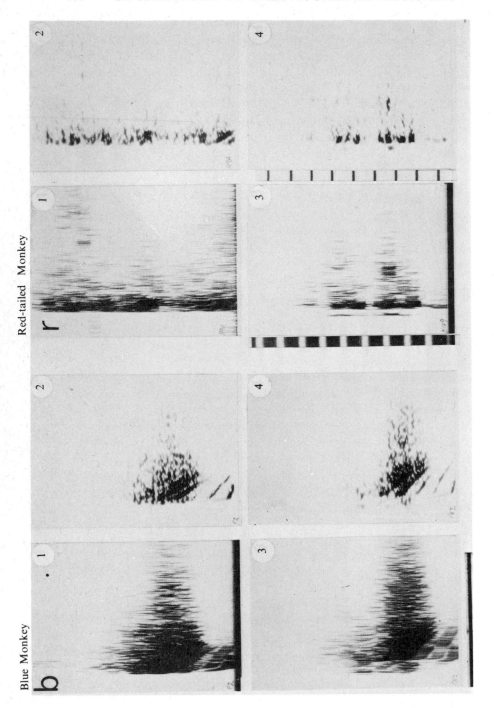

PLATE 12.1. A comparison of the adult male loud calls of the Blue Monkey, *Cercopithecus mitis*, and the Red-tailed Monkey, *Cercopithecus ascanius*. Two examples of each are given, and each example is given in both wide-band (300 Hz on the left) and narrow-band (40 Hz, right) analyses. The frequency scale is marked in 500 Hz intervals and the time marker is 0·5 seconds in duration.

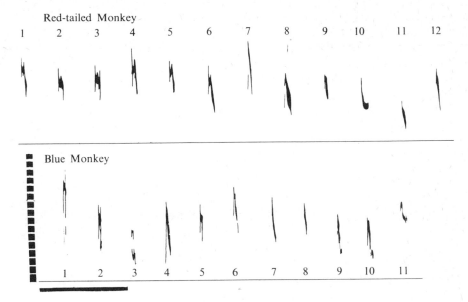

FIG. 12.1. 'Chip' calls of the Red-tailed Monkey and of the Blue Monkey (bottom row) in Uganda. This is an assemblage of calls from many individuals. Frequency scale is in 500 Hz intervals and the time marker is 0·5 seconds.

A comparison of functions of the adult male 'pyow' and 'hak' calls suggests a different history. These monkeys live in groups of about 10–15 animals with a single adult male (Aldrich-Blake 1970). At least in Blue Monkeys, there is evidence that each group maintains a territory, the adult male probably taking the initiative in its defence. Two functions have been suggested for the loud calls of the adult males: (1) that they serve to maintain distance between groups, and (2) that they serve to rally the group prior to movement. In both cases the interactions are primarily intra-specific. Confusion would result if the calls of the two species were hard to discriminate. I suggest that they have been selected in the course of evolution for divergence, resulting in a more rapid rate of evolutionary change than has occurred with the 'chip' calls. Here, of course, there are implications for taxonomy, but I am more concerned with patterns of ontogeny. Although we know nothing of the details in this case, one might presume that, since the differences are restricted to adult males, there was at some stage in the history of separation of these two species an ontogenetic change in the pattern of responsiveness of a target tissue to testicular androgens, perhaps growing tissues in their throats. Special sacs seem to endow the calls of adult male *Cercopithecus* monkeys with their distinctive tonal quality (Gautier 1971). The ontogeny of the 'chip' calls has evidently been left largely unchanged.

The point I am working towards is that a new need for signal diversity can only be met by a change in patterns of behavioural development. I am thinking of signal diversity here within a complex of closely related and sympatric species. A similar notion is relevant to signal diversity within a species, or within an individual. Here I want to focus on the circumstances that will call for relatively rapid evolutionary change in signal morphology to meet the demands of specific distinctiveness between sympatric species. In these monkeys the demand is not particularly great, for relatively few monkey species live within earshot of one another, and other organisms that are present make different types of sounds. I suggest that more extreme selective pressure for signal diversity in a community can call for more drastic revisions of developmental strategy, leading, for example, to an increase in the dependence on learning. With the monkeys there is no evidence that the adult male calls are dependent on learning in development, any more than the other sounds in the repertoire, although it must be admitted that there is little evidence to go on.

The contrast I have in mind is with birds. Again there is evidence that alarm calls may diverge relatively slowly as species evolve, with a gain thereby in mutual inter-specific benefit. The converse tends to be the case with the male song, where reproductive isolation and conspecific territorial defence place a premium on specific distinctiveness. But, whereas *Cercopithecus* monkeys may have 3 or 4 close relatives in sympatry to generate selective pressure for change in the male calls, the adaptive radiation of song-birds is such that, with from 30–50 vocal relatives evolving and co-habiting in the same habitat, there must have been an extraordinary demand for a rapid rate of evolutionary change in the male songs.

I believe that this may explain the revolution that took place at some point in the history of oscine birds, such that learning from adults came to make a major contribution to the ontogeny of the song (Marler and Mundinger 1971). If we see this occurring in birds rather than in non-human primates I think that this is a result of varying demands for signal diversity in various organisms and not to any basic difference in intelligence. I shall suggest later that the demand for signal diversity that emerged in man as he began naming objects and naming individuals with different roles in his societies might have led to a somewhat analogous revolution in ontogenetic strategy.

12.2. Avian vocal learning

Again what I have to say about bird-song development begins in the field. It is no accident that two of the prime subjects in study of bird-song learning—the chaffinch and the White-crowned Sparrow—were already known as classical examples of song dialects, such that, even within a restricted area, adjacent local populations can be distinguished by consistent variations in some properties of the song (Marler and Tamura 1962).

This led to their selection as subjects for experimental study, confirming what had been hinted at long before in the avicultural literature, that a male raised in social isolation develops a song unlike any to be heard from that species in the field.

It is important to realize how unusual this result is. Most of the basic patterns of motor co-ordination that characterize the natural behaviour of animals develop normally in social isolates. This remarkable fact still deserves more emphasis than it receives, notwithstanding its almost axiomatic status in ethological theory for many years. In this perspective the abnormality of song in socially isolated birds assumes a special interest. The abnormalities can take several forms, and I shall illustrate some of them briefly from studies in which my colleagues and I have been involved.

Fig. 12.2(a) shows a typical example of male song from the *nuttali* race of

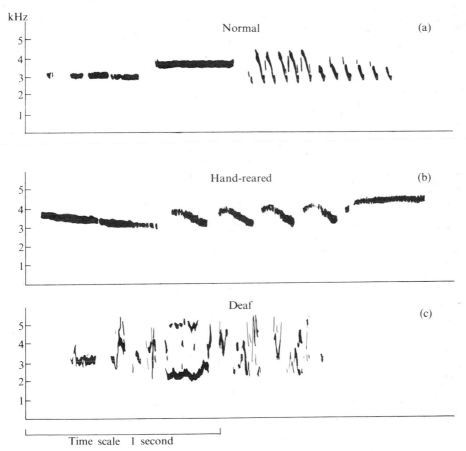

FIG. 12.2. Examples of the songs of male White-crowned Sparrows developing under three conditions: (a) as a wild bird; (b) hand-reared in acoustical isolation from normal song from about 4 days of age; (c) deafened in youth. (After Konishi 1965.)

the White-crowned Sparrow. This particular bird lived around Berkeley, to judge by the structure of the syllables in the second part of the song. The spectrogram in Fig. 12.2(b) is a typical example of a male White-crowned Sparrow song raised in social isolation. This individual was taken at 5 days of age and raised in a large sound-proof room with 9 other males of similar age, from several areas. The song of an individually isolated male is similar. The second or trill portion of the song is virtually absent. The sustained tones that characterize the introduction to the song are present, however,

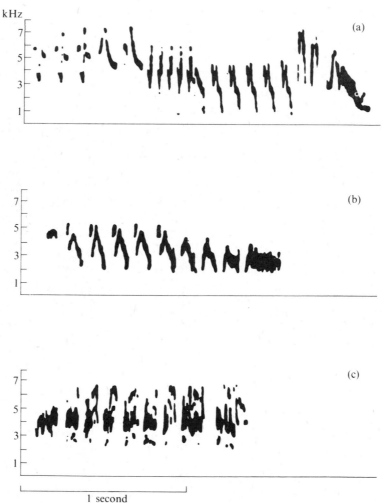

Fig. 12.3. Songs of male chaffinches developing under three conditions; (a) in the wild; (b) hand-reared in acoustical isolation after a few days of age, (after Thorpe 1958); (c) deafened at about 90 days of age (after Nottebohm 1968). The time marker indicates 1 second and the frequency scale is marked in 1 kHz intervals.

with a normal spectral composition. To the extent that a second portion is generated, it results from the rough fragmentation of the tones. The result is a song unlike any to be heard from a White-crowned Sparrow in nature (Marler 1970). However, there are definitely some normal traits, a point to which I will return.

We know from the work of W. H. Thorpe (1958, 1961) that the result of this experiment with the European Chaffinch is rather similar (Fig. 12.3). The dialects in the male song are characterized by the details of syllabic structure, especially that in the terminal phrase. In a social isolate, here a male taken from the nest at about 4 days and raised in individual isolation thereafter, a simpler song is typical. The syllable morphology is greatly simplified, the number of types is reduced, and the stepwise changes in pattern through the course of the song are lost. The result is a much simpler pattern than the natural song of European Chaffinches. Nevertheless, as in the White-crowned Sparrow, there are some normal traits. The duration is of the right order, and there is a tendency to terminate the song with a distinct and often strident syllable, reminiscent of the terminal phrase. To the ear at least, there is something normal in the tonal quality. This again is a point to which I will return.

Now a slightly different example from a recent study on song development in the Red-winged Blackbird (Plate 12.2). At the top are two examples of the major patterns we found in the normal song of male Red-wings in the New York area. We call one the slow type and one the fast type, with a rapid frequency modulation pulse train that gives the song its distinctive tonal quality. Below are some examples of the very variable songs produced by male birds taken from the nest at 5 days, kept in groups with brothers and sisters until 40 days, and then individually isolated. Five males that we raised in this way each developed a repertoire of 3 song types, 15 different patterns in all. With two possible exceptions, none showed the typical division into a short simple introductory phrase followed by a longer trill. The frequency–time envelope of many is conspicuously different from the wild sample which was quite homogeneous in this regard. Their average frequency usually increased from the first to the second part of the song and was never seen to decrease in our sample as occurred in several of the isolates. The over-all patterning of the isolates' songs is thus simpler than that of their wild counterparts (Marler, Mundinger, Waser, and Lutjen 1972).

In spite of the abnormal and variable song patterns of these socially isolated Red-winged Blackbirds, as in the other species, we can discern some normal traits. Each of the 5 birds had at least one song with a rapid frequency-modulated vibrato, giving the twangy quality of the natural song. A few of their songs resemble somewhat the slow type of normal trill. In general, it seems that a male raised under these conditions has the

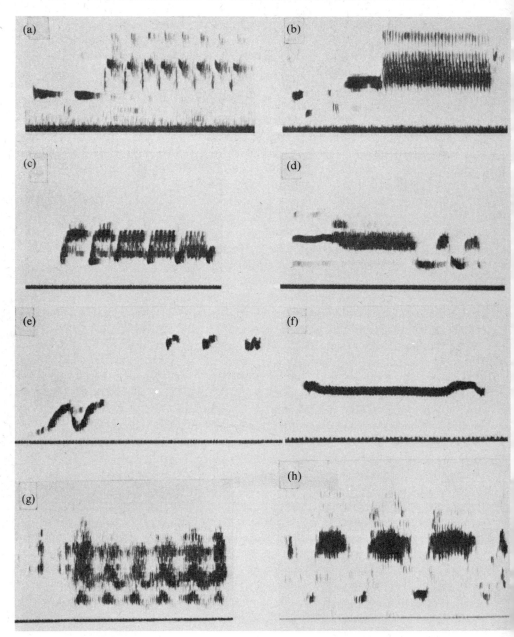

PLATE 12.2. Songs of male Red-winged Blackbirds developing either in the wild ((a) and (b)), illustrating two types of natural song in the New York area, or after hand-rearing in a group from about 5 days to 40 days of age and auditory isolation thenceforth (c)–(f). The latter illustrates a small selection from the many song types produced by birds raised with this degree of auditory isolation. (g) and (h) illustrate songs of birds deafened in youth.

capacity to produce the basic ingredients for normal song, certainly the fast trill, in some cases the slow trill as well, and some sounds that resemble the introduction. But they show no preference for assembling these ingredients in a normal manner. In addition, highly abnormal sound patterns some- times intrude.

One point of phylogenetic interest in these experiments is the possibility that the song of a social isolate might be sufficiently changed as to come to resemble the song of another species. There is some hint of this in the White-crowned Sparrow, where the rather simple song of a social isolate perhaps comes to resemble that of the White-throated Sparrow, for example. The juncos illustrate this rather better.

We have done a small number of experiments on song development in juncos which turned out to be interesting and deserving of more study. Field-work showed that the song of the Arizona Junco both in Arizona and in Mexico is elaborate. We were not able to detect any dialects. Neighbours did not seem to share many characteristics and the structure of the compo- nent syllables, of which there were always several, is quite complex (Fig. 12.4). This is different from the song of some of the other junco species, particularly another we studied, the Oregon Junco, where the song typi- cally consists of repetitions of a single syllable (Konishi 1964a). Only very rarely do songs contain more than one syllable, so there is almost no overlap between the Arizona Junco and the Oregon Junco in this regard. In the Arizona Junco there are almost always several syllable types, and in the Oregon Junco there is almost always only one.

On the right of Fig. 12.4 are illustrated 6 song types developed by a male Arizona Junco taken as a nestling and raised in individual isolation with no further exposure. Only one of these seems normal in the sense that it falls within the array of patterns that we encountered in nature. Half of them—3 out of 6—consisted of repetitions of a single syllable type and are really very similar to songs of the Oregon Junco. This then is another kind of abnormality which is found in social isolates, removing them largely or entirely from the normal range of patterns and in this case bringing them rather close to the song of a related species (Marler 1967).

The last case I want to mention comes from the work of Mulligan on song development in the Song Sparrow, a close relative of the juncos and the White-crowned Sparrow. Although the number of subjects is small, the behaviour of an acoustically-isolated Song Sparrow seems to be quite different. Mulligan raised some from the egg, inducing canaries to serve as foster parents, so the Song Sparrows had no opportunity to hear adult song. Two such birds developed remarkably normal songs. It is true that the song repertoire was reduced, 8 song types in one and 9 in the other, as compared with 12–20 in wild birds. But the number of different syllable types per song were similar to those in wild birds and, from Mulligan's reports, they

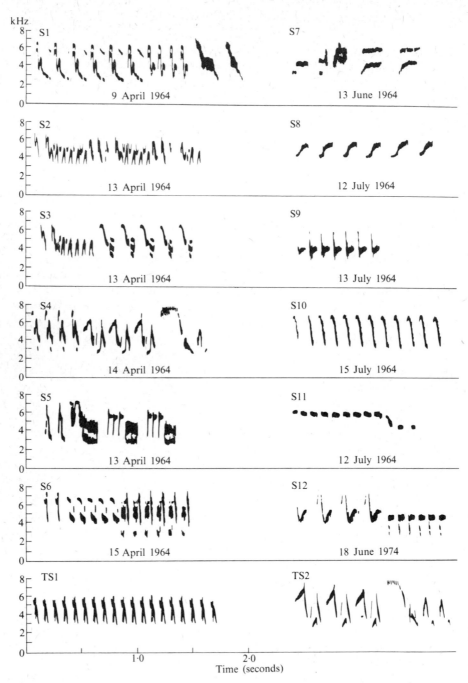

FIG. 12.4. A comparison of 6 song-types developed by a male Arizona Junco reared in auditory isolation from a few days of age (S7–S12) with the 6 song types produced by another male (S1–S6) after playback with recordings of a normal Oregon Junco song (TS1) and an Arizona Junco song (TS2). For details see the text.

seem to be acceptable as normal Song Sparrow song. A similar result was obtained with one male taken as a nestling at 6 days and raised in individual isolation. His 5 song types were again normal (Mulligan 1966). An ornithologist would unhesitatingly identify the songs of these socially-isolated Song Sparrows as coming from that species. So here is a quite different result from a bird so close to the White-crowned Sparrow that some have suggested they should be congeneric.

12.3. Use of the term 'innate'

It is interesting to consider briefly the proper approach to this range of results from essentially the same kind of isolation experiment. The classical ethological approach would be to characterize the song of the Song Sparrow as innate and the song of the other species, the junco, the Red-winged Blackbird, the chaffinch, and the White-crowned Sparrow as non-innate, acquired or learned. I interpret the position of Konrad Lorenz on this kind of question to be somewhat that of a systematist, as in much of his writing on the fixity of motor patterns and their applicability to taxonomic problems. It is clearly essential if one is contemplating the use of traits in systematic and phylogenetic analysis to distinguish between those which are variable within a species and those stable enough to be thought of as characterizing the species.

If, however, one's interest is not so much in taxonomic significance but rather in patterns of behavioural development—in understanding which events taking place in the course of ontogeny are responsible for the differences which finally result in the behaviour of the adults of two species—then this approach may be less appropriate, as Tinbergen (1963, 1968) has indicated.

To my mind the results of the experiments on bird-song development illustrate the problem rather well. In the first place, if we label one pattern of behaviour as innate and another as acquired, this distinction seems to imply a rather fundamental, far-reaching difference in the ontogenetic histories of the two traits. This would be puzzling in the present case, where we are comparing such close relatives as two species of sparrows. This is the first kind of discomfort that I feel about using the distinction between innate and acquired behaviour as a means of entry into problems of ontogeny.

A second point is that once you begin to look into particular cases it becomes difficult to decide on which side of the fence a given trait should be placed. Consider the Arizona Junco (Fig. 12.4). Social isolates develop a song which is much simpler than that of the wild individuals, often a trill composed of a single syllable type, rather reminiscent of the song of the Oregon Junco. One would infer on the basis of this experiment alone that the song of the Arizona Junco is not innate and is, presumably, acquired. The question then (as in so much of experimental biology) is what must you, as an experimenter, add to the environment of a socially isolated bird to restore the normal pattern of development? Obviously one presents, to the socially-isolated male, playback of a recording of a natural song. In Arizona Juncos it is found that this is sufficient by itself to re-establish normal development, as is illustrated on the left-hand side of Fig. 12.4. The bird studied here was given playback each day for 2 months with 23 minutes of song, starting about 3 weeks after fledging; half Arizona Junco song, half Oregon Junco song was used. The bird developed 6 song types, all of which were more or less normal. There were none of the simple, one-syllable type that made up half of the repertoire of the untrained isolate.

We might infer that normal song is indeed acquired. However, if you look closely at this result, there is little detailed resemblance in syllabic structure between most of the songs developed and the model presented.

We are reminded of the rarity with which wild neighbours have syllabic structure in common in the Arizona Junco. This suggested a further experiment in which the condition of a social isolate was again changed, but in a much simpler fashion, by putting the young males together in a group. Again none of them had any opportunity to hear adult song after they were taken at 5 or 6 days. But the songs that they developed were quite different from those of individual isolates. Five such birds developed 21 song types, only one of which was a simple trill. The patterning of all of the others would be acceptable as normal wild songs. The correspondence was quite remarkable. Again there was little evidence of detailed imitation of one bird by another (Marler 1967).

The inference I draw from this result is that social stimulation is indeed important in song development in the Arizona Junco. But it does not necessarily take the form of direct imitation of adults by young. It seems to be a matter of more generalized stimulation, evoking a greater degree of vocal invention, or improvization, as I am tempted to call it. Exposure to adults is not necessary to achieve this level, and an inexperienced age-mate will suffice.

Have we then sustained the hypothesis that the song of the Arizona Junco is acquired? If you raise a young male in individual isolation the answer is yes. If you raise the young males in isolation from adults, but in a group then the answer is no. We could presumably refine the classification by establishing levels of innateness or acquiredness, and this may be necessary as comparative ontogenetic studies proceed. However, before this is possible, we need to address attention to the actual processes involved in ontogeny, and attempts to classify the patterns of behavior may be premature at this stage.

12.4. Selective learning in birds

To persist with this topic just a little longer, if we play recordings of normal song to a socially-isolated White-crowned Sparrow, chaffinch, or Red-winged Blackbird. in sufficient quantity and at the appropriate phase of the life-cycle, this treatment will serve to restore normal development. The song that the young males produce will show evidence of actual imitation of the model presented. We would conclude then that the song is acquired. But a further experiment makes us hesitate all over again about this conclusion. There is some evidence for the chaffinch, and the White-crowned Sparrow, that if the young male is presented at the appropriate time with recordings not only of his own species song but also that of another species, he will selectively learn the conspecific song. This was demonstrated with two male White-crowned Sparrows, one of which was presented with both White-crowned Sparrow song and, for example, Song Sparrow song. The bird learned only the former. Three males presented with Song Sparrow song alone during the same period rejected the models and developed in much the same way as social isolates, without training. Here then is an interesting situation; a bird which cannot generate normal song of its own accord and yet has the capacity to identify the appropriate model to copy, without prior experience of such a model itself. Does this make the song any less learned? One might think of the song as being acquired, through a process that is innately selective, perhaps an illustration of Lorenz's 'innate schoolmarm' principle (Lorenz 1969).

As a further example, let me mention the result of such an experiment with Red-winged Blackbirds. As an illustration, two males were taken as nestlings, individually isolated at about 40 days and given playback of half normal Red-winged Blackbird song and half normal Baltimore Oriole song. The latter was selected, because this species can be heard at the nests where the young Red-winged Blackbirds were taken. They were given this exposure during the second month of life from about 40 days to 70 days of age. In songs recorded some 10 months later, one bird had developed 7 song types. Two resembled parts of the oriole song, one was a rendition of the Red-winged Blackbird song, and the remainder resembled those of untrained isolated birds. The second bird developed 4 songs. One resembled the oriole song closely, another rather less closely, two were acceptable Red-winged Blackbird patterns, and one resembling the Red-winged Blackbird model. We have confirmed this result in a number of other experiments on males raised in pairs, males raised together with a female, and so on—some 15 all told.

In this situation, with sound presented through a loudspeaker, Red-winged Blackbirds learn Baltimore Oriole song as readily as the song of their own species. We know that this does not happen in nature, and this raises the question of what might be the basis for what is presumably another case of selective learning. The obvious hypothesis, now the subject of an experimental programme we are just developing, is that naïve Red-winged Blackbird males are responsive to the conspicuous and specific plumage colours typical of their family. Perhaps selective responsiveness to other kinds of social stimulation, non-vocal concomitants of song—especially presentation of the red epaulets of the Red-winged Blackbird in displays—will prove to constrain the learning to a particular situation. This would of course have the same effect of restricting learning to a biologically appropriate set of models.

In the White-crowned Sparrow selection seems to be manifest simply with sounds presented through a loudspeaker. In either case we have to postulate a selective mechanism that is endogenous to the individual bird. An endogenous capacity for selective learning could be regarded as innate, though I would prefer to think of it as a function of earlier events in the organism's ontogeny, without any commitment as to the nature of those events until we know more about them. I would like now to describe some further experiments which begin to explain the nature of the actual ontogenetic machinery that is involved in the development of bird song and also give us a framework with which we can begin to make some sense of the species differences that I have described in developmental terms—an assignment that I feel is important at this time—and is not being undertaken by behavioural geneticists as far as I can tell.

The experiments were designed to explore the role of auditory feedback—the capacity of the bird to hear its own voice. I shall confine myself to just one kind of study which asks what is the effect of cutting off auditory feedback at a stage of development prior to the accomplishment of full song? When Mark Konishi (1965) did this with male White-crowned Sparrows the result was very interesting. A male deafened in youth, in this case at about 40 days of age, developed a song much more abnormal than anything generated by an intact isolated bird (Fig. 12.2(c)). The highly variable songs consist in large part of irregularly modulated sounds which succeed one another rapidly, often approximating clicks. Tones fluctuate in frequency in an irregular fashion. The net impression is of a more noisy, scratchy, unstructured sound than that of either normal song or that of an intact social isolate.

Bearing this in mind, compare the result from a similar experiment on chaffinches by Fernando Nottebohm (1968). Here, in a male deafened at 90 days of age, we see a somewhat similar result—amorphous song with little regular internal structure and with much of the normal tonal quality lacking (Fig. 12.3). Instead it has a rasping noisy quality. One has an impression too that the songs of these two deafened birds, a White-crowned Sparrow and a chaffinch, differ from each other less than those of either normal or intact isolates of their species. Konishi (1964b) obtained a

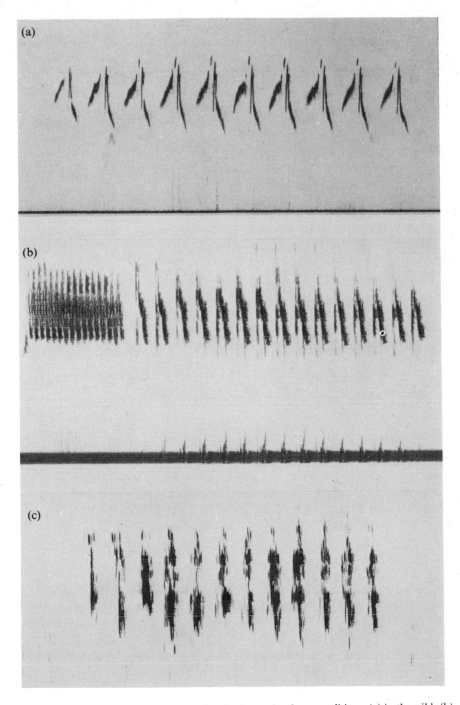

PLATE 12.3. Songs of Oregon Juncos developing under three conditions. (a) in the wild; (b) hand-reared in acoustical isolation from early youth; (c) deafened in youth. (After Konishi 1964*b*.)

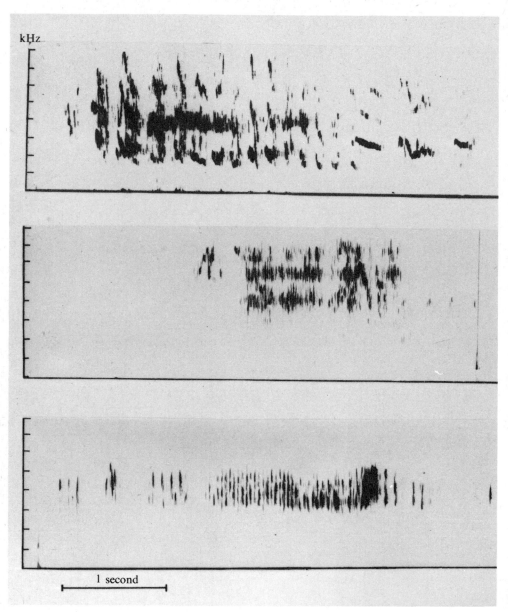

PLATE12.4. A comparison of songs of (a) an early-deafened Song Sparrow, (b) a White-crowned Sparrow, and (c) an Oregon Junco. Frequency scale in kilohertz intervals. (After Mulligan 1966; Konishi 1964*b*, 1965.)

similar result with juncos, in this case the Oregon Junco. Although some syllabic structure persists, we see again this erratic internal morphology and noisy spectral organization (Plate 12.3).

Finally, consider the result obtained by Mulligan (1966) in one of his Song Sparrows deafened at the age of 70 days (Plate 12.4). It is certainly less easy to distinguish from the songs of a deaf White-crowned Sparrow or of a deaf junco than the normal songs of their species. This is a remarkable result, when you recall the very different results of the isolation experiment in these species.

12.5. Sensory templates and vocal development

What conclusions can be drawn from this result? We have noted that, in species which must learn the normal song, social isolates are nevertheless able to retain some normal characteristics of the song; as a result something of the species differences still persists. Deafening seems to eliminate still more of these species differences. This finding has led us to hypothesize that at least some of the species differences that emerge during ontogeny arise from the exploitation of what we might think of as a sensory template, lying somewhere at points in the auditory pathway, to which the bird matches its vocal output by auditory feedback.

Provided that the feedback loop is intact, the information embodied in this auditory template is accessible for purposes of motor development. In some cases, as in the Song Sparrow, the auditory template may suffice to generate a normal song. In the White-crowned Sparrow, the template alone is less adequate, and we perhaps see in the songs of intact socially isolated White-crowned Sparrows a picture of what the template of a naïve bird embodies.

Given this notion of the template we can perhaps extend it further to explain what happens in the course of learning. Take the White-crowned Sparrow for example. Song-learning occurs in Nuttal's White-crowned Sparrow between about 10 days and 50 days of age. Singing begins some months after this. This temporal separation of learning and singing enabled Konishi to deafen males after learning was completed but before they began to sing. Their subsequent utterances were no different from those of a bird deafened earlier without training. Thus deafening a trained bird makes the information that has been acquired inaccessible as far as motor development is concerned.

We can hypothesize that, during the learning period, the male White-crowned Sparrow listens to adult song, and as it does so learns some of the characteristics. These become incorporated in the auditory template, which is then more highly specified than in the naïve bird, incorporating not only the general characteristics we see in a social isolate but also those of fully normal song and of the particular dialect to which the bird has been exposed. When song begins, the male will proceed to match his vocal output to these more highly specified dictates of the auditory template, and a copy of the model will result.

One further point we have still to explain is the basis for selective learning in the male White-crowned Sparrow. I postulate that the crude template for the species song possessed by the naïve bird, although inadequate to define fully normal song, is nevertheless sufficient to serve as a kind of filter that will pass White-crowned Sparrow song and reject songs of other species.

In this way I believe we can begin to derive a general explanation which helps in understanding the processes of development in several species; some like the White-crowned Sparrow requiring modification of the auditory template for normal development; and others such as the Song Sparrow already possessing an adequate template.

Although something like an auditory template seems to play a role in the song development of many species, as work proceeds it becomes clear that it is not sufficient to explain all of the species differences one finds. As already shown, the tendency in some early-deafened finches is to revert to a very elementary pattern of sound with little structure. But in other species this is less obvious. The Red-winged Blackbird is such a case. Two males were taken from the nest at 7 days of age and deafened at 9 days. The 6 song patterns that they generated between them were by no means lacking in internal structure, although they had much of the instability that we associate with singing of early-deafened birds (Plate 12.2). In some respects their structure is reminiscent of the slow type of normal song. In canaries we find a more extreme case of the persistence of structure after deprivation of auditory feedback.

One deficiency of most of the experiments I have described thus far is that they were conducted with birds taken as nestlings. There is always the uncertainty that exposure to natural song during the first few days might have had some effect. There is positive evidence that the effect is minimal (Marler 1970), and we should remember that, unlike precocial birds such as chicks and ducklings which can hear well before they hatch (Gottlieb 1971), these are altricial birds and are still in a primitive developmental state at this time. Nevertheless, the worry remains. Another difficulty concerns the deafening technique, which has the drawback of being irreversible.

As a way of overcoming both of these difficulties we have been experimenting with loud masking noise as a method for reversible blocking of auditory feedback, with the additional possibility of breeding birds in noise to control for the effects of early exposure. The initial results of this technique with canaries are encouraging. They will breed in chambers with 100 decibels of white noise broadcast within, so that young birds are prevented from hearing both their own voice and that of their parents and siblings. Comparisons with surgical deafening show that the masking is indeed effective. Although they suffer considerable permanent deafening, they seem still to be able to hear their own voices afterwards (Marler, Konishi, Lutjen, and Waser 1973). This has made it possible to raise birds totally deprived of auditory feedback from singing, by keeping male canaries in noise to 40 days of age and then deafening them surgically before removal from the noise.

Guided by the other song-bird studies, my expectation was that the song in these canaries would be if anything still more elementary than that of our other deafened birds. To our suprise the results came out rather differently. Controls were taken from the noise at the same age and then placed either in individual isolation in one class, or in pairs in another class. By comparison with these the songs of the deaf birds were simpler in the sense that they included fewer syllable types. There was nevertheless a remarkable degree of structure of songs of 5 birds so treated. The number of syllables ranged from 1 to 8, with an average of 5, and the syllables were clearly organized in bouts in much the same way as in normal song. There are again signs of the instability associated with all early-deafened birds, but the remarkable result is the degree of structure they achieved. Although somewhat odd in tone, the resemblance to normal canary song is obvious.

It seems clear then that the auditory-template hypothesis is not sufficient by itself to explain the developmental basis of species differences, even within a restricted group such as the finches. What alternative method is being exploited by these canaries, and perhaps in a lesser degree by the Red-winged Blackbird (Plate 12.2), is not clear. It may be that species differ in the extent to which the patterning of song depends, on the one hand, on refined modulation of the properties of the vibrating syringeal membranes, and on the dynamics of respiratory flow through the sound-producing equipment, on the other. The latter type of control might be especially important in species like the canary and some other cardueline finches where the song is a very long drawn-out affair, with its pattern resulting from long-term changes in bout organization. The respiratory machinery may play a larger role in this case, perhaps under endogenous motor control from the central nervous system. This could be a step towards (or back to) the condition that Konishi (1963) and Nottebohm and Nottebohm (1971) have described in chickens and doves, where birds deafened even at a very early age are capable of perfectly normal vocal development.

Taken as a whole, the results of these studies of bird-song development provide evidence for something of a revolution in developmental strategy, such that learning plays a more radical role in shaping the adult behaviour than is typical of most natural patterns of animal behaviour. As we learn more of the details of development in particular cases, the most general principle emerging is that of the imposition of constraints on learning, manifest here as the template hypothesis. Such a notion may have relevance to other studies, including those on man. Only in man do we see evidence of such a radical ontogenetic revolution in the basis of motor development, perhaps even involving some of the same principles. It is interesting that this kind of change does not seem to have taken place in nature in any other primate so far as we know. I have suggested that the extraordinary demand for signal diversity in the oscine birds called forth this change in developmental strategy. One could think of early human society as placing analogous demands, in this case for diversity, not in a community of many species, but rather within a population and its individual members. The radical increase in the subdivision of labour that took place with the expansion of tool-using in early human society, such that members of the same sex and age-class assumed many different roles, must have created a need for dramatic increase in signal diversity to specify those roles—to name objects and the tasks to be performed. This alone might have sufficed to induce the revolution. Once accomplished for the satisfaction of relatively simple biological requirements, not so different in principle between birds and our own pre-language ancestor, the greatly increased potential for signal diversity thus created would pave the way for the much more remarkable changes in signal design and coding that eventually made the language of man a unique phenomenon.

References

ALDRICH-BLAKE, F. P. G. (1970). Problems of social structure in forest monkeys. In *Social behavior in birds and mammals* (ed. J. H. Crook). Academic Press, New York.

GARCIA, J., McGOWAN, B. K., and GREEN, K. F. (1972). Biological constraints on conditioning. In *Classical conditioning II: current theory and research* (eds A. H. Black and W. F. Prokasy). Appleton–Century–Crofts, New York.

GAUTIER, J-P. (1971). Etude morphologique et fonctionnelle des annexes extralaryngées des cercopithecinae; liason avec les cris d'espacement. *Rev. Biol. Gabonica* **7**, 230–67.

GOTTLIEB, G. (1971). *Development of species identification in birds.* University of Chicago Press, Chicago.

KONISHI, M. (1963). The role of auditory feedback in the vocal behavior of the domestic fowl. *Z. Tierpsychol.* **20**, 349–67.

—— (1964a). Song variation in a population of Oregon juncos. *Condor* **66**, 423–36.

—— (1964b). Effects of deafening on song development in two species of juncos. *Condor* **66**, 85–102.

—— (1965). The role of auditory feedback in the control of vocalization in the white-crowned sparrow. *Z. Tierpsychol.* **22**, 770–83.

LORENZ, K. (1969). Innate bases of learning. In *On the biology of learning.* (ed. K. Pribram). Harcourt, Brace, and World, New York.

MARLER, P. (1967). Comparative study of song development in sparrows. *Proc. Int. orn. Cong.* **14**, 231–44.

—— (1970). A comparative approach to vocal development: Song learning in the white-crowned sparrow. *J. comp. Physiol. Psychol. Monogr.* **71**, 1–25.

—— (1973). A comparison of vocalizations of Red-tailed Monkeys and Blue Monkeys, *Cercopithecus ascanius* and *C. mitis* in Uganda. *Z. Tierpsychol.* **33**, 223–47.

——, KONISHI, M., LUTJEN, A., and WASER, M. S. (1973). Effects of continuous noise on avian hearing and development. *Proc. natn. Acad. Sci. U.S.A.* **70**, 1393–6.

——, MUNDINGER, P. (1971). Vocal learning in birds. In *Ontogeny of vertebrate behavior.* (ed. H. Moltz). Academic Press, New York.

—— MUNDINGER, P., WASER, M. S., and LUTJEN, A. (1972). Effects of Acoustical Stimulation and Deprivation on Song Development in Red-winged Blackbirds (*Agelaius phoeniceus*). *Anim. Behav.* **20**, 586–606.

—— TAMURA, M. (1962). Song dialects in three populations of white-crowned sparrows. *Condor* **64**, 368–77.

MULLIGAN, J. A. (1966). Singing behavior and its development in the song sparrow, *Melospiza melodia. Univ. Calif. Publs. Zool.* **81**, 1–76.

NOTTEBOHM, F. (1968). Auditory experience and song development in the chaffinch (*Fringilla coelebs*). *Ibis* **110**, 549–68.

—— NOTTEBOHM, M. E. (1971). Vocalizations and breeding behavior of surgically deafened ring doves. (*Streptopelia risoria*). *Anim. Behav.* **19**, 313–27.

SELIGMAN, M. E. P. and HAGER, J. L. (eds) (1972). *Biological boundaries of learning.* Appleton–Century–Crofts, New York.

THORPE, W. H. (1958). The learning of song patterns by birds, with especial reference to the song of the chaffinch, *Fringilla coelebs. Ibis* **100**, 535–70.

—— (1961). *Bird song: the biology of vocal communication and expression in birds.* Cambridge University Press.

TINBERGEN, N. (1951). *The study of instinct.* Clarendon Press, Oxford.

—— (1963). On aims and methods of ethology. *Z. Tierpsychol.* **20**, 410–33.

—— (1968). On war and peace in animals and man. *Science, N.Y.* **160**, 1411–18.

13. Conservatism of displays and comparable stereotyped patterns among cephalopods

M. MOYNIHAN

THE EVOLUTION of ritualized behaviour patterns (most notably displays, specialized signals, and signal-like patterns) has been a major interest of ethologists. See, for instance, Heinroth (1911), Whitman (1919), Lorenz (1941), and Tinbergen (1951, 1964). Attention has been paid to a variety of aspects of the subject, for example, the derivation of displays, their changes in form and physiology, and the selection pressures involved.

It may also be useful to calculate rates of evolution of behaviour—to plot the stability or instability of patterns against an actual long-term time-scale. This is not always easy, however. Signals, in particular, do not leave a fossil record directly (see Tinbergen 1963). The only way to tackle the problem is by comparing the ritualized patterns of living species, and then deducing and extrapolating (logically it is hoped) from whatever is known of the histories of the species themselves and their habitats.

Some components of displays can be deduced to be relatively young. Thus, for example, the degree of morphological differentiation among populations, and the evidence of Pleistocene and post-Pleistocene climatic changes (with inevitable movements or extinctions of populations) would suggest that the diagnostic peculiarities of many 'song dialects' of such species as *Zonotrichia capensis* and other passerine birds (Nottebohm (1969) and appended references) are unlikely to have developed more than a few thousands or tens of thousands of years ago.

Other components of displays, especially basic outlines or fundamental structures, can sometimes be traced farther back. Such ritualized signals as are shared by living viverrids and felids but not canids, mustelids, or procyonids (Ewer 1968; Kaufmann 1962; Kaufmann and Kaufmann 1965; Kleiman 1966, 1967) probably originated at some time in the late Eocene or earliest Oligocene when the ancestral miacids were proliferating to give rise to the modern Carnivora (Romer 1968). The basic, distinctive, and partly ritualized communication system of all living platyrrhine monkeys (Moynihan 1967) must have come into existence by the middle of the Miocene when the radiation of the group was in full swing or perhaps already accomplished (Stirton 1951). The original forms of many of the more elaborate displays (those unlikely to have been evolved by parallelism or convergence) of many species of certain single families or orders of reptiles and birds, for example, Iguanidae (Carpenter 1967), Pelecaniformes (Van Tets 1965), Anatidae (Johnsgard 1965), and Columbidae (Goodwin 1967), may date back to the Palaeocene or late Cretaceous (records of the relevant groups are summarized in Fisher (1967) and Appleby, Charig, Cox, Kermack, and Tarlo (1967)).

The living cephalopods of the subclass Coleoidea provide several instances of even greater conservatism.

The three major surviving orders of the group are Teuthida (squids), Sepiida (cuttlefishes and their relatives), and Octopida (octopuses and argonauts). Their fossil records are not good (very inferior to those of the related nautiloids and ammonites), but sufficient to indicate that they must have separated from one another in the earlier Mesozoic, possibly late Triassic or early Jurassic (Jeletzky 1966).

General surveys of the known social reactions of the living species are contained in Lane (1957), Wells (1962), Clarke (1966), and Packard (1972). Although many notes are fragmentary or anecdotal, there are useful accounts of the fast-swimming, pelagic or semi-pelagic squids of the genus *Loligo sensu lato* (Drew 1911; Stevenson 1934; Tinbergen and Verwey 1945; McGowan 1954; Arnold 1962; Waller and Wicklund 1968) and detailed analyses of three littoral and more sedentary species, the squid, *Sepioteuthis sepioidea* (Arnold 1965; Moynihan and Rodaniche, in preparation†), the common octopus of warmer European waters, *Octopus vulgaris* (Cowdry 1911; Young 1964; Packard and Sanders 1971), and the cuttlefish, *Sepia officinalis* (Tinbergen 1939; Holmes 1940).

Many ritualized behaviour patterns are remarkably similar in different species of cephalopods. The resemblance extends to two or three different functional types of patterns.

This brings up a semantic problem. There has been a tendency among students of animal communication (myself included) to use the terms 'ritualized' and 'display' as if they were interchangeable, to refer to patterns which have become specialized in form or frequency to subserve a signal function, to transmit information from one individual to another. Unfortunately, the physical features which are characteristic of most displays, that is, stereotypy and exaggeration of form, are equally characteristic of some behaviour patterns which are adapted to promote crypsis (*sensu stricto*), to make a performing individual effectively invisible or at least inconspicuous (for example, the 'eucryptic' specialized resting postures of many orthopteroid insects discussed by Robinson (1969*a*, *b*). Such

† Unless otherwise noted, all subsequent references to *Sepioteuthis* are based upon personal observation by Moynihan and Rodaniche.

patterns are designed especially *not* to transmit information. They must be considered to be ritualized (presumably the basic physiological processes are similar in all cases—see also Crane (1949) and Blest (1961), but it would be confusing—absurd—to give them the same name as patterns which are signals. Perhaps it would be convenient to call them 'anti-displays' instead.

The category is not entirely discrete. Anti-displays may be effective only against certain backgrounds. They may be conspicuous against other backgrounds. In the latter circumstances, with their exaggeration and stereotypy, they may appear to be just another kind of display. It is evident, in fact, that some ritualized patterns normally function as displays at some times and as anti-displays at other times (see also below).

The ritualized patterns of cephalopods include postures and movements and a whole host of colour-changes which can be combined and re-combined almost *ad infinitum*. Among the more widespread performances are the following.

1. *The 'dymantic' or 'black spot' display*. Packard (1972) has already remarked that this display is performed by *Sepia officinalis*, *Octopus vulgaris*, and *Sepioteuthis sepioidea*. In all three species it is a hostile or agonistic pattern, often or usually an inter-specific signal, a 'startle' display, provoked by and directed toward a potential predator or some other alarming stimulus. It takes slightly different forms in the different species (see Fig. 13.1). A cuttlefish in an extreme dymantic has a pale body with two dark blotches (false eye spots) on the back toward the rear and sides, black rings around the sides and lower edges of the (real) eyes, and black borders along the outer edges of the fins (Holmes 1940). An octopus in the equivalent pattern is pale with dark circles around the eyes and dark edges to the horizontally re-curved arms, in appearance the border of the body mass (Wells 1962; Packard and Sanders 1969). A *Sepioteuthis* can produce false eye spots in combination with a variety of colour patterns, including a pale background and/or dark stripes along the edge of the body.

2. *'Zebra' stripes*. This pattern has been observed in the same three species by the same observers. It is characterized by bold, dark, transverse stripes across the body and all or most of the arms which may be spread simultaneously (Fig. 13.2). It also is sometimes or always hostile, but primarily an intra-specific signal and often quite aggressive.

3. *Upward V-curls*. Packard and Sanders have described and illustrated a 'flamboyant' posture of *Octopus vulgaris* in which the foremost two arms are raised, divergently, and curled backward over the head. *Sepioteuthis sepioidea* may raise all or several pairs of arms, closely appressed in two divergent bundles to form a V, either straight upward or upward and

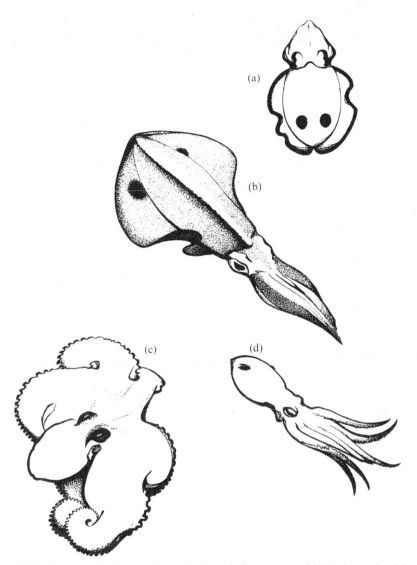

FIG. 13.1. Dymantic displays and associated or similar patterns. (a) the dymantic of *Sepia officinalis* (after Holmes, *op. cit.*). (b) the false eye spots of *Sepioteuthis sepioidea*, superimposed upon a medium dark coloration and a trace of longitudinal striping. This may be considered a modified form of dymantic. More extreme forms are much more reminescent of *Sepia*. (*Note.* the false eye spots are shown on the fins. This the usual position. Packard (*op. cit.*) shows spots on the sides of the body. This is not typical.) (c) The dymantic of adult *Octopus vulgaris* (from Wells, *op. cit.*). (d) A pattern of young *O. vulgaris*. A branchial heart shows through the skin to produce a dymantic-like effect.

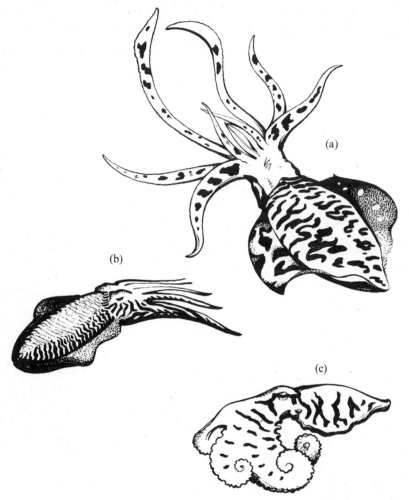

FIG. 13.2. Zebra stripes. (a) *Sepioteuthis sepioidea.* (b) *Sepia officinalis* (after L. Tinbergen, *op. cit.*). (c) *Octopus vulgaris* (from Packard and Sanders 1971). Note that there is some appreciable spreading of the arms in each of the performances.

backward over the head (see Fig. 13.3). In both species, these patterns are essentially infantile or juvenile. The flamboyant seems to be performed only by young individuals; never by large adults (in which it is replaced by the dymantic). The upward Vs of *Sepioteuthis* are performed much more frequently by young than by adults. In both species, the patterns are primarily anti-predator adaptations, indications of alarm. The arm movements are combined with several colour arrangements: uniform dark, mottled, chevron and other stripes in the octopus; plain brown or transverse and more-or-less blotchy bars (broader and less numerous than

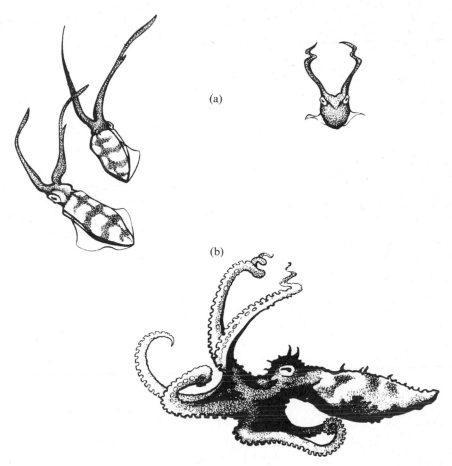

(a)

(b)

FIG. 13.3. Upward V-curls. (a) Three views of Vs by young *Sepioteuthis sepioidea*. Two of the individuals are barred. The third is in a uniform dark coloration. (b) The flamboyant of young *Octopus vulgaris*. (From Packard and Sanders 1969.)

zebra stripes) in the squid. The effect is supposed to be cryptic in the octopus. It can be cryptic or mimetic in the squid. An animal with V and bars resembles a small piece of drifting weed, especially *Sargassum*. It may pass unperceived or be mistaken for an inedible object. This would seem to be a valid example of a behaviour which can be either an anti-display or a display to convey misleading, false information depending upon background, that is, the abundance and distribution of weed in the environment at the time. (I have recently seen a young individual of another squid, almost certainly '*Doryteuthis*' *plei*, perform a curled upward V quite like that of *Sepioteuthis*, but in a generally pale or whitish coloration. This must, at least, have served to break up the usually streamlined shape of the animal, and may perhaps have reduced its attractiveness to predators by making it look less typically squid-like.)

4. *Longitudinal streaks.* Another reaction to alarming stimuli is the production of two dark longitudinal stripes down the back. In *Sepia* (Holmes 1940), the principal back stripes are accompanied by two more along the sides of the body at the bases of the fins. *Sepioteuthis* has a similar 'double streak–fin stripe' complex. It differs from the cuttlefish perform-ance in that the back stripes may be continued onto the arms and tentacles. The display of *Octopus* which Packard and Sanders call 'longitudinal stripes' would seem to include the same basic components as the squid complex, only superficially distinctive in form because the body of the animal has a different shape (see Fig. 13.4). In all three species, the pattern appears to be hostile with a strong escape component. The *Sepioteuthis* version is used in many social situations but rather more frequently during inter-specific encounters than intra-specific disputes.

Resemblances on this scale can hardly be the result of convergence. In each of the four cases (the four kinds of patterns), the performances are so elaborate, distinctive, and similar (in causation and functions as well as forms) in the different species as to indicate that they cannot have evolved entirely independently in each phyletic stock. They must be descended from patterns which were already ritualized in something like their present shapes in the common ancestor of the living species.

There is further miscellany of less complicated but still ritualized postures and movements (arm-spreads, downward and upward pointing, rocking, etc.), general colour-changes (dark and pale flushes), and other stripe, spot, and bar patterns, some or all of which appear in much the same forms in many species, including *Loligo* spp. (references as on p. 277), '*Doryteuthis*' *plei* (Moynihan and Rodaniche, personal observation), the peculiar dwarf squid *Pickfordiateuthis pulchella* (Moynihan and Rodaniche, personal observation), *Octopus horridus* (Young 1962), and *O. cyanea* (Wells and Wells 1972), in addition to *O. vulgaris, Sepia officinalis,* and *Sepioteuthis sepioidea,* and which might also be homologous in the strictest sense. Many or most of these simpler patterns seem to be further expressions of alarm and are often provoked by the approach of a potential predator. Some may be warning signals to conspecifics; others may be semi-cryptic or semi-mimetic; still others may inform a predator that it has been discovered (some predators are reluctant to attack alerted prey).

Ritualized performances which can be traced back to the common ancestor of the three living orders of coleoids must have originated about 190 million years ago (a conventional date for the beginning of the Jurassic) or earlier. This is a great age, perhaps greater than any previously

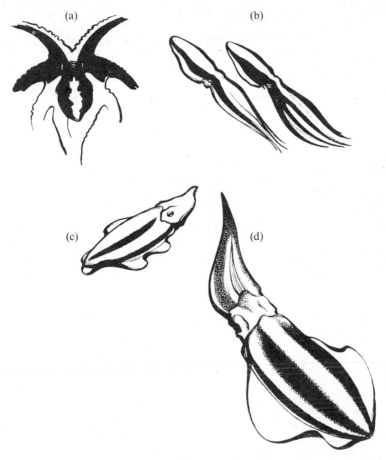

FIG. 13.4. Longitudinal streak patterns. (a) Top view of streak-like pattern of *Octopus vulgaris* (from Packard and Sanders 1971). (b) Side views of two versions of streaks in the same species (after Cowdry, *op. cit.*). (c) Longitudinal streak plus fin stripe in *Sepia officinalis* (after Holmes, *op. cit.*). (d) The same patterns in *Sepioteuthis sepioidea.*

suggested for particular ritualized patterns of other groups. Such stability is, or should be, surprising, especially in the case of displays.

   Displays are not immortal. They must change with time. Like all characters, they respond to alterations of both external and internal environments, mutations, and other accidents. It has been suggested that they might even be unusually vulnerable or susceptible to change, transformations, and decay and eventual disappearance because of the very nature of communication processes. The argument for this is set out at length in Moynihan (1970). Some of the main points can be summarized as follows. Displays are only worthwhile if perceived. They should all, with the possible or probable exception of the mimetic patterns, be emphatic to some extent. They must attract and fix the attention of appropriate receivers—the individuals that they are designed to influence—in order to be effective. But they must often tend to become progressively less effective with time. Competition among performing individuals will favour increased repetition or exaggeration of the displays. But exaggeration cannot continue indefinitely and repetition may lead to gradual 'attenuation' of the information conveyed. Sooner or later, the receivers probably will become blasé and react less strongly or not at all, and some other actual or potential signal pattern will be selected to replace or supplement the attenuated performance. (A passage from the earlier paper (Moynihan 1970) may be quoted directly. '. . .this outline of the probable evolution of displays is closely comparable to what is known to happen to certain human signals. A particularly "strong" word or phrase; for example an obscenity or striking new technical term, is apt to be very impressive and effective when it first begins to be used. Simply because it is effective, it tends to be used more and more frequently. And, unless it is constantly reinforced by new variations or further elaboration, it eventually becomes essentially meaningless.'—At which stage, some other word or phrase should be used instead.)

This argument is admittedly hypothetical as applied to non-human animals, but it may be attractive insofar as it can help to account for certain non-obvious but widespread phenomena of animal communication systems. It is relevant to cephalopods and will be referred to in subsequent discussions. I should stress, however, that neither acceptance nor rejection of the hypothesis can render the conservatism of some cephalopod patterns less puzzling. Some of the ritualized patterns of cephalopods *do* seem to have remained more conservative than many or most of the corresponding reactions of other organisms (it is not usually possible to determine homologies among the more complex performances of species of different orders of other classes and phyla). Equally significantly, some of the other ritualized patterns of cephalopods have been far from stable (see also below).

It would seem to be pertinent, therefore, to ask why those patterns that are conservative have been so comparatively slow to change.

Part of the answer may be intellectually trivial. Certain aspects of the external environment may also have been slow to change, not only physical substrates but also some of the organic features which provide backgrounds for crypsis and models for mimicry. Organic reefs of corals and algae, components of which are imitated by some living cephalopods, began to assume their modern forms in the middle Triassic (Newell 1971). The frondose algae and higher plants which provide other models and backgrounds probably have been around for a long time or were preceded by others of similar forms and colours (algae may tend to be relatively conservative—see Banks, Chester, Hughes, Johnson, Johnson, and Moore (1967), and Black, Downie. Ross, and Sarjeant (1967)). This, in itself, might be enough to explain the stability of both cryptic anti-displays and mimetic misleading displays. It cannot explain the persistence of non-misleading intra-specific and inter-specific displays such as the dymantic and zebra stripes. The latter are conspicuous and emphatic against their usual present backgrounds. The problem is to understand why they have remained conspicuous in their own characteristic ways. It is possible to conceive of other patterns which could be equally conspicuous against the same backgrounds.

Cephalopods do not seem to display unusually infrequently. Nor do they have access to unique remedies for whatever attenuation may occur. They do not have more kinds of basically different ritualized signals than other animals, or so many as to avert or greatly reduce the incidence of boredom or satiety among receivers (their ability to combine colours must be helpful, but only to a limited extent). Their conservative displays are exaggerated, but not much more so than those of many vertebrates and arthropods.

Granted that the non-mimetic patterns of cephalopods should not be intrinsically (internally) more resistant to transformation or decay, it must be supposed that some of them have been exposed to weaker selection pressures in favour of change than have most displays of other organisms. Or, perhaps more precisely, that the counter-selection against change has been relatively greater in the case of the relevant cephalopod patterns. (There are always advantages and disadvantages to any character: its form and fate are always the result of conflicting pressures.)

Why should there have been such counter-selection?

The persistence of the non-mimetic displays definitely cannot be corre-lated with any general stability of environments. Attenuation should occur as frequently and proceed as rapidly, on the average, in stable as in fluctuating circumstances—perhaps more rapidly. Many parameters of the marine habitats to which cephalopods are more or less confined are quite variable in the long run. Corals and algae may have continued, but other elements have come and gone in the Mesozoic and Cenozoic. There have been variations in physical parameters such as temperature and distribu-tion of land and water masses (see summary and comments in Newell (1971). Biotic changes have also been numerous and are known to have involved important competitors, predators, and preys of coleoids. Thus, for instance, among the bony fishes (which impinge upon cephalopods in many different ways), the sub-holosteans and holosteans have been largely replaced by teleosts (Romer 1966). Several groups of marine reptiles, some of which may have been significant predators of cephalopods (see remarks by L. B. H. Tarlo appended to Cox (1967), have radiated and disappeared, to be succeeded by marine birds and mammals which may feed and depend upon cephalopods to an equal or greater extent (Moynihan, in preparation). Other cephalopods which must have competed with squids, cuttlefishes, and octopuses, that is, the ammonites and belemnites, first flourished and then became extinct within the same vast stretch of time (Donovan, Hodson, Howarth, House, Tozer, and Wright 1967; Jeletzky 1966).

It should also be mentioned that the living cephalopods which have retained conservative ritualized patterns have diverged in social organization, feeding behaviour, and activity rhythms. *Sepia officinalis, Octopus vulgaris*, and *Sepioteuthis sepioidea* are all predatory and more or less littoral, but otherwise heterogeneous in many respects. The octopus is essentially solitary except when breeding, and usually active at night, primarily bottom-dwelling, crawling, resting, and hiding in holes and crevices among rocks and feeding on crabs, by preference (Lane 1957; Wells 1962). The cuttlefish is also primarily nocturnal or crepuscular and probably not very gregarious, but is more of a swimmer, although often buried in sand when resting; perhaps feeding on shrimp by preference and certainly hunting them actively (Hardy 1956; Boycott 1958). The squid, which recalls the cuttlefish in body and fin shape (Boycott 1965), is even less closely tied to the bottom, often occurring high in the water column or at the surface, usually over turtle grass (*Thalassia*) or coral, alert or at least awake during most of the day as well as all or part of the night, feeding on fishes of small to medium size and small crustaceans such as copepods, highly gregarious during the day when groups wait for prey to come to them, and less gregarious at night when individuals scatter to search for the same or other prey.

Perhaps the solution to the problem of the conservatism of some displays may be revealed by comparing them with the others which have not been conservative. It is highly suggestive that among cephalopods, the sexual displays, at least the patterns performed by males immediately before copulation, are rather varied. Male *Sepia officinalis* assume a version of zebra striping (Tinbergen 1939). Male *Sepioteuthis sepioidea*, if they perform any display at all, are most likely to flutter their fins and turn pastel-pink or lavender in colour. Male *Loligo pealii* may spread their arms, raising the dorsal pair, and develop dark spots at the bases of the second and third pairs (Drew 1911). *Octopus vulgaris* males show the white suckers at the bases of the 'lateral' arms (Wells and Wells 1972). There are even differences between closely related species. A Pacific squid, *Loligo opalescens*?, photographed by Cousteau and Cousteau (1970), would appear to develop a red flush on all or most of the arms and tentacles. Male *Octopus cyanea* show a pattern of longitudinal black stripes and erect horn-like papillae on the head (Wells and Wells 1972). Male *O. horridus* can combine longitudinal stripes with dark patches or bars (Young 1962). It will be noted that some of these patterns are similar in form to some of the hostile displays cited above. They may well be partly hostile or of mixed import in the sexual context. 'Courtship' is apt to be ambivalent. But the use of these patterns as pre-copulatory signals cannot be conservative in all cases. Other patterns may have been purely sexual from their origin. And the patterns used are obviously not homologous in all species.

The variety of pre-copulatory displays would not necessarily have been expected. In many other groups of animals, for example, birds of the family Laridae (Moynihan 1962) and of the order Anseriformes (Johnsgard 1965), the immediately precopulatory patterns have tended to less mutable than most other kinds of displays. The situation among cephalopods is all the more remarkable in that individuals of many or all species of the group develop comparatively rapidly (Wells 1962; Packard 1972; Clarke 1966) and typically reproduce in 'big bangs' (Gadgil and Bossert 1970), that is, only once or a few times in their lives (Wells 1962; Clarke, 1966; Mangold-Wirz 1963). These features must tend to limit opportunities for both evolutionary and ontogenetic 'experimentation', to make it more difficult to conduct 'trial runs' of new patterns at leisure and in favourable circumstances.

The fact that the pre-copulatory patterns of cephalopods are, nevertheless, diverse is a further indication that their displays can be subject to changes like those of other animals, and also that their environments have not been so stable in all respects as to preclude selection for change in general.

This again serves to emphasize the peculiarity of the displays that have remained conservative. Many or most of the latter may be purely hostile. Hostility should not be more conservative than ambivalence or sex *per se* (namely, the Laridae and Anseriformes cited above). But there is one way in which the pre-copulatory signals of cephalopods do differ from all or most of their purely hostile signals *quite apart from the specific nature of the information encoded and transmitted*. This is the *number and variety* of receivers.

Pre-copulatory displays may be adapted to influence only a few individuals. Almost certainly, they are adapted to influence only a single category or class of receivers—individuals of the opposite sex in breeding condition. Many purely hostile displays, on the other hand, must be designed to affect several different categories, different age, size, and sex classes of the same and/or other species, as well as many more individuals, probably over a greater time-span in many cases.

It is easy to imagine how diversity of receivers could favour conservatism of signals. Any change in a pattern adapted to influence different kinds of receivers will usually or always have to satisfy all of them, or at least satisfy some of them without dissatisfying others very greatly. Of course, there may be general improvements which increase the efficiency of a signal in most social circumstances. They will be selected for strongly. But there must be many other changes which enhance or refine the impact of a signal in one type of interaction with one kind of receiver but do not enhance— may even impair—its effectiveness in other interactions with other receivers. The probability of a change having mixed effects must increase with the diversity of receivers, with the variety of their sensory equipment, ways of life, and behavioural systems (appetites and aversions, flexibility, etc.). The more conflicting the effects, the less likely is the outcome of competition among selection pressures to be strongly unidirectional, the more likely is the actual selection to be stabilizing.

Some of the hostile displays of cephalopods would appear to be extreme examples of this sort of stability because they are often inter-specific signals, observed by a heterogeneous assemblage of potential predators, birds, and mammals as well as many fishes. (This, in turn, may explain several evolutionary contrasts. The hostile displays of Laridae and Anseriformes which have been progressive and changeable are primarily or exclusively intra-specific. Cephalopods may have had to take more precautions against predators because they are both unusually attractive as prey and particularly vulnerable to predation. There is, at least, evidence that they tend to be more attractive and potentially more vulnerable than most other organisms of comparable size and complexity occurring in the same environments at the present time (see Moynihan 1975).)

Speaking of the cryptic patterns of cephalopods, Packard (1972) suggests that they are 'a reflexion of the visual discrimination of the predators'. The same might be said of some other patterns. The crux of the matter is that they have had to be stable because they have had to reflect such a wide diversity of discriminators.

The correlation itself may be a general rule among all animals, including those in which the specific details are very different. All other things being equal, the more widely reflected or broadcast a signal, the more conservative it will be; the more narrowly reflected or broadcast, the more likely it is to change rapidly.

Summary

Cephalopods have a variety of ritualized, stereotyped, behaviour patterns. Some are displays, designed to convey information, true or false (mimetic), from one individual to another. Others promote crypsis and might be called 'anti-displays'. Some can be cryptic in some circumstances, signals in other situations. Several of the more elaborate patterns, of different functional types, occur in essentially the same forms in one or more species of each of three orders (Sepiida, Teuthida, Octopida) which are supposed to have diverged at some time in the early Mesozoic, probably around the end of the Triassic or the beginning of the Jurassic, something like 190 million years ago. They would appear, therefore, to have been conservative during evolution. This may be ascribed to two factors. The cryptic patterns probably have remained the same because the backgrounds (corals, plants) that they must 'match' have not changed much in colour or form. The other conservative patterns may have been stable because they are designed to influence a diversity of receivers; different age, size, and sex classes of the same species and/or individuals of other species, especially potential predators. There is reason to believe that signal patterns adapted to many kinds of receivers should tend to change less frequently or more slowly, on the average, than signals adapted to only one kind of receiver.

References

APPLEBY, R. M., CHARIG, A. J., COX, C. B., KERMACK, K. A., and TARLO, L. B. H. (1967). Reptilia. In *The fossil record* (ed. W. B. Harland C. H. Holland, M. R. House, N. T. Hughes, A. B. Reynolds, M. J. S. Rudwick, G. E. Satterthwaite, L. B. H. Tarlo, and E. C. Willey), pp. 695–731. Geological Society of London.

ARNOLD, J. M. (1962). Mating behaviour and social structure of *Loligo pealii*. *Biol. Bull. mar. biol. Lab., Woods Hole* **123**, 53–7.

—— (1965). Observations on the mating behaviour of the squid *Sepioteuthis sepioidea*. *Bull. mar. Sci. Gulf Carribb.* **15**, 216–22.

BANKS, H. P., CHESTER, K. I. M., HUGHES, N. F., JOHNSON, G. A. L., JOHNSON, H. M., and MOORE, L. R. (1967). Thallophyta. 1. In *The fossil record* (ed. W. B. Harland), pp. 163–80. Geological Society of London.

BLACK, M., DOWNIE, C., ROSS, R., and SARJEANT, W. A. S. (1967). Thallophyta. 2. In *The Fossil Record* (ed. W. B. Harland), pp. 181–209. Geological Society of London.

BLEST, A. D. (1961). The concept of 'ritualisation'. In *Current problems in animal behaviour* (ed. W. H. Thorpe and O. L. Zangwell), pp. 102–23. Cambridge University Press.

BOYCOTT, B. B. (1958). The cuttlefish—*Sepia*. *New Biol.* **25**, 98–118.

—— (1965). A comparison of living *Sepioteuthis sepioidea* and *Doryteuthis plei* with other squids, and with *Sepia officinalis*. *J. Zool. Lond.* **147**, 344–51.

CARPENTER, C. C. (1967). Aggression and social structure in iguanid lizards. In *Lizard ecology: a symposium* (ed. W. Milstead), pp. 87–105. University of Missouri Press.

CLARKE, M. R. (1966). A review of the systematics and ecology of oceanic squids. In *Advances in marine biology* (ed. F. S. Russell), pp. 91–300. Academic Press, New York and London.

COUSTEAU, J.-Y. and COUSTEAU, P. (1970). *The shark: splendid savage of the sea*. Doubleday, Garden City, New York.

COWDRY, E. V. (1911). The colour changes of *Octopus vulgaris*. *Univ. Toronto Stud. biol. Ser.* **10**, 1–53.

COX, C. B. (1967). Changes in terrestrial vertebrate faunas during the Mesozoic. In *The fossil record* (ed. W. B. Harland), pp. 77–89. Geological Society of London.

CRANE, J. (1949). Comparative biology of salticid spiders at Rancho Grande, Venezuela. Part IV. An analysis of display. *Zoologica* **34,** 159–214.

DONOVAN, D. T., HODSON, F., HOWARTH, M. K., HOUSE, M. R., TOZER, E. T., and WRIGHT, C. W. (1967). Mollusca: Cephalopoda (Ammonoidea). In *The fossil record* (ed. W. B. Harland), pp. 445–460. Geological Society of London.

DREW, G. A. (1911). Sexual activities of the squid, *Loligo pealii* (Les.). *J. Morph.* **22,** 327–52.

EWER, R. F. (1968). *Ethology of mammals.* Logos Press Ltd., London.

FISHER, J. (1967). Fossil birds and their adaptive radiation. In *The fossil record* (ed. W. B. Harland), pp. 133–154. Geological Society of London.

GADGIL, M. and BOSSERT, W. H. (1970). Life historical consequences of natural selection. *Am. Nat.* **104,** 1–24.

GOODWIN, D. (1967). *Pigeons and doves of the world.* British Museum, London.

HARDY, A. (1956). *The open sea—its natural history. The world of plankton.* Collins, London.

HEINROTH, O. (1911). Beiträge zur Biologie, namentlich Ethologie und Psychologie der Anatiden. *Ver. 5. Int. orn. congr.* **5,** 589–702.

HOLMES, W. (1940). The colour changes and colour patterns of *Sepia officinalis* L. *Proc. zool. Soc. Lond.* **A110,** 17–36.

JELETZKY, J. A. (1966). Comparative morphology, phylogeny, and classification of fossil Coleoidea. *Univ. Kans. paleontol. Contrib.* (Mollusca) **7,** 1–162.

JOHNSGARD, P. A. (1965). *Handbook of waterfowl behavior.* Constable, London.

KAUFMANN, J. H. (1962). Ecology and social behavior of the coati, *Nasua narica*, on Barro Colorado Island, Panama. *Univ. Calif. Publs. Zool.* **60,** 95–222.

—— KAUFMANN, A. (1965). Observations on the behavior of tayras and grisons. *Z. Säugertierk.* **30,** 146–55.

KLEIMAN, D. G. (1966). The comparative social behavior of the Canidae. *Am. Zool.* **6,** 335.

—— (1967). Some aspects of social behavior in the Canidae. *Am. Zool.* **7,** 365–72.

LANE, F. W. (1957). *The kingdom of the octopus.* Jarrolds, London.

LORENZ, K. (1941). Vergleichende Bewegungsstudien an Anatinen. *J. Orn., Lpz., Suppl.* **89,** 194–294.

McGOWAN, J. A. (1954). Observations on the sexual behavior and spawning of the squid, *Loligo opalescens*, at La Jolla, California. *Calif. Fish Game* **40,** 47–54.

MANGOLD-WIRZ, K. (1963). Biologie des Céphalopodes benthiques et nectoniques de la Mer Catalane. *Vie Milieu, Suppl.* **13,** 1–285.

MOYNIHAN, M. (1962). Hostile and sexual behavior patterns of South American and Pacific Laridae. *Behaviour, Suppl.* **8,** 1–365.

—— (1967). Comparative aspects of communication in New World primates. In *Primate ethology* (ed. D. Morris), pp. 236–66. Weidenfeld and Nicolson, London.

—— (1970). Control, suppression, decay, disappearance and replacement of displays. *J. theor. Biol.* **29,** 85–112.

—— (1975). Some distinctions of cephalopods; their causes and evolutionary consequences. (In preparation.)

NEWELL, N. D. (1971). An outline history of organic reefs. *Am. Mus. Novit.* **2465,** 1–37.

NOTTEBOHM, F. (1969). The song of the Chingolo, *Zonotrichia capensis*, in Argentina: description and evaluation of a system of dialects. *Condor* **71,** 299–315.

PACKARD, A. (1972). Cephalopods and fish: the limits of convergence. *Biol. Rev.* **47,** 241–307.

PACKARD, A. and SANDERS, G. D. (1969). What the octopus shows to the world. *Endeavour* **28**, 92–9.

—— —— (1971). Body patterns of *Octopus vulgaris* and maturation of the response to disturbance. *Anim. Behav.* **19**, 780–90.

ROBINSON, M. H. (1969a). The defensive behaviour of some orthopteroid insects from Panama. *Trans. R. ent. Soc. Lond.* **121**, 281–303.

—— (1969b). Defenses against visually hunting predators. *Evol. Biol.* **3**, pp. 225–259.

ROMER, A. S. (1966). *Vertebrate paleontology*. University of Chicago Press.

—— (1968). *Notes and comments on vertebrate paleontology*. University of Chicago Press.

STEVENSON, J. A. (1934). On the behaviour of the long-finned squid (*Loligo pealii* (Leseur)). *Can. Fld Nat.* **48**, 4–7.

STIRTON, R. A. (1951). Ceboid monkeys from the Miocene of Colombia. *Univ. Calif. Publs. geol. Sci.* **28**, 315–56.

TINBERGEN, L. (1939). Zur Fortpflanzungsethologie von *Sepia officinalis*. *Archs. néerl. Zool.* **3**, 323–64.

—— VERWEY, J. (1945). Zur Biologie von *Loligo vulgaris*. *Archs. néerl. Zool.* **7**, 213–86.

TINBERGEN, N. (1951). *The study of instinct*. Clarendon Press, Oxford.

—— (1963). On aims and methods of ethology. *Z. Tierpsychol.* **20**, 410–33.

—— (1964). The evolution of signaling devices. In *Social behavior and organization among vertebrates* (ed. W. Etkin), pp. 206–30. University of Chicago Press.

VAN TETS, G. F. (1965). A comparative study of some social communication patterns in the Pelecaniformes. *Ornithological Monographs*, Vol 2, pp. 1–88. Allen Press, Lawrence.

WALLER, R. A. and WICKLUND, R. I. (1968). Observations from a research submersible—mating and spawning of the squid *Doryteuthis plei*. *Bioscience*, **18**, 110–11.

WELLS, M. J. (1962). *Brain and behaviour in cephalopods*. Heinemann, London.

—— WELLS, J. (1972). Sexual displays and mating of *Octopus vulgaris* Cuvier and *O. cyanea* Gray and attempts to alter performance by manipulating the glandular condition of the animals. *Anim. Behav.* **20**, 293–308.

WHITMAN, C. O. (1919). The behavior of pigeons. *Publs. Carnegie Instn.* **257**, 1–161.

YOUNG, J. Z. (1962). Courtship and mating by the coral reef octopus (*O. horridus*). *Proc. zool. Soc. Lond.* **138**, 157–62.

—— (1964). *A model of the brain*. Clarendon Press, Oxford.

14. The evolution of predatory behaviour in araneid spiders

M. H. ROBINSON

NIKO TINBERGEN has repeatedly (1963, 1968, for instance) emphasized the fact that ethology is a branch of biology, and is properly concerned with explaining behaviour in terms of function, causation, ontogeny, and evolution. Those ethologists who are fortunate enough to work in the tropics are confronted with a rich variety of exciting animals. Even the most cursory initial observation on any part of this rich fauna serves to draw attention to problems belonging to each of Tinbergen's four categories of analysis. Tropical spiders are no exception to this generalization, and orb-weavers are particularly conspicuous in many areas—they (the araneids) are often represented by a large number of species, and individuals, that are present for a longer period of each year than are their temperate zone relatives. In these circumstances opportunities for comparative studies approach the state of an embarrassment of riches.

Araneid spiders frequently have complex and interesting behavioural repertoires associated with web-building, predator-avoidance, courtship and mating, and prey-capture. The totality of behaviours associated with the process of prey-capture is here called predatory behaviour. Over the past 5 years, in conjunction with a number of co-workers, I have studied the predatory behaviour of araneids in Panama, Africa, India, and New Guinea. These studies have been carried out in depth on some species and in less detail on others. So far, they have encompassed 23 species from 8 genera. Data are available from other sources on several species of 2 additional genera. The studies have been concentrated on questions about the function and causation of behaviour, and through the comparative work, on its evolutionary origins. Studies on the ontogeny of predatory behaviour, in selected species, are currently in progress.

In this essay the results of the comparative studies are used (alongside functional interpretations derived from the studies of particular species) to develop a scheme outlining the possible stages in the evolution of araneid predatory behaviour. This scheme is largely derived from a detailed analysis of variations in the process by which spiders enswathe their prey in silk (wrapping behaviour). To provide a background for assessing the roles of wrapping behaviour, and for the interpretations involved, a general account of the main features of araneid predatory behaviour is given at this stage.

14.1. General features of the predatory process

Adult female spiders of all the species that have been studied possess general patterns of predatory behaviour which they repeat with a high degree of predictability (males, in all but a few species, are diminutive in size and difficult to study). Essentially all araneid spiders deal with their prey in a series of steps. The prey item is located, attacked, and immobilized, removed from the web at the capture site, and then transported to the feeding site. The entire process involves a sequence of behaviour units. Spiders of the same species use different behaviour sequences to deal with different types of prey. From this it can be inferred that they discriminate between these types of prey. Sequences may differ in both the number, nature, and order of the behaviour units that comprise them. Different species may possess different repertoires of both predatory sequences and units of predatory behaviour.

All the araneids bite the prey during a predatory sequence. The bite involves the injection of venom and may occur at different stages in the predatory process and differ in the accompanying behaviours, and duration, according to the stage at which it occurs. In wrapping behaviour the spiders use multi-strand silk to enswathe their prey. The silk is deposited on the prey by movements of the fourth pair of legs. The exact details of the process may vary in the nature of the leg movements involved, and also in the orientation of the spider in relation to the prey item. These variations are considered later. There are also differences in the stage (or stages) at which wrapping occurs within a predatory sequence. These differences are of particular evolutionary interest.

Transportation of prey from the capture site to a feeding site is a basic feature of the predatory behaviour of araneids. The feeding site is usually the hub of the web (the point of origin of its radial elements) but, in some cases, may be a nearby retreat which is connected to the hub by a signal line. In the following sections the feeding site is simply referred to as the 'hub', the qualification 'or retreat' being omitted.

The spiders carry prey to the hub in two ways. It is either carried held in the chelicerae (*in jaws*), or suspended on a silk thread behind the spider (*on silk*). In each case it may be supported by one or more legs in addition to the primary support. Insects are removed from the web, prior to transportation, either by being pulled off the adhesive part of the structure or by being cut free of entangling web elements. With these details as a background the roles of wrapping behaviour can now be considered.

14.2. Wrapping behaviour in the predatory process

Early studies of araneid predatory behaviour, at least those of an exhaustive and detailed character, were confined to spiders of temperate regions. These studies, such as those of Peters (1931, 1933a, b) on *Araneus diadematus* showed that spiders could wrap prey at the capture site as either the first or second stage of attack and immobilization, and also could wrap or re-wrap prey after it had been transported to the hub from the capture site. *Araneus diadematus* attacked non-vibrating prey by wrapping and then biting (wrap–bite-attack couplet) and vibrating prey by biting and then wrapping (bite–wrap couplet). Studies of the tropical (to sub-tropical) araneid *Argiope argentata* (Robinson 1969; Robinson and Olazarri 1971) revealed that this spider had three attack-behaviour couplets. Very small prey (less than a hundredth of the spider's weight) were seized in the jaws and pulled from the web without any prior wrapping (seize–pull-out). All other prey, except lepidopterans, were wrapped and then bitten (wrap–bite), and the greater proportion of lepidopterans was bitten first and then wrapped (bite–wrap). Lepidopterans formed only 2·9 per cent, by numbers, of the total yearly catch of a sample population of this species in Panama (Robinson and Robinson 1970). Thus the bite–wrap couplet may receive a minor amount of usage. *Argiope argentata* wrapped prey at the hub in two sets of circumstances: if they had been attacked by seize–pull-out, or if they had been attacked by either of the two methods but were then carried to the hub in the jaws.

To summarize, the data from studies of *Araneus diadematus* and *Argiope argentata* show that wrapping can occur at the following stages in the predatory process.

At the hub:

(1) after the prey has been carried there in the jaws of the spider.

At the capture site:

(2) after a biting attack but before the prey has been freed from the web;

(3) as the initial means of attack, followed by a bite.

Fig. 14.1(a) illustrates this situation diagrammatically.

The use of wrapping behaviour at these three stages has been tacitly assumed to be involved in the predatory behaviour of all araneids (Eberhard 1967). A study of the predatory behaviour of *Nephila clavipes* showed that this was not the case (Robinson, Mirick, and Turner 1969; Robinson and Mirick 1971).

This species attacked all types of prey by biting; attack wrapping was completely absent from its predatory repertoire. Interestingly, in addition, the behaviour of wrapping after a biting attack was found to be nothing like as predictable as it is in the case of *Argiope argentata*. Large prey were bitten and the spider then made attempts to pull the prey from the web at the capture site (bite–pull-out). Many kinds of prey are freed from the web by this method, and then carried to the hub where they are then wrapped for the first time. Other types of prey are not consistently freed by pull-out attempts and are then wrapped *in situ* and cut from the web before transportation (bite–pull-out-failure–wrap).

Robinson, Mirick, and Turner (1969) attached threads to the bodies of prey (crickets) that the spider could normally pull-out; by pulling on the thread in the opposite direction to the spider they were able to prevent the spider from freeing the prey. In these circumstances the spider wrapped the prey *in situ* after making prolonged pull-out attempts. Some bulky prey that *Nephila clavipes* succeeded in removing from the web by pulling were subsequently wrapped beneath the web before transportation (free-wrapping: Robinson, Mirick, and Turner (1969)). Wrapping after the failure of pull-out attempts and free-wrapping are details of importance to the assessment of the functional significance of wrapping at the capture site (see §14.3, *Wrapping at the capture site following a biting attack*).

Later studies showed that attack-wrapping is absent from the predatory repertoires of three other *Nephila* species: *N. turneri*, *N. constricta*, and *N. maculata*, as well as *Nephilengys cruentata*, which is placed in the genus *Nephila* by some authors, (Robinson and Robinson 1973). All these *Nephila* species have the bite–pull-out attack couplet and wrap prey *in situ* after the failure of pull-out attempts. Fig. 14.1(c) illustrates the behaviour sequences of *Nephila* species in diagrammatic form.

Attack-wrapping is also absent from the behavioural repertoires of all the species of the genus *Micrathena* that have been studied, and from that of *Herrenia ornatissima*.

Table 14.1 details the types of wrapping behaviour, classified by the stage at which they occur in predatory sequences, for all the araneids that I have studied, and for those others for which complete descriptions exist. The sample is fairly large, and the spiders fall into three categories:

(1) those with wrap–bite and bite–wrap attack couplets (the latter often used for attacks on lepidopterans) plus wrapping at the hub;

(2) those with only the bite–wrap attack couplet plus wrapping at the hub;

(3) those with bite–pull-out and bite–wrap couplets, which also wrap at the capture site after the failure of pulling-out attempts, these also wrap at the hub.

Fig. 14.1 summarizes these categories.

Before going on to consider whether the absence of attack wrapping in categories (2) and (3) represents a primitive condition, it is useful to consider the detailed form of the various types of wrapping behaviour.

14.3. The form of wrapping behaviour

Wrapping at the hub

Prey that are wrapped at the hub have been freed from the web, and, at the commencement of wrapping, are held in the jaws. They may or may not

Table 14.1.
Wrapping behaviour of araneid spiders

Type	Species	Data source	Where studied
Wrap at capture site after biting attack; also after pull-out failure	Nephila clavipes (Linnaeus)	Robinson and Mirick (1971)	Panama
	Nephila maculata (Fabricius)	Robinson and Robinson	New Guinea
	Nephila constricta Karsch	Robinson and Robinson	Africa
	Nephila turneri Blackwall	Robinson and Robinson	Africa
	Nephilengys cruentata Simon	Robinson and Robinson	Africa
Wrap at capture site after biting attack	Herrenia ornatissima (Doleschall)	Robinson and Robinson	New Guinea
	Micrathena clypeata (Walckenaer)	Robinson (unpublished)	Panama
	Micrathena schreibersi (Perty)	Lubin (in preparation).†	Panama
	Micrathena sexspinosa (Hahn)	Lubin (in preparation).†	Panama
Attack-wrapping: Wrap at capture site following biting attack	§Caerostris tuberculosa (Vinson)	Robinson (in preparation)	Madagascar
	§Gasteracantha cancriformis (Linnaeus)	Muma (1971)	U.S.A.
	§Cyrtophora citricola (Forskal)	Lubin (1972)	Africa
	§Cyrtophora moluccensis (Doleschall)	Lubin (1972)	New Guinea

§*Cyrtophora cylindroides* (Walck.)	Lubin (1972)	New Guinea
§*Cyrtophora monulfi* (Chrysanthus)	Lubin (1972)	New Guinea
Argiope argentata (Fabricius)‡	Robinson (1969), etc.	Panama
Argiope savignyi Levi†	Robinson (unpublished)	Panama
Argiope florida Chamberlain & Ivie‡	Robinson (unpublished)	U.S.A.
Argiope aurantia Lucas‡	Robinson (unpublished)	U.S.A.
Argiope aemula (Walck.)‡	Robinson and Robinson (in preparation)	New Guinea
Argiope aetheria (Walck.)‡	Robinson and Robinson (in preparation)	New Guinea
Argiope picta Koch‡	Robinson and Robinson (in preparation)	New Guinea
Argiope species F, new‡	Robinson and Robinson (in preparation)	New Guinea
Araneus marmoreus Clerck‡	Robinson (unpublished)	U.S.A.
Araneus rufipalpis (Lucas)‡	Robinson (unpublished)	Ghana
Araneus diadematus Clerck	Peters (1931), etc.	Europe
Eriophora fuliginea C. L. Koch‡	Robinson et al. (1972), etc.	Panama
Eriophora nephiloides (P-Cambridge)‡	Robinson (unpublished)	Panama

† Wrap at hub.
‡ Spiders bite–wrap lepidopterans.
§ Spiders bite–wrap many insects.

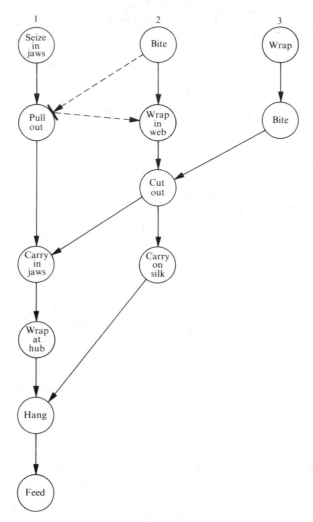

FIG. 14.1. Additive diagram of araneid predatory behaviour. Circles represent behaviour units, arrows connect these in sequences. Where more than one arrow leaves a circle alternate pathways exist. Sections 1+2+3 constitute Fig. 1(a), sections 1+2 constitute 1(b), and sections 1+2 and the loop in broken lines constitute 1(c). Section 1 corresponds to Stage 1 of the evolutionary scheme outlined in §14.5. Sections 1+2 plus loop in broken lines correspond to Stage 2(a), these sections minus broken lines loop correspond to Stage 2(b). Sections 1+2+3 correspond to Stage 3.

have been wrapped prior to this (at the capture site: by wrap–bite, bite–wrap, or during free-wrapping).

At-the-hub wrapping always starts with the spider casting skeins of silk forwards, under its body, onto the surface of the prey. The silk is picked up from the spinnerets on the tarsi of the fourth legs and is moved forwards by accurately oriented leg movements. The long fourth legs are strongly flexed during the silk-casting. (Very few of the webs of spiders are exactly perpendicular, most of them slope to a greater or lesser extent, and the spider operates beneath the slope.) In wrapping at the hub the spider hangs beneath the web in a head-upwards position and only faces head-downwards after completing prey-wrapping. During the process of turning from the wrapping position to the predatory stance the spider attaches the silk line from the prey package to several points on the hub silk. The prey package is thereby securely anchored to the web.

Spiders that build nearly horizontal webs hang below these, suspended from their drag line, during the wrapping process. The leg movements involved in wrapping at the hub are generally similar in all species of araneids. Major differences do, however, exist in the speed of the movements. Spiders that do not have attack-wrapping in their repertoires make slow alternate movements of the fourth legs ('bicycling'), whereas the others make very rapid movements that become nearly synchronous. The leg movements of *Nephila* species, for instance, can be counted without recourse to motion picture analysis, whereas those of *Argiope* species cannot be so counted. The prey is held in the jaws until a variable amount of silk is deposited and then is picked up on the short third legs and released from the jaws. The third legs of most araneids are markedly shorter than all the others. They are thus able to manipulate the prey beneath the spider while the longer legs are used to support the spider on the web. This adaptation of the third legs is not confined to araneids, but is most marked in their case. Wrapping continues when the prey is held by the third legs which move it from side to side during silk deposition. As an indication of the amount of effort involved in wrapping at the hub the following data can be cited. In 50 presentations of orthopterans (weight range 100–170 milligrams) to *Nephila maculata* the number of leg movements, per presentation, involved in wrap-at-hub ranged from 38–108. The spider weighed at least 20 times as much as these prey. At the most conservative estimate a minimum of 2 metres of silk was used for wrapping each of these relatively small insects. (Data from Robinson and Robinson 1973.)

The process of attaching the wrapped prey to the web is of considerable importance for the functional interpretation of wrapping at the hub. In all cases, a silk line from the packaged prey is fastened to the web by special movements of the spinnerets. There are differences in the attachment process that can be correlated to differences in hub structure. Thus some spiders, for example, *Eriophora fuliginea* (Table 14.1) have an open hub, that is, at the end of web construction the spider bites out the silk of the hub region, leaving a hole. It subsequently reinforces the edge of this hole and spans it with its body when it is in its normal predatory stance. Spiders with such open hubs have very elaborate movements during the process of hanging prey. The silk line from the prey package is attached to both sides of the silken circumference of the hole, and in several places. This contrasts with the situation in spiders that have closed hubs which merely attach the prey to one side of the hub region. It seems probable that the open hub is a point of structural weakness and the complex hanging behaviour results in an even distribution of the load around this. Spiders of the genus *Cyrtophora* which have very fine-meshed non-adhesive snare and platform webs (see later) have very simple prey-attachment behaviours (Lubin 1972).

Wrapping at the capture site following a biting attack

This resembles wrapping at the hub in that it is initiated when the prey is in the jaws (since it is being bitten). As in wrapping at the hub the first movements of the fourth legs carry silk forwards under the spider and onto the prey. Subsequent stages, however, may differ from those seen in wrap-at-hub. These differences, in part at least, can be attributed to the fact that the prey is not free of the web, since it is enmeshed to a greater or lesser extent according to the effect of its struggles. The web therefore acts as a barrier to the free passage of silk around the prey, and the spider, after the stage of in-jaws wrapping, releases its cheliceral hold on the prey and cuts away intervening web elements in a gradual and progressive manner. Wrapping thus continues intermittently. At this stage, in the case of spiders with attack-wrapping potential, the prey is frequently rotated about its long axis, on web attachments that act like spindles, while silk is laid down bobbin-fashion. In the case of very large prey the spider may run around the prey laying down silk directly from its spinnerets rather than pushing it onto the prey with leg movements. In *Nephila* species, on the other hand, as the prey becomes more and more freed from the web by cutting activities, the spider adopts a wrapping stance that is essentially similar to that assumed at the hub (see Fig. 14.2). Thus the additional behavioural complexity of this type of wrapping is greatest in those spiders that rotate their prey and least in *Nephila* species where the only major new component is the intermittent cutting of web attachments.

Attack wrapping

Attack wrapping differs from the other forms of wrapping in that it is initiated when the prey is held only by the web. Swathes of silk are cast onto the prey and (frequently) partly onto the web, so that the prey is sandwiched between the web and layers of wrapping silk. Fabre (1903) gives a superbly vivid description of this process. The spider may stand above the prey and cast silk directly onto it, or may face the prey and cast silk forwards onto it. Some spiders approach large prey, then turn to face

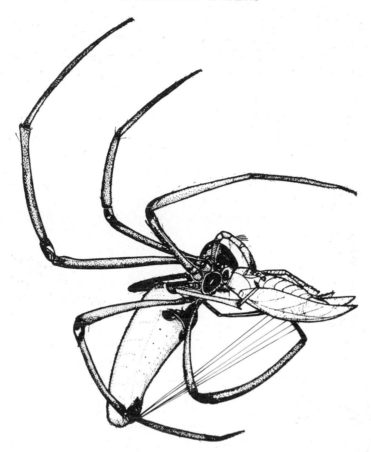

FIG. 14.2. *Nephila maculata* wrapping an orthopteran after a biting attack. The prey is still held in the jaws and silk is being passed onto the prey on the tarsus of left fourth leg while right fourth leg is moving back to pick up silk from the spinnerets. Note that the prey is gripped in third legs (short) and that the pedipalps are touching the prey package. Based on frames of 16 mm film. (From Robinson and Robinson 1974, by courtesy of Smithsonian Institution Press.)

away, out of contact distance, while they cast silk backwards (Robinson, Robinson, and Graney 1972). Later stages of attack-wrapping resemble the later stages of wrapping following a biting attack, involving intermittent cutting. The early stages of attack-wrapping presumably involve a more complex aiming component than that involved in other types of wrapping. When the spider is touching the prey and casting silk onto it the cues used in directing the silk could be basically similar to those used in other kinds of wrapping but when it is not touching the prey (or even facing away from it) the cues must be qualitatively different and the orientation of the silk-throwing movements presumably requires more complex control.

It seems reasonable to conclude that the increase in behavioural complexity involved in the step from wrapping after a biting attack to wrapping as an attack technique is greater than that involved in the step from wrapping at the hub to wrapping at the capture site following a biting attack.

14.4. Functional aspects of wrapping behaviour

Eberhard (1967) argued, on the basis of comparisons between the predatory behaviour of web-building spiders of different families, that attack wrapping probably evolved from wrapping at the feeding site. At the time he had access to data from only a small number of araneid species. Robinson, Mirick, and Turner (1969) supported the general hypothesis by a more detailed argument from the limited data then available from within the family Araneidae. It is now possible to add more weight to their views. It is logical to examine the adaptive function of the various types of wrapping behaviour and then outline their possible course of evolution.

Wrapping at the hub, as Eberhard (1967) suggested, clearly functions to allow the spider that has already got prey in its possession to make subsequent attacks without the danger of losing this prey. Wrapping makes it possible to anchor the prey to the web at the feeding site where it is secure against loss. This is suggested by the elaborate nature of the attachment part of at-the-hub wrapping (§14.3). If the spider had to carry already-caught prey into subsequent attacks these would be *at the very least* seriously impared in efficiency, and in some cases, impossible to effect. This was demonstrated in experiments carried out by Robinson, Mirick, and Turner (1969). By presenting *Nephila maculata* with small active prey (stingless bees) at 30-second intervals these authors contrived a situation in which the spider received a new prey before it had time to wrap and hang the old prey (at the hub). In these circumstances most of the spiders omitted the wrapping behaviour and rushed into a new attack carrying the 'old' prey in their jaws. Most of the spiders that did this lost the first prey, which was inactive by then, as they tried to bite the second bee. Those that succeeded in the second attack made the third attack carrying two prey in their jaws and losses increased accordingly. If the spider delayed the wrapping and hanging behaviour until it detected the presence of a new prey (in more normal conditions) it would lose time before attacking, and since some prey can rapidly escape from spider's webs (Robinson, Mirick, and Turner ibid.; Lubin 1972) this time loss loss could be critical.

Wrapping at the capture site after a biting attack has several possible functions. The most important of these may be that suggested by the behaviour of *Nephila* species which can be induced to wrap prey at the capture site if they are not allowed to remove it by pulling movements (see p. 295). Removing prey from the web other than by pulling it out inevitably involves cutting entangling web members. The spider cannot use its chelicerae for cutting silk and still use them to hold the prey. Once it starts to cut it must run the risk of dropping and losing the prey unless it can attach the prey to its body. Wrapping at the capture site enables the spider to do this. Once the prey is partly wrapped it is possible to quickly burn out all the web connections with a hot needle and prematurely free the insect from the web. This does not result in loss of the prey since it simply drops to the end of the silk line and swings below the spider hanging from the spinnerets.

There is an interesting difference in the pulling-out behaviour of *Nephila* species when it is applied to removing plant debris from the web rather than prey. This difference reinforces the conclusion that this stage of predatory behaviour is one at which the spider is in danger of dropping the prey. When pulling-out long prey items, the spider may tease one end away from the web and then move its jaw-hold to the other end. When it does so it always holds the prey with one or more legs. When dealing with leaves and other plant debris the spider does not do this. It may pull at one end, release its jaw-hold, pause, and then tackle the other end without holding the plant material between successive pulls. It does not, therefore, constantly restrain an object that it is about to reject.

The fact that spiders continue wrapping at the capture site *after* the prey has been secured against dropping suggests that the further wrapping may have an additional (and secondary?) function. This is also suggested by the fact that *Nephila* species wrap some prey *after* they have been freed from the web by pulling. The effect of this wrapping is to truss the prey so that it becomes a compact package. In the process long projections such as the thin abdomens of dragonflies, and the long jumping legs of orthopterans, become folded into the package (Robinson and Olazarri 1971). Such packaging must facilitate transportation, since the spider most frequently has to ascend a sloping, or nearly perpendicular, plane. Spiders walking forwards up the web with prey in their jaws occasionally get the prey stuck. Wrapping at the capture site also permits the spider to use a method of transportation that would not otherwise be available to it. It can carry the prey suspended on a silk thread behind it. In this case the prey swings, like a pendulum, away from the plane of the web and possible entanglement. This method is consistently used for the transportation of heavy prey (Peters 1933b; Robinson 1969) which are normally bulky and would not be easily transportable by other means. *Nephila maculata* transports some heavy prey by this method and carries others in the jaws by backing up the web. Backing-up results in a load distribution that is more favourable than walking forwards with prey in the jaws but is more time consuming than transportation on silk (see Robinson and Robinson (1973) for details).

Attack-wrapping confers a number of advantages in addition to sharing some of the functions of capture-site-wrapping after a biting attack (facilitating removal of prey from the web and trussing). It allows the spider to:

   (1) attack large and/or potentially dangerous prey without the dangers involved in the intimate contact of a biting attack;
   (2) reduce the time spent away from the hub of the web in predatory sequences.

The first function can be deduced from details of attack-wrapping behaviour, which does not require bodily contact between the spider and its prey (see §14.4: *Attack wrapping*), and also from observations on the biting attack of *Nephila* species on potentially dangerous prey. In fact, *Nephila* species attack large prey by using a special kind of biting attack (Robinson and Mirick 1971; Robinson and Robinson 1973). This special attack technique has been called 'bite and back-off behaviour'. The spider closes to the attack with its anteriorly directed legs reflexed back over its body, darts in for a very brief bite, and then immediately retreats out of contact range. The dart forward, brief bite, and retreat may be repeated many times before the spider makes a sustained bite. Such attack cycles can last many minutes, during which time the prey may struggle free of the web (12 out of 25 large acridiids presented to *Nephila maculata* escaped— Robinson and Robinson 1973). Spiders with attack-wrapping behaviour can attack proportionately large prey without hesitation, quickly, and without initial bodily contact (see below). Attack-biting also exposes the spider to the full blast of chemical defences when they attack insects such as pentatomids, and we have seen *Nephila* species retreat to the hub and groom themselves intensively after such counter attacks. *Argiope argentata*, on the other hand, by using attack-wrapping, can effectively capture pentatomids (Robinson and Olazarri 1971).

Economy of time spent in predatory sequences may be of great adaptive significance for two reasons. Many spiders have protective devices at the hub which conceal them from predators; others operate from a retreat where they are both concealed and enclosed. When they are away from these protections, on the surface of the web, they are almost certainly more vulnerable to predation; a reduction in time spent in predatory sequences must reduce this risk. Secondly, when the spider is away from the hub coping with a prey it inevitably has a reduced ability to detect the arrival of subsequent prey. A quick return to its web-monitoring position must therefore be advantageous. Evidence for the time-saving effect of attack-wrapping comes from two sources both of which, by themselves, are not entirely convincing. Taken together there seems little doubt that attack-wrapping does effect time-economy. Thus it is possible to compare the times involved in bite–wrap and wrap–bite attack couplets both between species and also within species.

TABLE 14.2

Times spent in total predatory sequence (attack to feed) by the spiders Argiope aemula *and* Nephila maculata.

Argiope aemula			Nephila maculata	
Weight 0·5–0·9 grams		Prey Melolonthid beetle 0·5–0·8 grams	3·0–4·0 grams	
Times			Times	
minutes	seconds		minutes	seconds
4	21		7	75
6	40		10	58
4	13		8	17
5	15		31	10
9	18		28	41
8	47		12	15
4	05		9	18
3	24		13	08
6	39		11	32
11	32		10	29

From Robinson and Robinson (1973), by courtesy of Smithsonian Institution Press.

Table 14.2 gives the durations of predatory sequences for *Argiope aemula* and *Nephila maculata* when dealing with living melolonthid beetles. The *Argiope* uses the wrap–bite attack couplet and the *Nephila* either bite–wrap or bite–back-off couplets. Note that *Nephila maculata* is more than 3 times as heavy as *Argiope aemula*, and is attacking prey that are much smaller and lighter than itself, whereas the *Argiope* is attacking prey of approximately its own weight and bulk. Despite this difference the *Argiope* predatory sequences are significantly shorter than those of *Nephila*. Comparisons of the time-expenditure within species that have both attack strategies are complicated by the fact that both strategies are not normally used for the same type of prey. However, although bite–wrap is used almost exclusively in attacks on lepidopterans, a small proportion of these are subjected to wrap–bite attacks (Robinson (1969) regarded these cases as mistaken discriminations). In these cases the durations of wrap–bite attacks is significantly less than those of bite–wrap attacks (see Robinson and Olazarri 1971, p. 27). Peters (1931) pointed out that the bite given in a biting attack was a long bite, whereas that following a wrapping attack was a short one; comparisons in this case were complicated by the fact that insects that were bitten first were vibrating, whereas those that were wrapped were not. In the case of *Argiope argentata* bites given after wrapping a given class of prey are consistently shorter than attack-bites, irrespective of the state of activity, or weight, of the prey. In fact the long bite may be long simply because, in addition to injecting venom, the spider is using its chelicerae to *hold* the prey and must continue to do so until the venom takes effect. The bite following a wrapping attack is not used to hold the prey, since it is already enswathed in silk. This point means that the spider's presence at the capture site is not essential, and, if there was a strong selective advantage to a speedy return to the hub (as suggested above), the spider could interrupt the predatory sequence and defer transportation to a later stage. *Argiope argentata*, in fact, does this, and does it most frequently following a wrap–bite attack (Robinson 1969).

Wrapping after a biting attack also allows the spider to interrupt the predatory sequence after the prey has been secured by wrapping. It is interesting to note that *Nephila* species do not do this, but *Microthena* species will do so if they already have prey at the hub. This difference is discussed later.

Thus attack-wrapping confers the advantages of capture-site-wrapping, and, in addition, allows the spider to attack prey with reduced risk to itself and to effectively attack large prey with a reduction in the risk of the prey escaping; it also allows the spider to achieve an economy in the time that it spends away from the hub of the web.

In view of these advantages the retention of the bite–wrap couplet by spiders capable of attack-wrapping must be explained. In fact bite–wrap is used exclusively against prey that have a high escape potential—either against lepidopterans or small rapidly vibrating prey, or both. Lepidopterans are able to escape rapidly from the adhesive elements of spider's webs (Eisner, Alsop, and Ettershank 1964). In effect, the loose wing-scales stick to the adhesive and are left behind as the insect struggles free. (Robinson and Robinson (1970) scored more lepidopteran escapes than captures in their census of the prey of *Argiope argentata*.) Small, rapidly vibrating prey may contact only a few adhesive droplets, and have a high escape potential as a consequence; Barrows (1915) reported that flies remained in the orbs of *Araneus sericatus* for only 5 seconds, on average. Clearly, the spider must restrain such prey as rapidly as possible and this is most readily achieved by immediate seizure in the jaws. The extent to which small rapidly vibrating prey become entrapped must vary with the mesh size of the web, and their escape potential will also depend on its strength. In this respect it is interesting to note that *Eriophora fuliginea*, which builds a web with at least twice the mesh size of *Argiope argentata*, bite–wraps prey of a size that the latter attacks by wrap–bite (Robinson, Robinson, and Graney 1972).

The fact that some araneids use the bite–wrap couplet to deal with lepidopteran prey, while attacking other prey of similar size by wrapping, was first revealed by the study of *Argiope argentata* (Robinson 1969). This piece of adaptive behaviour depends on a remarkably rapid discrimination between butterflies and moths, on the one hand, and all other insects. The discrimination does not depend on the vibrational state of the insect; it is made even if the lepidopterans are dead, and even when they are wingless (Robinson 1969; Robinson *et al.* 1972). Subsequent studies have shown that *Argiope argentata* is not alone in possessing this 'taxonomic sense'. Table 14.1 lists the species of *Argiope*, *Araneus*, and *Eriophora* that make the discrimination. It is clear that the phenomenon may well be widespread among araneids.

14.5. The influence of web structure and ecological factors on predatory patterns

The webs of araneids differ considerably in area, shape, mesh-size, inclination (from nearly perpendicular to horizontal), and siting. In some genera the sticky trap is even replaced by a non-sticky, knock-down device. The differences in the prey-trapping strategies involved in particular types of webs are clearly a potential source of differences in the nature of attack and other, predatory behaviours that must be taken into account in developing an evolutionary scheme. One example has already been quoted in detailing the relatively minor difference between the behaviour of *Eriophora fuliginea* and *Argiope argentata*. At present, two other examples are sufficiently clear-cut to be cited. Both involve spiders with attack-wrapping as part of their predatory repertoire.

The spider, *Caerostris tuberculosa*, builds large, relatively fragile, wide-meshed webs at heights of 5–7 metres above the ground, near streams and lakes. Its prey must be largely composed of day-flying insects (Robinson, in preparation). Orthopterans and other low-flying insects of the herb layer must be relatively unusual prey. This spider uses attack-wrapping for relatively few prey items (see Table 14.1). The preponderance of attack-biting can be related to the need to restrain active prey before they damage the fragile web, and its long supporting members, or struggle free of the widely spaced viscid element.

Studies of spiders of the araneid genus, *Cyrtophora*, carried out recently by Lubin (1972) are of great interest. These spiders build non-adhesive snare-and-platform webs. The spider attacks prey from beneath a domed platform that is close-meshed, very strong, and more-or-less horizontal in orientation. Since this platform has no adhesive components its prey-restraining efficiency is low compared with that of typical orbs of compar-able area (Lubin 1972). Very rapid restraint of the prey that strike the platform is essential if they are not to escape. All four of the species of *Cyrtophora* that Lubin studied are capable of attack-wrapping, but use it less consistently in their attacks than do spiders with typical sticky orbs (Table 14.1). Lubin argues convincingly that the fact that *Cyrtophora* species bite-attack a wider variety of prey than do comparable-sized *Argiope* species may be due to the fact that the close-meshed platform constitutes a barrier between the spider and its prey. At one and the same time it allows the spider to bite prey with reduced risk of injury to itself and also interferes with the wrapping process. Nevertheless, it is noteworthy that *Cyrtophora moluccensis* uses wrapping most consistently on very large insects (potentially dangerous) and also on pentatomids (which have a vigorous chemical defence). *Cyrtophora* species do not interrupt predatory sequences after the completion of wrap–bite attacks by returning to the hub (see p. 306). This may be due to the fact that wrapping cannot be accomplished, given the close-meshed platform, without freeing the prey from the web during the intermittent cutting stages.

14.6. Possible steps in the evolution of wrapping behaviour

The pitfalls of evolutionary interpretation are well known and are much greater in extent for the ethologist than for the comparative morphologist. Tinbergen (1959, p. 321) has said that compared with the morphologist the ethologist 'aiming at evolutionary interpretation has a very narrow and restricted inductive basis to operate from'. The comparative method assumes that 'differences between contemporary related forms are consequences of divergent changes in time' and that 'in order to translate differences between contemporary forms into changes of time, one further step is required: an interpretation of the direction of change' (Tinbergen 1959, p. 322).

In the case of araneid predatory behaviour this step must presently rely entirely on an interpretation of the form and function of behaviour, since (as far as I am aware), there is no generally agreed phylogenetic scheme for relationships within the family that is based on morphology. Nevertheless there is, I feel, a good inductive basis for assuming that wrapping at the hub is primitive. It is the type of wrapping behaviour that is most simple in form and probably the most basic in function. The latter point derives from the fact that it allows the spiders to safely attack further prey after they have subdued one prey, other forms of wrapping behaviour essentially allow the spider with this basic ability to extend the *range* of organisms which can be effectively attacked. Furthermore, as one would predict if the adaptation is basic, this behaviour is 'retained' in the more complex repertoires of more 'advanced' spiders. The behaviour occurs as the only form of wrapping behaviour in web-building spiders of the family Diguetidae (Eberhard 1967). This seems to be a case of convergence; the diguetids face the same functional problem in the operation of their traps as do the araneids. The assumption that attack-wrapping is the most advanced condition has also a strong inductive basis; it is behaviourally the most complex form of wrapping, requiring a complex steering or aiming component, and confers advantages that greatly extend the range of prey that the spider can efficiently subdue. On this basis the following scheme is proposed.

Stage 1. All prey are overcome by biting and are then pulled from the web, carried, and wrapped at the hub. No araneid yet studied relies exclusively on this method but *all* species utilize it for some prey items.

Stage 2(*a*). All prey overcome by biting, but in addition to Stage 1 behaviour the spider wraps some prey *in situ* at the capture site, *but only after the failure of pull-out attempts*. Again this stage is not found exclusively in any species. It is, however, the preponderant strategy of all the species of *Nephila* that have been studied. Evolution of this type of behaviour would not necessarily increase the range of prey that could be *attacked* but would increase the range of prey that could be removed from the web, and transported, after an attack. It would, therefore, effectively increase the range of prey that the spider could cope with.

Stage 2(*b*). All prey overcome by biting, but in addition to the use of behaviours of Stages 1 and 2(a), the spider can omit pull-out behaviour and wrap prey immediately after a biting attack. *Nephila* species are at this stage, and *Micrathena* species have gone further in the elimination of the association between pull-out attempts and capture-site wrapping. Again this system involves no enhancement of the initial attack but does extend the spiders handling capacity for prey. If *Nephila* species are at a less advanced stage than the *Micrathena* species (because of the greater use of pull-out units) this might also explain why the capacity to interrupt predatory sequences after bite–wrap attacks is not found in this group.

Stage 3. Attack-wrapping added to the repertoire of predatory be- haviours found in Stage 2(b). This stage is found in all the species of *Argiope, Araneus, Eriophora, Caerostris*, and *Cyrtophora* that have so far been studied (Table 14.1). It represents a major advance in attack potential.

It should be noted that this scheme is essentially additive and that new units and sequences are used as alternatives to the 'pre-existing' ones in the extension of the range of prey. Fig. 14.1 illustrates this point.

There is no implication in this scheme that contemporary spiders that are at Stage 2 are ancestral to those at Stage 3. It is only suggested that their predatory behaviour may represent a surviving form of the stage through which the Stage 3 forms have passed. Of course this scheme could be read backwards giving the interpretation that the *Nephila* and *Micrathena* species have lost the capacity for attack-wrapping. The studies of the behaviour and ecology of these species, and particularly the functional analyses, provide, however, no basis for such a view.

Kaston (1964) used data on the structure and modes of construction of spider's webs to build a scheme of evolutionary trends. He places the *Cyrtophora* type of snare and platform web as basal to the araneids, with the *Nephila* web as the most primitive of the true orbs. This placing of *Nephila* accords well with the above conclusions based on predatory behaviour. The placing of *Cyrtophora* does not. In fact the interpretation of the web-structure of *Cyrtophora* species is difficult. Kaston (1964) quotes Marples (1949) to the effect that the *Cyrtophora* web can be regarded as either a specialized derivative of an orb-web or a step towards the orb-web. In a case like this where the evidence from web-structure is ambiguous the evidence from the study of predatory behaviour may be critical. Lubin (1972) regards the *Cyrtophora* web as derived from the typical orb rather than the source of typical orbs. Kaston (1964) also regards the *Micrathena* group as possessing more advanced webs than the *Argiope* species since they have an open hub and no barrier web. This interpretation does not necessarily conflict with the scheme based on predatory behaviour, since the divergence in web structure could have occurred before the divergence in predatory behaviour. This point could possibly be resolved by morphological studies.

In conclusion it is necessary to point out that despite the increasing attention being given to araneid spiders by students of behaviour comparative studies are at an early stage. The field is wide and exciting. The ideal has been summarized by Niko Tinbergen, 'the keys to a full understanding of the effects of selection therefore are (1) functional study, (2) study of all functional systems an animal possesses, and, last but not least, (3) studies of interactions of all kinds.' (1959, p. 327–8). The application of these 'keys' to the study of spiders cannot fail to extend our appreciation of the intricacy and beauty of the evolutionary process.

14.7. Summary

This paper concentrates on an examination of the role of prey-wrapping behaviour in the predatory sequences of orb-web spiders. Comparative data are used to reconstruct the possible course of evolution of araneid predatory behaviour.

All araneids attack and immobilize prey and then transport it to a feeding site. This is either the hub of the web or a nearby retreat that is connected to the hub. Prey that are carried to the hub in the spiders' jaws are wrapped and secured to the web, with silk, on arrival. Wrapping at the hub thus functions to secure the prey to the web and enables the spider to make subsequent attacks without the danger of losing this prey. Prey that have been wrapped at an earlier stage and are carried suspended on silk behind the spider are not usually wrapped at the hub but *are attached to the web on arrival*. Some prey are also wrapped, by all species, after biting attacks and before transportation. This wrapping functions to allow the spider to remove the prey from the web without the risk of dropping it and also has a trussing effect that facilitates the transportation of bulky prey. It also enables the spider to carry prey that are heavier or larger than it could possibly carry in its jaws. Such prey are packaged in silk and then carried suspended on a line behind the spider. In this position the package swings away from the web and thus away from possible entanglement.

Some, but not all, araneids use wrapping as an initial attack weapon and only bite the prey, in this case, after the wrapping attack. Attack-wrapping confers two major adaptive advantages in addition to serving the functions of wrapping after a biting attack. It permits the spider to attack large and otherwise dangerous prey without the dangers involved in the intimate contact of a biting attack. It also achieves an economy in the time that the spider spends away from the hub during an attack sequence. (Spiders that are capable of attack-wrapping continue to use attack-biting against certain prey that have a high escape potential—particularly against lepidopterans.)

It is argued that wrapping at the feeding site after transportation is functionally a basic adaptation and is the evolutionary forerunner of the other types of wrapping. In terms of behavioural complexity the step from this type of wrapping to wrapping after a biting attack is a relatively small one. The functional advantages of this step include an extension of the range of prey with which the spider can cope. Some araneids have either not evolved beyond this stage or have regressed to it. The latter interpretation is regarded as improbable. Attack-wrapping involves a greater step in behavioural complexity but one which confers great functional advantages. Present-day spiders with attack-wrapping are considered to possess the most advanced patterns of predatory behaviour. The evolutionary scheme proposed herein is illustrated in Fig. 14.1.

References

BARROWS, W. M. (1915). The reactions of an orb-weaving spider, *Epeira scolopetaria* Clerk, to rhythmic vibrations of its web. *Biol. Bull.* **29,** 316–32.

EBERHARD, W. (1967). Attack behavior of diguetid spiders and the origin of prey wrapping in spiders. *Psyche* **74,** 173–81.

EISNER, T., ALSOP, R., and ETTERSHANK, G. (1964). Adhesiveness of Spider Silk. *Science N.Y.* **146,** 1058–61.

FABRE, J. H. (1903). L'Epeire fasciée. *Souv. entomol.* **8.**

KASTON, B. J. (1964). The Evolution of Spider Webs. *Am. Zool.* **4,** 191–207.

LUBIN, Y. D. (1972). *Behavioral ecology of tropical tent spiders of the genus Cyrtophora (Araneae, Araneidae)*. Ph.D. thesis, University of Florida at Gainesville.

MARPLES, B. J. (1949). An unusual type of web constructed by a Samoan spider of the family Argiopidae. *Trans. R. Soc. New Zealand* **77,** 232–3.

MUMA, M. H. (1971). Biological and behavioral notes on *Gasteracantha cranciformis* (Arachnida, Araneidae). *Fla Ent.* **54,** 345–51.

PETERS, H. M. (1931). Die Fanghandlung der Kreuzspinne (*Epeira diademata* L.) Experimentelle Analysen des Verhaltens. *Z. vergl. Physiol.* **15,** 693–748.

PETERS, H. M. (1933a). Weitere Untersuchungen über die Fanghandlung der Kreuzspinne (*Epeira diademata* Cl.) *Z. vergl. Physiol.* **19,** 47–67.

—— (1933b). Kleine Beiträge zur Biologie der Kreuzspinne *Epeira diademata* Cl. *Z. Morph. Ökol. Tiere* **26,** 447–68.

ROBINSON, M. H. (1969). Predatory Behavior of *Argiope argentata* (Fabricius). *Am. Zool.* **9,** 161–74.

—— MIRICK, H. (1971). The predatory behavior of the golden-web spider *Nephila clavipes* (Araneae: Araneidae). *Phyche* **78,** 123–39.

—— —— TURNER, O. (1969). The predatory behavior of some araneid spiders and the origin of immobilization wrapping. *Psyche* **76,** 487–501.

—— OLAZARRI, J. (1971). Units of behavior and complex sequences in the predatory behavior of *Argiope argentata* (Fabricius) (Araneae: Araneidae). *Smithson. Contr. Zool.* **65,** 1–36.

—— ROBINSON, B. (1970). Prey caught by a sample population of the spider *Argiope argentata* in Panama: a year's census data. *J. Linn. Soc., Zoology* **49,** 345–57.

—— ROBINSON, B. (1973). The behavior and ecology of the giant wood spider *Nephila maculata* (Fabricius) in New Guinea. Smithson. *Contr. Zool.* **149,** 1–76.

—— —— GRANEY, W. (1972). The predatory behaviour of the nocturnal orb-web spider *Eriophora fuliginea* (C. L. Koch) (Araneae: Araneidae). *Rev. Peru. Entomol.* **14,** 304–15.

TINBERGEN, N. (1959). Behaviour, systematics and natural selection. *Ibis* **101,** 318–30.

—— (1963). On aims and methods in Ethology. *Z. Tierspsychol.* **20,** 410–33.

—— (1968). On war and peace in animals and man. *Science, N.Y.* **160,** 1411–18.

15. Functional aspects of behaviour in the Sulidae

J. B. NELSON

15.1. Nature and scope of discussion

A SHAG, *Phalacrocorax aristotelis,* can move the tips of its mandibles much more rapidly than a gannet, *Sula bassana,* whose bill exerts a far greater gripping force. The former, diving from the surface and pursuing fish by swimming, must be able to counter the lightening twists of its small and agile quarry by faster corrections of its bill movements. The plunge-diving gannet uses hit-or-miss tactics and its problem is to maintain a grip on prey which is often large and muscular. The rest of the body is equally well adapted and as a result imposes its own inevitable restrictions on the types of activity open to the animal. These, in turn, mesh with ecological attributes such as breeding regime, clutch size, choice of breeding habitat, and social behaviour. Clutch size, for example, is related to the frequency with which parents can return with food. The shag, fishing close inshore, can return frequently and so rear three or four young. The gannet, hunting for shoaling fish often far from the breeding colony, returns much less frequently and lays but a single egg.

Thus, anatomical details are deeply related to many ostensibly discrete dimensions of the animal's life. The same is true of behaviour. Ritualized behaviour, in particular, when interpreted in terms of its place in the species' over-all adaptedness, demonstrates many and often devious cross-links with aspects of the animal's life other than that of the primary context in which it evolved and now occurs. The narrow, primary context of a greeting ceremony may be 'pair-meeting on the nest-site' but, in addition, its form, motivation, and total function may be deviously related to the nature of that site. Thus, the meeting ceremony of Abbott's Booby, *Sula abbotti,* is an elaborate display conducted between mates at a distance because, one may suppose, typically sulid contact behaviour such as sparring or jabbing would be dangerous in this species' unique jungle-top environment, where a slip to the forest floor is inevitably fatal. Furthermore, the internal threshold for intra-specific aggressive behaviour in Abbott's Booby has become so high that even blatant intrusion from a potential rival does not elicit attack or indeed any form of contact behaviour—a situation without parallel in this notably quarrelsome and assertive family. Finally, and in the widest sense, the total function of the meeting ceremony in Abbott's Booby is to help form and maintain a pair-bond in that specific jungle environment. In marked contrast, the gannet, even though it, also, has an important and elaborate meeting ceremony, differs diametrically in all the points mentioned above.

In this essay therefore, I will look at the social behaviour of the sulids, asking, what is the function or adaptive value of any particular behaviour pattern (and its homologues or analogues within the group); and how does it fit into a more total (it would be presumptuous to say 'the' total) picture of the species' adaptive web? For in this kind of study, to do more than simply indicate the 'immediate' function of particular behaviour patterns, one must link them with major adaptive systems,† such as the timing of breeding. This inevitably leads to speculative interpretations but the alternative is a motley assortment of facts which hardly help one to understand how (under what selection pressures) the observed behaviour has arisen and how it is now orchestrated. Many of these observations and comments have been published before in the context of the species concerned, but I have not previously discussed the comparative aspects of sulid behaviour in any detail.

15.2. Some basic intra-familial differences due to adaptive radiation

Wherever, geographically, the ancestral sulid began to diverge, the seven sulid species, though still closely related, now occupy a range of feeding and breeding habitats. Three (the Red-footed Booby, *Sula sula*; the White or Blue-faced Booby, *Sula dactylatra*; and the Brown Booby, *Sula leucogaster*) are widely distributed in the warm, blue-water, pan-tropical belts. Two (the Peruvian Booby, *Sula variegata* and the Blue-footed Booby, *Sula nebouxii*) are to differing degrees associated with cold, green-water areas. The Peruvian Booby occurs only in Peru and Chile in the region of the Humboldt Current, whilst the Blue-foot, overlapping with the Peruvian in Northern Peru, continues north and west with the Humboldt to Ecuador and the Galapagos and turns up again in the Revillagigedos and the Gulf of California, also near to green-water areas. The remaining species, Abbott's Booby, is entirely restricted to the Indian Ocean Christmas Island and may conceivably be a relict population which became associated with upwellings in the eastern Indian Ocean, adopted a tree-breeding habit and through subsequent destruction of forested islands, which are now extremely rare, may have arrived at its present status.

† A system is defined as a number of events which are interconnected because each contributes to fitness by relating to the same survival factors. In the above case, the many factors which affect timing of breeding are parts of the system; the timing of breeding, and *ipso facto* the events affecting it, has survival value.

The gannets, three forms which are probably best grouped as a super-species under *Sula (sula) bassana*,† the North Atlantic Gannet, *S. (s.) capensis*, the South African or Cape Gannet, and *S. (s.) serrator*, the Australasian Gannet, are cool–temperate water birds with breeding stations in, respectively, the North Atlantic and North Sea, on islands off the south-west tip of South Africa in the region of the cool Benguela Current, and (mainly) off New Zealand.

A major aspect of this radiation has been that the different species now live with drastically different climatic and food situations and in very different habitats. The blue-water species often suffer massive loss of young by starvation and in their breeding biology have evolved a battery of adaptive features to mitigate the effects of irregular food scarcity. The cold- and temperate-water species have a much more assured, indeed often abundant, supply (although there may be complicating factors, as for the Peruvian Booby, whose foraging areas occasionally become utterly devoid of catchable fish, with catastrophic consequences).

The habitats of the different species (and one includes social aspects) range from cliffs through to bare, flat ground and trees and from huge and densely packed colonies to almost solitary pairs. Gannets and Peruvian Boobies nest densely in huge colonies, often on cliffs; Brown, White, and Blue-footed Boobies (in that order) nest in several habitat types but in general on progressively flatter ground and much more widely dispersed; and Red-footed and Abbott's Boobies nest in trees, the former often in large colonies and the latter often solitarily.

From this point onwards, my account must become comparative, considering each behavioural phenomenon in the context of all the sulids and their circumstances. Within this framework a chronological sequence of events within the breeding cycle is useful.

15.3. Behaviour functional in site acquisition and maintenance

Fighting in the Sulidae

From a functional point of view the behaviour involved in acquiring or re-acquiring a nest-site need be no more and can be no less than the establishment and maintenance of the site in the face of potential competition. Competition, however, is very different in the different sulids and so is the corresponding development of site-oriented behaviour.

In the first phase, locating a site, there is little difference, since all species prospect on the wing, and on foot if habitat permits. However, the chosen site must then be defended, and here there are dramatic differences in behaviour between the species. The most extreme method of

† All remarks in this essay refer to this allo-species.

site-defence—that of frequent, damaging, and prolonged fighting—is found only in the gannet and, even there, mainly on flatter ground. In this situation, and on present knowledge, gannets habitually fight more damagingly than any other sea-bird, or perhaps any other bird. Other sulids conform much more closely to the general pattern of bird-fighting—relatively short-lived encounters which, though they involve actual gripping, jabbing, and wing-flailing, do not harm the antagonists. In actual order of severity and frequency of fighting, the Peruvian Booby certainly comes next to the gannet, then (more arguably) the White and Brown Boobies, the Red-foot, the Blue-foot, and finally and again certainly, Abbott's Booby, which does not fight at all.

One may be tempted to conclude from this and the vast, crowded gannetries that suitable gannet sites are so scarce that there is strong competition for them and hence a premium on the strongest form of aggression. *But at least for* Sula (bassana) bassana *topographically suitable sites are certainly not scarce and the observable competition is concerned with the social quality of the site,* for reasons to be suggested later. The Peruvian Booby, unlike the gannet, shows every evidence of being forced to nest in huge, dense colonies, because a prodigious food supply attracts vast numbers of sea-birds (several species potentially use similar sites) to relatively few islands. In all other sulids, however, there is little site competition, since (1) sites are not scarce and (2) due to blue-water ecology (a seasonal and relatively irregular or scarce food) there is no advantage to be gained from seasonally timed breeding, a phenomenon which is facilitated by social (behavioural) factors, as in the gannet (Nelson 1966, 1970), and which, therefore, can lead to competition for 'socially acceptable' sites.

Territorial (site-ownership) behaviour in the Sulidae

Once the site has been established, in all species by the male, both sexes perform certain highly stereotyped displays on it. Analysis of their form, context, and effect shows that these are mainly aggressively motivated (all of them incorporate biting directed to the substrate) performed in response to real or potential intrusion and communicating the site-defending propensities of the displaying bird (Nelson 1965, 1967a, 1969, 1970, Kepler 1969).

Apart from behaviour which is obviously derived from low-intensity attack—mainly threatening with open mandibles and jabbing, more or less ritualized according to the species (and commonest, as one would expect in the dense-nesting gannet and Peruvian Booby)—one is faced with a range of apparently heterogeneous 'site-ownership' displays. As with overt fighting, their characteristics might be expected to bear some relationship to the value placed upon the site by natural selection.

The gannet. Thus, in the gannet, the site-ownership display—bowing—is highly conspicuous, incorporating widely spread wings, vigorous head-movements, and loud calling, all repeated several times. It is also frequent, particularly after returning to the nest (effectively re-establishment each time) and continues throughout the very long breeding season, during which the male bows some 5000–10 000 times. Furthermore, females, too, bow throughout the season. Thus, the importance of the site to the gannet, strongly suggested by the associated fighting, is endorsed by the marked development, high frequency, protracted seasonal duration, and performance by both sexes of the site-ownership display.

Some predictions which one might make on the strength of the above interpretations are fulfilled. First, males bow more often for more of the season and at higher intensity than females; their wings are held out further and they perform a greater number of 'dips' down to the ground (ritualized, re-directed biting). This is to be expected since males establish the site, have a stronger attachment to it as measured by duration of seasonal attendance and length of attendance spells throughout the breeding season, and are more overtly aggressive than females as measured by their tendency to fight and to attack their mate and young. Secondly, males that have newly established their sites bow more frequently than old-established males, a difference which may even persist in the second year of ownership. This is to be expected, since new males are observably less 'secure' than old ones (more likely to be challenged) and must constantly proclaim their site-owning status.

So one may summarize the male gannet's situation by saying that it fights to the uttermost to get its site and then displays almost unceasingly in its defence.

The boobies. In the White and Blue-footed Boobies territories are not only large, but may also be multiple, a pair holding two, three, or even four simultaneously before finally laying in one. Moreover, their boundaries are blurred, and indeed the defended area of the White Booby shrinks to almost a quarter of its original size as the season progresses (Kepler 1969). In this species fights are rare and, when they do occur, brief. Close-range hostile behaviour, as at boundaries, is a somewhat stereotyped jabbing and wing-flailing, whilst the site-ownership display (a highly ritualized pattern of horizontal and vertical head-movements) is largely confined to dawn and dusk sessions in the prelaying phase of the breeding cycle, with a brief recrudescence after the chick has fledged. Also, it is less clearly confined to defending the site against intrusion and is more likely to occur in a mosaic of agonistic behaviour in the pair context. Thus, it differs markedly from the gannet's bowing. Nevertheless, the White Booby's territorial display seems to repel potential intruders; in 30 out of 33 cases in which a male displayed 'at' a potential intruder and then walked away, the intruder did not enter the displaying bird's territory (Kepler 1969).

The situation is similar in the Blue-footed Booby, except that fights are even rarer and the site-ownership display is somewhat different, having lost the horizontal component of the head-nodding but retained a similar context and function. The Peruvian Booby much resembles the Blue-foot (having probably diverged from stock which resembled the present day Blue-foot more than the present day Peruvian). It fights more, because it is more crowded, but its site-ownership display is merely a slowed down, and exaggerated version of the Blue-foot's head-nodding.

The close-range agonistic behaviour of the Red-footed Booby is jabbing. In its long-range (site-ownership) display the forward-leaning head is swept down to the twigs on each side. Often the bird seizes a branch, perhaps even plucking a spray and continuing the display whilst holding it. Throughout it calls rapidly and abrasively. Whilst one of the contexts in which it is typically performed by the male—namely, by a territory owner after a hostile encounter with a neighbour—shows that it is a territorial display, it most frequently occurs in both sexes as post-landing behaviour. The Red-foot has nothing comparable to the frequent bouts of ownership display shown by the White Booby, much less to the gannet's bowing. Nests are often widespread, but then so are those of the White Booby. The arboreal habitat, however, precludes territorial trespass on foot, whilst this is easy in White Booby terrain.

In Abbott's Booby, the most solitary of sulids, there is little or no agonistic behaviour involving contact between the antagonists, but there is a head-jerking display which, from context and associated behaviour, is recognizably aggressive (Nelson 1972). Thus, it involves quick darts of the head towards the other bird and downward movements to touch twigs, which may be bitten and plucked. It also incorporates violent wing-movements similar to wing-flailing in other boobies. It is (like aggression in other sulids) very frequently a part of pair interactions in contexts which, one knows from comparison, elicit aggression between mates, principally immediately after change-over as one partner moves away from the other. However, more than any other sulid, Abbott's Booby lacks any display which could justifiably be categorized as principally territorial, directed against potential intruders other than the mate. In fact, one looks to Abbott's Booby in vain for a territorial display even after landing—the one situation in which all other sulids perform site-ownership behaviour.

Thus, there seems here to be a positive correlation between the type of territory, particularly its 'intruder accessibility' (a measure compounded of spacing and topography) and the possession of territorial fighting and display.

15.4. Behaviour in pair formation and maintenance, in relation to habitat

Besides territory and its associated behaviour, the behaviour concerned with forming and maintaining pairs and the nature of the pair relationship differs between sulids.†

Before they can copulate, partners must find each other and interact to synchronize their respective motivational states sufficiently to permit mating. In species which depend on successful co-operation between mates over relatively long seasonal periods, this minimum requirement may be greatly increased, and behaviour subserving pair relations can be very important. This is particularly so in species like the gannet, in which mates remain faithful over many years or for life.

In colonial species, finding a mate is presumably not difficult for the female, if receptive males advertize the fact sufficiently clearly. Prospecting females simply fly over the colony or land near or within it and are reacted to by territorial males. In the case of some sulids, this has been proved by observations of dyed birds.

Gannet 'advertising' and its relationship to aggression

The male gannet has one of the most inconspicuous sexual 'advertising' displays imaginable. It is a highly reduced version of bowing (the aggressive site-ownership display), in which the dip component (biting redirected to the ground) is reduced to a slight intention movement and the head-shake (a bill-cleansing movement incorporated into bowing because in redirected ground-biting the bill usually becomes soiled) is exaggerated. The wings are held closed (in bowing they are spread) and the display is silent (bowing birds call loudly). By thus eliminating or reducing those components which largely give bowing its character and which signify the aggression involved, the gannet has produced an advertising display which accurately communicates ambivalent motivation. The approach of the female gannet bearing all the appropriate releaser characters (the sexes are almost identical) elicits aggression which is just inhibited, presumably by simultaneous activation of sexual tendencies. The resultant display satisfies two requirements. It employs, albeit at low intensity, the established circuitry and motor movements which such a stimulus 'normally' and adaptively elicits, and it conveys to the female that she is unlikely to be treated as a full territorial rival. The fact that the result is an oddly inconspicuous movement is presumably unimportant, given the topographical context, which

† There are here good examples of several behavioural phenomena—for instance, the derivation of functionally equivalent displays from different sources, the progressive elaboration in different species of a homologous display which has acquired specifically distinct functional overtones, and emancipation, so that a display which is manifestly of the same origin in the different sulids has acquired totally different motivation and function in different species.

normally allows a prospecting female to perch temporarily near to a large number of potentially receptive males and hence readily to perceive even slight movements. Indeed, the very contrast between the male who is behaving territorially (bowing) and one who is behaving sexually (advertising) enhances the effectiveness of the latter. In that context, it is the *difference* that is important.

I said above that aggression in the advertising male is only just inhibited. The most convincing evidence for this is that, when the female actually joins him, she may be strongly attacked. This is perhaps less surprising than the fact that after pair formation and continuing throughout the long association of the partners—often many years—the male bites the female's nape *whenever they reunite after a reasonably long separation* (hours rather than minutes). He also bites her nape during copulation, though this is quite unnecessary to retain balance (no boobies do it). The female responds by facing-away and, often, by soliciting copulation. Thus male gannets are quite exceptionally aggressive to their mates, and one cannot fail to note the correlation between this and their equally exceptional aggression in the territorial context. Before discussing the functional aspects of this phenomenon, however, it is necessary to describe other behaviour important in the pair relationship.

Gannet meeting behaviour and its relationship to aggression

Gannets have an elaborate meeting ceremony (mutual-fencing) which, like male advertising, is probably basically a modified form of the aggressive territorial display, bowing (details in Nelson (1965, 1972)). As in bowing, the male's version of mutual-fencing contains more dips (redirected bites) and a more extreme wing position than that of the female. The intensity and duration of mutual-fencing bouts shows a seasonal pattern of incidence and intensity similar to that of behaviour known to be aggressive—actual fighting, threat behaviour, and bowing itself. Also, the gannet has reciprocal allo-preening (modified biting rather than feather-care (for interpretation of which see Nelson (1965) and Harrison (1965)) of a conspicuously intense and prolonged kind. Next, it is important to mention that the gannet has a conspicuous display called sky-pointing, which functions between mates as a pre-leaving signal and plays an important role in change-over at the nest by ensuring that each partner 'understands' who is to leave and who to remain; the penalty for even momentary desertion of the nest often being the loss of egg or small chick. It is the homologue of this display which has a totally different function (sexual advertising) in the boobies (see below).

Finally, although advertising and the meeting ceremony are the most immediately and obviously functional behaviour in bringing the gannet pair together and forming the bond, there is other less obvious behaviour ('leave-and-return') involved, and this is of interest in the comparative context. The actual motor patterns are not ritualized, the behaviour consisting merely of repeated departures from and quick returns to the site and to the female (though the departures are preceded by the usual pre-movement ritualized posture—sky-pointing—and the return is followed by nape-biting and mutual-fencing).

The 'leave-and-return' procedure is undoubtedly important in conditioning the apprehensive female of a newly formed pair to remain on the nest-site in the male's absence. Then, when he finally goes off on a long foraging trip, the female remains guarding the site, and from then on it is never unguarded throughout the season. Until then, the male takes extremely short foraging trips and spends long periods on the site, reaching his lowest weight at this time. Were the male of a new pair to depart for long *without* going through this procedure, the female would not remain until his return; for some time after initial pair formation they are extremely wary and prone to leave. This leave-and-return behaviour occurs in all sulids and is probably the origin of pair-flighting in those species which show it.

Booby 'advertising'

The boobies present a very different picture. Booby advertising, or sky-pointing, unlike that of the gannet, seems not to have been derived from an aggressive display but, as mentioned above, from a pre-movement signal. Nor does any male booby bite his female upon meeting at the nest or during copulation. As for meeting ceremonies, again excluding Abbott's Booby, these are brief and relatively crude though nonetheless clearly ritualized interactions—mainly mutual jabbing and bill-touching. Thus, there is a contrast between the different sulids—particularly the gannet versus most boobies—in the nature of the pair relations and the behaviour used in this context, again as it relates to their social and physical habitats. Perhaps over-simply, one could say that where there is most territorial aggression there is more aggression between mates and/or the most highly developed pair interactions, probably of a bonding nature.

Advertising, though different in form in the different boobies, is homologous in them all. Its main features are the upward stretching of the head and neck, bill pointing skywards, the busking of the wings (distinct from normal spreading) and the accompanying call. It would be too much to suppose that by chance six species could each develop these same three components in combination. Nor can the similarity be attributed to convergence, since this is hardly likely to be environmentally shaped behaviour and in any case the environments differ radically.

There are two kinds of differences in booby sky-pointing: first those due to elaborating some components or adding new ones and secondly differences in context, some boobies displaying unilaterally (male sky-points at or to the female, never vice versa), others reciprocally, and two species simultaneously or mutually.

The White Booby has the least conspicuous sky-pointing and it is unilateral, performed by the male in his territory and oriented towards females which are fairly near or flying over. Then comes the Brown Booby, the male of which has evolved the additional feature of sky-pointing in flight. Again, the display is unilateral but, more often than in the preceding species, performed when the pair are close together. The Red-foot raises its wing tips to a greater extent and the display has become prolonged and reciprocal. The Peruvian and the Blue-foot have exaggerated the wing-movement to a bizarre extent, twisting the wings grotesquely, and the display is both reciprocal and mutual. Lastly, the male Peruvian has incorporated one-foot raising.

Before attempting to comment on these differences one should link up the associated meeting behaviours and other pair interactions. This takes some time, and one should keep in mind that the topic being discussed is how the many differences in pair behaviour, whose function in all cases is to form and maintain adequate pair relations, may be related to habitat and social circumstances.

The White Booby has a vigorous and unmistakably ritualized mutual jabbing or sparring display in which the partners' open bills are flung into or against each other's with a head-shaking motion and loud calling. This behaviour occurs in bouts, periodically flaring up and dying away, when the pair are together on the territory and usually slightly off the centre of the site. White Boobies possess, in mutual jabbing, a ritualized behaviour pattern which, like the gannet's mutual fencing, enables the mates to 'express' aggression in the pair context.

The Brown Booby, too, has a close-range ritualized pushing or sparring, and the same comments apply. The aerial advertising is probably an unimportant extension of context, associated with this species' aerial agility (the males are very small and light) and its predilection for nesting at cliff edges or on steep ground, from which the pair often take off to fly round and return together.

The Red-foot lacks mutual, close-range meeting behaviour. Mates may jab each other, but it is difficult to detect any difference between this and jabbing as functional repelling behaviour; indeed, females do not usually participate as they do in White and Brown Boobies, but retreat or show signs of fear and avoidance by tremoring, feather-ruffing, and facing-away.

The Peruvian Booby has a well-marked and clearly ritualized mutual jabbing which commonly occurs in indistinguishable form between rivals. The situation is in this respect similar to that of the Red-foot, but comparable withdrawal is of course impossible in the Peruvian.

Blue-foot mates jab each other relatively rarely, having neither a vigorous and ritualized version nor a hostile, unritualized unilateral one.

The male Abbott's Booby lacks a special advertising display and mates, upon meeting, usually do not engage in contact behaviour but display at a distance. The mutual meeting ceremony, which is highly ritualized and prolonged, consists of an impressive, slow wing-beating, with sweeping downward movements of head and neck and loud shouting. As it ends and the pair are about to come closer, elements of a second display, head-jerking (mentioned earlier), begin to obtrude. Head-jerking is the main agonistic display.

Reciprocal allo-preening, often after meeting behaviour, is found in White and Peruvian (commonly), Brown (sometimes), and Blue-foot (rarely), whilst unilateral allo-preening occurs in the Red-foot.

Omitting details, it may be noted that males of all species show the leave-and-return; Brown and Blue-footed Boobies show pair-flighting over the breeding area, the male of the latter landing with a beautiful 'salute', presenting the soles of his blue feet skywards immediately before touchdown. Peruvian Boobies show a less extreme salute. Once together on the territory Brown, White, and particularly Blue-footed Boobies make great play with locomotory and associated behaviour, parading around with flaunting steps and ritualized postures, the two main ones being an upward and sideways twisted head position (bill-up-face-away) and a bill-tucking position. Associated with bill-up-face-away is a brisk head-shake and, as a pre-flight signal, a ritualized wing-shake. Peruvian Boobies are often prevented by their cliff habitat from this sort of activity, but retain all the Blue-foot's behaviour either 'on the spot' or in reduced form. Red-footed Boobies, probably because of their arboreal habitat, also lack these locomotory and associated behaviour patterns, as do Abbott's Boobies, though both show the pre-flight wing-shake.

Interpretation of advertising, meeting, and associated behaviour in gannets and boobies

The preceding paragraphs have outlined the differences that occur in pair interactions. With this information one may attempt a functional interpretation of the advertising, meeting, and associated behaviour (or its lack) in the various species.

The Gannet's case seems clear; extremely aggressive in competition for territory, sexually isomorphic, and dense-nesting, its highly developed nape-biting and meeting behaviour enables the motor patterns which express territorial aggression (namely biting and bowing) to be elicited by the approach of and sustained contact with another gannet (withdrawal being impossible). Further, this basically aggressive interaction has become closely linked with copulation, and thus has probably acquired a sexual connotation and pair-bonding function. Correspondingly, partnerships are long-lived or permanent.

Immediately one steps over the phylogenetic divide into the booby camp one is spanning a period of time during which considerable changes have occurred.

The Peruvian Booby resembles the gannet insofar as it nests densely and often on cliffs and it is relatively aggressive. However, by every indicator—morphology, ecology, and protein structure—its closest relative is the Blue-footed Booby, which is a flat-ground and spaced nester. Therefore, one must expect a stronger behavioural resemblance between the Peruvian and other boobies, particularly the Blue-foot, than between the Peruvian Booby and the gannet. In fact this is so, even although the obvious competition for sites and its concomitant fighting in the Peruvian does (due to convergence) correlate with both a better-developed meeting ceremony than in other boobies (always excepting Abbott's), well-developed mutual preening and highly developed sexual display (sky-pointing), which not only incorporates a new element (foot-raising) but is frequent, prolonged, and mutual. Pair interactions based on locomotion are often impracticable but the pre-movement signals (bill-up-face-away and wing-rattle) are much used in connection with the leave-and-return behaviour used in pair formation and in change-over at the nest. Pair-forming behaviour is thus extremely well developed, but *pair-bonding* behaviour much less so. This seems consistent with the observation that (due to ecological factors) nesting sites, and almost certainly mates, are changed more frequently than in the gannet.

The significant features of the Blue-foot's pair relations are the relative lack of overt hostility (one may recall the lack of territorial fighting, the well-spaced nesting, and the use of multiple territories), the lack of a meeting ceremony and the rarity of mutual preening, and the extreme development of mutual advertising and of ritualized locomotion and the postures associated with this. Because the Blue-footed and Peruvian Boobies are so closely related, comparison is particularly apt and shows that the latter evinces more aggression between partners and has less highly developed flight and locomotion-based displays. These differences correlate with its greater territorial aggression and more restrictive habitat (densely packed and/or on cliffs). Both in the Peruvian and in the Blue-foot, there is a tremendous amount of pair-*forming* behaviour (reciprocal advertising, etc.) but no pair-*maintaining* behaviour comparable to that found in the gannet. Correspondingly, it is highly likely (though not proven) that pair fidelity is low.

In the White Booby advertising plays merely a restricted primary role and does not occur as a reciprocal or mutual display, but the species has well developed ritualized mutual jabbing and mutual preening. Locomotion-based displays (parading and symbolic nest-building involving the fetching of fragments) are well developed, which is consistent with spaced-out nesting on flat ground. What little information there is on mate fidelity indicates that it is fairly high. One might tentatively suggest a link between this and the fairly well-developed pair interactions.

Red-footed Booby pair relations are marked by a comparative paucity of pair interactions imposed by the difficulty of movement among twigs. There is little or no competition for sites—either topographical or social. Copulation and nest-building, the main areas of co-operation involving close bodily contact, are marked by protracted bouts of mutual advertising, which is thus functioning—one is tempted to suggest instead of a meeting ceremony—not merely to bring them together but, afterwards, to effect a temporary rapport close enough to permit copulation. The aggression which undoubtedly exists, particularly from male to female, and which elicits in the latter unmistakable signs of fear (flinching, tremoring, feather-ruffing, and facing-away) seems to effect withdrawal—a response which is both adequate and practicable but which partly obviates the development of ritualized pair interactions. One suspects relatively low mate fidelity, but there are no figures available.

Abbott's Booby, exceptional in many ways, apparently has no advertising display as such, but merely posts itself in conspicuous positions on the outer branches and reacts to visitors by one or other of its two main displays (mentioned earlier). The distant mutual greeting ceremony is so well developed in this species—birds will display for minutes on end—that it must be important. A possible functional interpretation is two-fold. First, it promotes a strong and permanent pair-bond, something which only gannets, with a comparable meeting ceremony, have within the family. In the case of Abbott's Booby, the advantages could be connected with the unusually long breeding season and the difficulties of forming a pair in its unusual environment. Secondly, the marked lack of contact pair behaviour is probably to be explained as a modification adapting it to canopy-nesting by reducing the likelihood of a fall.

Summarizing this section, one may suggest that the nature of pair relationships in the sulids reflects habitat—both physical and social (the latter principally via aggression developed primarily for site competition)— and perhaps, also, whatever ecological features dictate whether it is an advantage to have a strong and permanent pair-bond. With regard to the latter, there is just a hint that pair-forming and pair-maintaining behaviour are inversely correlated; when one is highly developed the other tends to be less conspicuous.

15.5. Parent–young relationships

There are significant differences between the sulids with respect to interactions between parents and their young.

First, the begging behaviour of young White, Brown, Blue-footed, and particularly Red-footed Boobies is frenzied, involving violent wing action and a physical onslaught on the parent. That of the young gannet and Abbott's Booby is restrained, with folded wings and, in the latter's case, often does not even involve bill contact. This must surely reflect the danger of the gannet's cliff and Abbott's tree-top site, from which it would be easy to fall in the mêlée attending frenzied food-begging.

Secondly, the manner and the timing of discrimination by adults between their own and strange young differs between species—gannets do not at any time reject strange or additional young if these 'appear' on the nest, but violently reject even their own young if these try to approach from a position off the nest. Ground-nesting booby adults discriminate between their own and strange young over a much wider area than merely the nest. The arboreal Red-foot adult discriminates between its own chick and a stranger only when the young concerned can fly. Thus only when it would be possible (and disadvantageous) to feed somebody else's young does discrimination occur.

Thirdly, and most intriguing of all, the degree of plumage difference between the fully feathered young and the adult differs. The two extremes are the gannet, in which the juvenile is totally different from the adult, being virtually black all over, and Abbott's Booby, in which the juvenile, whilst still unable to fly, is indistinguishable from the adult male even down to the colour of the beak. It is entirely consistent with the facts already presented, showing that the gannet has an extremely low internal threshold for attack on other gannets, including his mate, to suggest that the juvenile's black plumage acts as an external (morphological) check on the release of attack. Conversely, Abbott's Booby has been shown to possess an abnormally high internal threshold, which 'permits' the juvenile to assume adult male plumage without danger of eliciting attack. Presumably, in this case, the rapid adoption of adult plumage has positive value, though its nature is totally unknown. It is, however, unlikely to be mere coincidence that only Abbott's Booby has a juvenile which looks like an adult and only Abbott's Booby lacks the usual sulid type of aggressive behaviour between mates and between parents and offspring.

15.7. Discussion

Here one may comment on the two types of functional interpretation of behaviour mentioned earlier—first narrow, and secondly in the wider context of 'total' adaptedness. The immediate function of male advertising, for example, is to attract a female and of mutual-fencing, probably to establish and maintain rapport between pair mates—the pair-bond. The nape-biting accompanying meeting after advertising and before mutual-fencing and the form of the latter however, plainly reveal that both are functioning within a motivational framework evolved in connection with territorial behaviour. The male gannet indisputably recognizes the female as an individual (White and White 1970) yet, meeting after meeting, for years on end, he bites her. Possibly the most convincing interpretation is that by retaining actual physical attack upon any gannet entering his territory, rather than evolving special-contingency inhibitory mechanisms, the male is most effectively retaining the whole stimulus–response mechanism between intrusion and the consequent elicitation of attack. This harks back to the pre-eminence of territory and, in this interpretation, the associated need to evolve pair behaviour subject to this.

This species, like others, has nevertheless had to handle aggression in the pair context. But it seems that natural selection has done so, not by eliminating aggression from male to female, but by allowing it very considerable play and encouraging effective inhibitory responses in the female. In my view, natural selection has in fact gone much further even than this, by producing a species in which the male not only physically attacks the 'intruder' aspect of his mate, but the female is sexually stimulated by the nape-biting. This seems to have been achieved by associating nape-biting with actual copulation.

Mutual-fencing could function in allowing a pair bond to consolidate despite hypertrophied male aggression and despite mutual-fencing being itself derived from an agressive display (bowing). In a sense, the display 'accommodates' aggression and indeed (no surprise since it is itself basically aggressive) is longer and more intense precisely when, in both the long and short term, males are more aggressive as judged by the frequency of menacing and bowing. Thus not only is the male gannet 'allowed' to show overt attack to the female, but also to perform his site-ownership display 'at' her, and, contrary to what one might suppose, both activities strengthen, rather than weaken, the pair bond.

I have described some of the differences between sulids in the nature of their territorial behaviour, pair, and parent–young relationships, dwelling mainly on the narrow functional contexts of the behaviour patterns concerned. A most important functional question, however, is, 'how are the systems which are subserved by this breeding behaviour related to biological fitness?'. This question is on quite a different level, for it asks 'why is the gannet site so competed for?', and 'why is the pair bond important?'. The nuts and bolts of the behaviour employed to *secure* the site or *maintain* the pair have their place as the executive arm of the evolutionary strategy which the species is following, but the big unknown is the survival value of the systems that these behaviour patterns serve.

I have more than once (Nelson 1966, 1967, 1970) expressed the view that timing of breeding is very important in the North Atlantic Gannet, both in British and Canadian waters. This species alone within the Sulidae lacks post-fledging parental feeding. This lack is part of its adaptation to seasonal and super-abundant food, full utilization of which produces a chick with huge fat deposits which, once it has left the nest, cannot return and, for several cogent reasons, cannot be accompanied to sea by its parents. Lack of post-fledging feeding means, for the young, total dependence on the rapid acquisition of the species' skilful and highly specialized hunting behaviour which almost certainly hinges partly on weather. Theoretically, early chicks should survive better by experiencing calmer weather in the North Atlantic and North Sea during the critical transition period. But variability in these areas is so great that no period is guaranteed good weather (hence the great survival value of the observed spread in laying). I have become convinced that, in this species, the undoubted potential of social (behavioural) stimulation in advancing breeding (Nelson 1966) (which, with a fixed point for the earliest date and the bringing-forward of potentially later layers, means synchronization) has been utilized by natural selection, with the result that the nest-sites which confer exposure to such stimulation have become the objects of strong competition. Doubtless this 'sociality' has been superimposed on a colonial tendency initially imposed by site requirements and/or other advantages conferred by large local numbers—perhaps mutual benefit in locating food or because communal feeding helps each individual to catch more. However, once considerable survival value attaches to a socially stimulating site, not only will competition for such sites inevitably arise in an expanding population, but the properties which give the site its socially stimulating qualities will themselves be under selection pressure to intensify, until opposing factors intervene. If this explanation for the severe site competition observable in gannetries is rejected, it becomes difficult if not impossible to understand why competition should be so marked and hence why such aggression should have evolved.

In this connection, recent work on the Cape Gannet (Jarvis 1972) is of great interest. Despite much greater physical shortage of sites and denser nesting, aggression (the frequency and intensity of fighting, menacing, and aggressive site-ownership display or bowing) is demonstrably less than in the Atlantic Gannet. At first this seems odd, but in fact the Cape Gannets seasonal peak of egg-laying is not nearly so concentrated as that of the Atlantic Gannet. The timing of laying will not be subject to the same degree (or perhaps type) of selection pressures, which implies that the role of social stimulation in achieving a concentrated peak will be reduced, and hence competition for a socially adequate site will also be reduced in the Cape Gannet.

The Red-footed Booby is enormously instructive by comparison. In its extreme blue-water areas such as the Galapagos, exactly the opposite of the gannet's environmental conditions obtain. Food is not seasonally abundant but irregularly scarce; there is no marked seasonal change in weather which could influence the transition of young to independent fishing and these are fed by their parents long after fledging. There is thus little advantage in timing breeding, and the cycle (lengthened to cope with the irregular food by slowing the growth of the chick) can overrun the 12 months; it does not matter if a pair lay their egg in December of one breeding attempt and April of the next. There should thus be little incentive for behaviourally mediated seasonal timing of breeding and in fact one finds, correspondingly, that Red-foots in such environments are very sparsely scattered. This, in turn, affects much of their territorial and pair behaviour as described. Indeed, one can perhaps go further, for in some areas the Red-foot is in fact an annual and loosely seasonal breeder, and in such areas it probably tends to nest in larger, denser groups.

The remaining boobies fit equally well into such a scheme. In none of them is accurate, seasonally timed breeding important and none breed densely (except, for quite different reasons, the Peruvian). Some (like the Brown Booby on Ascension Island and the Blue-foot on Hood Island) have less than annual cycles; others (like the White Booby) have annual cycles but with an enormous spread; Abbott's Booby shows seasonal but biennial breeding.

Despite the almost entire lack of over-all synchronization or accurate timing in booby populations it is very noticeable that subgroup synchrony is marked in all species. This at one and the same time supports the suggestion that proximity enhances synchronization whilst leaving untouched the contention that over-all synchrony of a breeding population is, in some species, at best unnecessary and at worst maladaptive. The survival value of subgroup synchrony remains obscure, but it is such a general phenomenon among birds that some equally general advantage must be envisaged. This could very well be the minimizing of interference from conspecifics, which undoubtedly ensues when everybody is more or less at the same stage of breeding simultaneously. There is much scattered evidence that in many species the main cause of breeding failure is interference from conspecifics. This is especially true of all the Fregatidae.

Within an eco-ethological framework such as the one outlined, one can begin to see why gannets have developed their specific breeding regime, why they have become so territorially aggressive and how this has influenced the nature of their ritualized breeding behaviour. The function of the many behaviour patterns used in breeding can be understood for what they achieve in the short term (a site, or a mate, etc.), and in addition the long-term (evolutionary) strategy favouring the acquisition of such a site or such a pair-bond, etc. also seems understandable. Similarly, for the other sulids, the form and function of the mosaic of breeding behaviour looked at against the background of their ecological circumstances, makes composite sense in a way which it does not if one asks merely about causation and neglects function.

'A biological science which gives all its energies to the analysis of causal mechanisms underlying life processes and neglects to study with equal thoroughness how these mechanisms allow the animals to maintain themselves is a deplorably lop-sided Biology' (Tinbergen 1967).

References

HARRISON, C. J. O. (1965). Allopreening as agonistic behaviour. *Behaviour* **24,** 161–209.

JARVIS, M. J. F. (1972). The systematic position of the South African gannet. *Ostrich* **43,** 211–16.

KEPLER, C. (1969). Breeding biology of the blue-faced booby *Sula dactylatra personata* on Green Island, Kure Atoll. *Publs Nuttall orn. Club* **8,** 1–97.

NELSON, J. B. (1965). The behaviour of the gannet. *Br. Birds* **58,** 233–88, 313–36.

—— (1966). The breeding biology of the Gannet *Sula bassana* on the Bass Rock, Scotland. *Ibis* **108,** 584–626.

—— (1967a). The breeding behaviour of the White Booby *Sula dactylatra*. *Ibis* **109,** 194–231.

—— (1967b). Colonial and cliff nesting in the Gannet compared with other Sulidae and the Kittiwake. *Ardea* **55,** 60–90.

—— (1969). The breeding behaviour of the Red-footed Booby *Sula sula*. *Ibis* **111,** 357–85.

—— (1970). The relationship between behaviour and ecology in the Sulidae with reference to other sea birds. *Oceanogr. Mar. Biol.* **8,** 501–74.

—— (1972). Evolution of the pair bond in the Sulidae. *Proc. int. orn. Congr.* **15,** 371–88.

TINBERGEN, N. (1967). Adaptive features of the Black-headed Gull *Larus ridibundus* L. *Proc. int. orn. Congr.* **14,** 43–59.

WHITE, S. J. and WHITE, R. E. C. (1970). Individual voice production in gannets. *Behaviour* **37,** 40–54.

16. The evolution of duck displays

F. McKINNEY

THE SOCIAL signals of waterfowl have been well known to students of behaviour evolution since the pioneer studies of Heinroth (1911) and Lorenz (1941) revealed their spectacular diversity. Most of the special movements, postures, calls, and mechanical noises used by these birds in a variety of social situations are easy to study under captive conditions and these 'displays' have been especially well catalogued for the family Anatidae. The form of many, such as the 'grunt-whistle', is so distinctive that they can be used with confidence as taxonomic characters, and waterfowl displays provide especially convincing illustrations of the phenomenon of behavioural homology. As a result, the family has become famous as the classic example of the successful use of comparative behaviour studies in systematics (Lorenz 1941; Delacour and Mayr 1945; Johnsgard 1965).

Much less attention has been given to explaining the sources and significance of specific differences in display activities, however, and in many ways the waterfowl are ideally suited for study of the adaptive radiation of signalling methods. As shown by the now famous work along these lines in gulls (Tinbergen 1959, 1963) and weaver-birds (Crook 1964), and similar studies on sulids (Nelson 1970), icterids (Orians and Christman 1968), and tetraonids (Hjorth 1970), a blend of ethological, ecological, and evolutionary approaches is essential. A great deal of ecological information is available on waterfowl, especially on species of economic significance to wildfowlers, and although there have been many behaviour studies, the results have not generally been integrated and interpreted in an evolutionary framework.

Each species has to be examined separately as a product of a unique evolutionary history, its specializations being viewed as repercussions of certain major 'decisions' during its history. As argued by Hinde (1959), these are most likely to have been related to changes in habitat or food, as new 'niches' became available. As these changes were taking place, in many cases selection must have favoured new breeding strategies requiring different social signals. Thus while changes in feeding methods and habitat-preferences may have been of basic significance, the modifications in display repertoires that accompanied them have to be interpreted in the light of variations in mating systems, pair-spacing patterns, parental roles, and related phenomena. This crucial area of 'social systems' has not been given the attention it deserves by students of waterfowl behaviour.

In respect to the dabbling ducks (tribe Anatini, including 36 species of *Anas*), with which I will be primarily concerned here, two facets of the social system have remained enigmas since the beginning of this century. First, the significance of 'social courtship' (*Gesellschaftsspiel*), in which several males perform displays around a female, and especially its relationship to pair formation, have been unclear. Secondly, the aerial pursuits of males after females, during the breeding season, have been confusing for the mixture of aggressive and sexual motivation they seem to betray, and their relevance to male promiscuity and to the defence of territories has been much debated. Recent reviews of the controversies are included in McKinney (1969) and Weidmann and Darley (1971a). I think that the time is ripe for an attempt to formulate a series of working hypotheses embracing these aspects of duck behaviour and the most promising approach seems to lie in exploring the implications of natural selection at the level of individuals.

Three overlapping questions are discussed here: why do ducks need displays, why do displays have the characteristics they have, and what factors have been most important in producing specific differences in display repertoires? Displays are regarded as 'signalling devices', the performance of each being advantageous to the performer in dealing with his or her personal social relationships, primarily with conspecifics. Behavioural traits are assumed to be under ultimate genetic control and selection is viewed as operating through differential survival and reproduction of individuals. Not only is there co-operation between the members of a pair for breeding but there is also competition between the interests and investments of males and females, as argued by Trivers (1972).

The point of view presented in this discussion was developed while I studied the breeding behaviour of captive Green-winged Teal, *Anas crecca carolinensis*, in May and June 1973. Preliminary conclusions from this research are summarized here but the detailed results will be published separately.

16.1. Constraints and pre-adaptations

The 36 species of *Anas* occupy a variety of aquatic habitats around the world. Their structure and life-styles are basically similar, indicating that the ancestral stock from which they arose must have had pre-adaptive commitments to exploiting ecological niches associated with marshes, ponds, rivers, and lagoons. As adults and as ducklings they are dependent on shallow waters for their main foods, and they need water every day for drinking and bathing. Aquatic plants and invertebrates are taken at or near the water surface and, although a few species gather food on land or by diving, dabbling ducks spend their lives, first and foremost, surface-swimming.

The members of several duck tribes have specialized in diving for food (for example, Aythyini, Mergini, Oxyurini) and to varying degrees have sacrificed efficiency in walking through changes in body proportions and leg position. In contrast, the Anatini have remained adept walkers and associated with this ability is their potential for nesting far from water. Their agility out of water enables them to use a variety of nesting sites, including tree-stumps, and other elevated sites. But most nest on the ground, in vegetation, and this can entail hazardous trips away from the safety of water both for adults, when selecting the nest-site or for females while laying and incubating, and for ducklings, when leaving the nest or crossing between water areas.

Anatidae are built for strong, direct flight. They have high wing-loading and a requirement for steady flapping. This flying method evidently evolved in association with a life-style demanding rapid changes in location, often over long distances. Most living anatids travel extensively, making daily journeys between feeding grounds or between water and nest-site, and seasonal movements between areas used for breeding and moulting and other areas used outside the breeding season. For the dabbling ducks, in particular, the use of shallow fresh water habitats demands adaptability and mobility when conditions change. Feeding areas and brood-rearing habitat can be created or disappear almost overnight when rivers flood or ponds dry up, and these events can seldom be predicted with certainty from week to week or from year to year.

These specializations for feeding from the water surface while swimming and the use of flight primarily for changing location have had many repercussions on the signalling methods of dabbling ducks. Maintenance of feathers for waterproofing and flight are vitally important and these birds spend much time bathing, preening, shaking, and oiling their plumage. These activities often occur in close temporal association with social interactions (pair formation, hostile encounters, copulations) and many comfort movements have been adapted for signalling purposes (reviewed in McKinney (1965a)). The flying method, involving constant flapping, precludes soaring, and aerial manoeuvre such as. hovering or sailing are possible only momentarily. As a result, aerial displays are limited mainly to the giving of vocalizations and to the ritualization of complete short flights performed just above the water surface (Lebret 1958; McKinney 1970). On the other hand, rapid direct flight involves the risk of separation for mates or flock members and alarm calls, pre-flight signals promoting

synchronous take-off, and contact calls given when flying, are well developed in waterfowl. In turn, the attention-getting potential of these
signals has favoured their further adaptation for use in courtship situations.

Changes in the roles of the sexes in care of the eggs and young have
evidently been of fundamental significance in the evolution of waterfowl
social systems and signalling methods. The main tasks to be accomplished
are incubating the eggs, leading the young, brooding and oiling them in the
early days, warning them of danger, and protecting them from predators.
The young feed for themselves (for exceptions, see Kear (1970)) and they
can be cared for either by one or by both parents. Co-operative parental
care is highly developed in the swans, geese, and Whistling Ducks (sub-
family Anserinae) while in the sheld-ducks, sheld-geese, and all the other
duck tribes (subfamily Anatinae) the female plays the major role, males
never sharing in incubation.

Johnsgard (1962) and Kear (1970) have discussed evolutionary trends in
waterfowl breeding systems, basing their conclusions on a phylogenetic
arrangement derived from comparative studies of living forms. While there
is still much room for argument about the probable characteristics of the
two ancestral stocks, commitment in the Anatinae to incubation by the
female only seems to have had especially far-reaching consequences. This
system could have been favoured by a need for males to specialize in
defence (as Kear suggests) or by a need for females to specialize in
incubation. The latter could have been associated with anti-predator nest-
concealment where the danger of betrayal of the nest-site was greater if
two individuals had to make flights to and from adjacent water areas. In
this event, females would have been pre-adapted for the role of incubator
through their knowledge of local topography (and perhaps also the local
predators) through their experiences during the egg-laying period.

The divergence in the roles of males and females in breeding in Anatinae
has evidently involved differences in the vulnerability of the sexes to
predation and perhaps to other sources of mortality. Unbalanced adult
sex-ratios, with a preponderance of males, are well known in many duck
species (for example, Bellrose, Scott, Hawkins, and Low 1961) and are
generally attributed to heavier losses to predators among incubating
females, as reflected strikingly in a recent study of fox kills by Sargeant
(1972). In species where the male is very aggressive in defense of mate,
territory, nest, or young (for example, sheld-ducks and sheld-geese), on the
other hand, the risks entailed in advertisement and fighting could be
greater than those to which nesting females are exposed. In either case,
unbalanced adult sex-ratios may be expected to increase competition for
the sex partners in shortest supply. Such competition is likely to have been
the main driving force behind the evolution of conspicuous male plumage
features, elaborate male courtship displays, and the phenomenon of social

courtship itself in those duck species where males predominate. Conversely the type of courtship found in some sheld-duck, where females aggressively compete for mates (Heinroth 1911; Johnsgard 1965) has presumably been favoured by an adult sex-ratio with a preponderance of females.

Among the dabbling ducks, no species is known to be completely promiscuous and, apart from his role in copulation, the male's presence is evidently required by the female for other supporting activities. Most important are probably his roles in protecting the female in which he will invest or has invested his genes (1) from predators, (2) from stolen matings by other males, and (3) by defending a food source for her during the egg-laying and incubation phases. Each of these roles has favoured a pair-bond system and most of the social signals that have evolved in this group appear to be necessary to accomplish pairing, to maintain bonds, to protect females from predators and stolen matings, and to defend areas for the female's use.

However, the demand on a paired male's time required to safeguard his mate's breeding effort must vary between different *Anas* species. In some, the male stays close to his mate throughout incubation, but in others, paired males spend much time and energy exploiting a second method of reproduction—stolen matings. Specific diversity in male social behaviour seems to have many of its roots in the extent to which males have combined these two breeding strategies—pair-bonding and promiscuous rape.

Predation on eggs is heavy in *Anas* species and females regularly lay two, or even three, clutches in a breeding season before succeeding in bringing off a brood. Thus females may require the services of a male for copulation at various, unpredictable times and, in species where the pair bond breaks early, there may be many opportunities for males to inseminate females. Again, competition between males is inevitable.

Males are observed accompanying females with broods in some duck species (Weller 1968, 1972; Kear 1970; Siegfried 1974) but the significance of this behaviour is not yet clear. The male's presence may contribute in some way to survival of the young or the mate, or it may reflect persistence of the pair bond with benefits for the male in future breeding attempts.

16.2. Life-style and social systems

Many of the freshwater habitats used by dabbling ducks around the world have been modified by human activities and it is no longer a simple matter to reconstruct the ecological conditions to which each living species is primarily adapted. In southern Africa, for example, waterfowl now depend heavily on irrigation dams (Siegfried 1970) and artificial water areas are widely used in many countries. In the North American prairies, duck ecology has been influenced greatly both by agricultural practices (for example, pothole drainage, grain farming) and by wildlife management practices (local and seasonal control of hunting, preservation of refuge areas, creation of new breeding habitat). Nevertheless, specific differences in preferences for feeding areas, food, habitats, and nest-sites are well-known to field workers who have studied the ecology of several sympatric species simultaneously (see, for example, Sowls 1955; Keith, 1961; Hildén 1964). Although it is often difficult to decide which of these differences reflect ancestral predelictions and which are relatively recent adaptations to changing environments, much of the picture may still emerge through comparative eco-ethological work.

Major characteristics of the adaptive radiation of *Anas* are obvious, however, if species or groups of species with very distinctive morphological adaptations or habitat-preferences are considered. Specialization for plankton-straining in Shovelers and for grazing in wigeons are two of the clearest examples. The African Black Duck, *A. sparsa*, uses primarily river habitats. The Pintail *A. acuta*, is an open-country bird often nesting far from water in sparse cover, and ranging widely to use temporary water areas. Mallards (*A. platyrhynchos* and related species) use many different habitats, both wooded and open, but their agility in using wooded ponds, their ability to nest in trees, and their preference for more permanent water areas suggest quite a different historical background from that of the Pintail.

Specific differences in feeding habits and habitat preferences appear to have had basic repercussions on breeding strategies, as illustrated by the contrasting social systems seen in the Shoveler *A. clypeata*, and the Pintail *A. acuta*. Of prime importance here seems to be the concept of 'economic defendability' (Brown 1964; Brown and Orians 1970). Breeding Shoveler pairs have small, discrete territories, defended by the males, the pair bond is strong and long-lasting, and males spend little time attempting promiscuous matings with other females. In the Pintail system, on the other hand, males divide their time between accompanying the mate and trying to rape other females; paired males do not defend areas, and females suffer heavy harassment (R. I. Smith 1968). These differences seem to be related to the use of rich, localized food sources (plankton in relatively permanent ponds) by Shovelers and the exploiting of temporarily available and widely scattered feeding places by Pintails (see McKinney (1973) and subsequent discussion in the same symposium by G. H. Orians). Defence of an area is feasible in the former, and I suspect that it is advantageous in providing an assured, secure feeding site for the female of the pair during the pre-laying, laying, and incubation phases. Studies of feeding ecology and bioenergetics are needed to test this hypothesis, the suggestion being that plankton-straining is a time-consuming feeding method entailing special problems for breeding females.

My current observations on the breeding behaviour of captive Green-winged Teal have revealed a Pintail-like system involving a similar combination of pair bonding and male promiscuity, but they suggest that characteristics of feeding ecology are unlikely to provide a complete explanation for the evolution of this type of breeding system. Conflicts between the breeding strategies of males and females are evidently involved and the advantages to males of investing their time and energy in the reproductive effort of one female (through a pair bond) and the advantages of trying to inseminate as many females as possible (through raping) seem likely to vary in response to various other factors as well (for example, habitat characteristics, availability of females, vulnerability to predators). The importance of time and energy budgeting in European Teal, *A. crecca crecca*, while they are on their wintering grounds in France is well shown by Tamisier (1972) and his studies strongly reinforce the need to view time spent in courtship as an expense that individuals must balance against the dangers and effort involved.

The behaviour of Shoveler pairs and Green-winged Teal pairs, confined in the same densities under similar conditions in large flight-pens (described in McKinney (1967)) is dramatically different. Male Shovelers spend much of their time threatening, chasing, and fighting with neighbouring males, with the result that each pair comes to occupy more or less exclusive areas in the pen. Paired males do not perform courtship displays to other females and they spend little time actively trying to rape them; almost all of their activities centre round the mate and the territory. Male Green-winged Teal, in contrast, maintain similar strong pair bonds, but they spend very little time chasing other males and essentially there is no 'defended area'. From time to time, a male leaves his mate, approaches another female, and performs courtship displays to her; if his mate should approach, however, displays are then directed to her almost exclusively. Males also leave their mates to engage in vigorous pursuits after other females, frequently culminating in attempts to rape. After these pursuits, each male returns to his mate.

These rape attempts were so frequent and persistent in the Green-winged Teal that females deserted their eggs and spent much of their time hiding from males. Unfortunately we underestimated the seriousness of their plight and four of eight females died, from direct or indirect effects of harassment, before they could be removed from the pens.

Intermediate between these two systems is that adopted by the Mallard. Males are very aggressive and attempt to defend exclusive areas, but they do not usually maintain these areas through the incubation period, as Shovelers do. They become involved in rape attempts much more frequently than Shovelers, but their investment in territorial defense probably leads to somewhat less investment in raping activity than is the case in the Pintail or Green-winged Teal.

Raping activity is very widespread, perhaps universal, in the genus *Anas*, and specific variations in its frequency are probably dependent largely on whether paired males can afford to neglect their mates or their territories. This, in turn, must depend on the needs of females, though, in some species, there may be relationships between male behaviour in defence of an area and the subsequent survival of the brood. For most species, such an effect seems unlikely since broods are highly mobile, but defence of a brood-rearing area may prove to be important in species such as the resident, river-dwelling *A. sparsa* (Siegfried 1968).

In addition, many ecological and behavioural factors must influence the risks, rewards, and energy expenditures involved in rape attempts for the males of each species, and variation in male mating systems will not be fully understood until the 'compromises' concerned have been worked out.

16.3. Social systems and signalling needs

Signals promoting synchronous take-off and the co-ordination of a male and female in copulation are unlikely to be influenced greatly by variations in social systems and the movements serving these functions are essentially uniform throughout the genus. On the other hand, with differences in the character of the pair bond and varying needs for territorial defence, the associated signals might be expected to vary.

In the Blue-winged Ducks (Shovelers and their allies), conspicuous 'hostile pumping' movements are used both as long-range and short-range threat signals, and they appear to have evolved in response to the need of these species to defend territories. Non-territorial species such as the Pintail and Green-winged Teal lack a conspicuous long-range threat signal.

Orientation components have often been noted in the courtship displays of ducks (for example Fig. 16.1) but little attention has been given to the factors promoting these characteristics from species to species or from one display to another. The care with which males aim their displays at a specific female achieves its ultimate refinement in those species that use the grunt-whistle. At the peak of this movement, the male directs a spray of water droplets, flipped up by the tip of his bill, toward the female. In order to achieve this, the bird must first align his body laterally to the female, and clear signs of the male's preoccupation with 'aiming' can be detected in the ritualized shaking movements that precede the display (Simmons and Weidmann 1973). The importance of this precise orientation of male displays is apparent when the male has 'interests' in more than one female at a time and rapid alternation in the direction of signalling is required. Probably this display (and perhaps other highly directional displays associated with it in species such as the Green-winged Teal and Mallard)

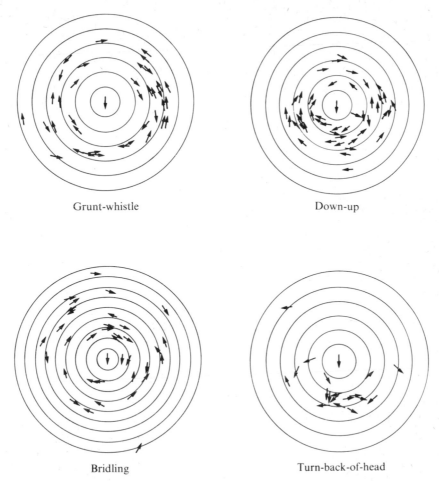

Grunt-whistle   Down-up

Bridling   Turn-back-of-head

FIG. 16.1. Positions of male Green-winged Teal in relation to the female (centre arrow) during performances of grunt-whistle, down-up, bridling, and turn-back-of-head displays. Note the precise lateral body orientation of males when performing the grunt-whistle, shorter distance from the female in the case of the down-up, variability in distance for bridling, and positioning in front of the female for turn-back-of-head. Grunt-whistle, bridling, and turn-back-of-head can occur when only one male is present, but down-up is performed only when a second male is present also. The distance between concentric circles is one foot; a swimming teal measures slightly less from bill-tip to tail-tip.

evolved in response to situations involving more than one female (see p. 337).

Arguments such as these could be carried further, but they are perhaps premature until more is known about *Anas* social systems. Rather than pursuing them further here, I will stress the different signalling needs of males and females, since these are basic to the interpretation of display repertoires in all duck species.

16.4. Signalling needs of males

A great deal can be inferred about the information that is probably being transmitted by displays through observation of the situations in which they are given, their characteristics, and the effects they appear to have on other animals present. These methods have been used by many ethologists and the possibilities of this functional approach to signals have become clearer recently (see W. J. Smith 1968). In some cases, experimental testing of hypotheses is possible, as has been done for only one duck display, the Mallards's 'decrescendo call' (Lockner and Phillips 1969; Abraham 1974). For interpreting displays occurring in complex social interactions, however, we must rely on indirect evidence and natural experiments.

Courtship and establishment of bonds with females

In order to maximize their reproductive effort, it must be advantageous for both males and females to choose their sexual partners with care. Except for individuals that were paired in a previous year, females must be using indirect methods of evaluating the potential of males as breeding partners. The most likely criteria would seem to be (1) vigour, skill, and persistence in courtship, (2) healthy physical condition, as reflected in quality of plumage, (3) attentiveness, compatibility, and constancy in re-affirmation of the bond, (4) success in competition with other males, and (5) efficiency in copulation. Female ducks could well make use of all these qualities in choosing among potential mates, and Johnsgard (1960c) has given some suggestive evidence in respect to plumage features, but experimental evidence is lacking.

The most likely method of achieving efficient mate-selection would seem to be through a series of trial liasons, and the earliest bonds among Mallards (Weidmann 1956) and captive Shovelers (personal observations) are temporary. In northern-hemisphere *Anas* species, courtship begins several months before the birds move to the breeding area, and individuals that begin testing potential mates early must, on average, be at an advantage. Among the captive Shovelers I observed, adult males came into breeding plumage and began giving displays earlier than yearlings. Adults are heavier and more experienced and on average may be expected to be superior to yearling males in all the qualities used by females for assessment.

To establish a bond with a female, a male (1) must attract her attention to him and let her know that he is interested in her; (2) he must compete successfully with other males in holding her interest until she allies herself with him and follows him; (3) he must discourage rival males by threatening and, if necessary, fighting them; (4) he must overcome the female's aggressive tendencies toward him. To accomplish all of these things, preferably simultaneously, and to be able to maintain them whether the female is swimming, walking or standing on land, or flying, requires a complex series of signals. All four accomplishments are important, each requires different kinds of signals, and none can be over-emphasized at the expense of the others. In general, the male's behaviour relating to the female is incompatible with his hostile behaviour toward other males, and the two activities must either alternate rapidly or be combined in behaviour patterns meaning different things to female and males.

It must be to the advantage of unpaired males to explore on-going courtship activities, since these indicate that at least one other male has discovered a female worthy of his attention and the chances are good that, even is she has established a bond already, she has been judged to offer 'possibilities'. Thus courting groups (which are noisy and conspicuous because of the competing males' attention-getting signals) tend to attract other males. But each male must have to make decisions, throughout the pairing season, on which female to devote his attention to. As with males, females must vary in 'quality' and, by judging behaviour and appearance, males may be expected to become expert in reading the signs governing the expenditure of their courting effort.

By studying the situations in which each male courtship display is performed, the effects they appear to have on other birds, and the form of the displays, it is possible to classify them in relation to their probable functions. I have made a preliminary attempt to do this for the Green-winged Teal (Table 16.1), and the same principles should apply to all ducks although many variants may be expected.

The discussion above obviously concerns the phenomenon known in the literature as 'social courtship', but I have avoided reifying this concept as has often been done in the literature. It seems better to regard these group activities as the results of simultaneous attempts on the part of individual males to court one female, rather than some sort of organized performance with a *raison d'etre* of its own. The activities of the group are explainable, I believe, on the basis of the individual interests and responses of the birds present.

A major problem in understanding social courtship has arisen with the observation that paired male Mallard will leave their mates to join such a group (Weidmann 1956). This has lead to the impression that the activity is 'infectious', that participation may be important in stimulating 'interest' in pairing, or that there are psychosomatic effects related to gonad maturation and breeding synchronization. Also, since 'pairs' of ducks can be seen in autumn before social courtship has really begun (Lebret 1961) there has been some discussion of whether group activity is 'necessary' for pair formation. It seems more profitable to work from the assumption that a male does not participate in social courtship unless he

TABLE 16.1

A simplified attempt to classify Green-winged Teal displays according to their probable functions; displays of ducklings, juveniles, and females with broods are not included (see McKinney, (1965b) for descriptions and photographs of many of the displays)

	Male	Female
(A) *Pre-flight intention*	Head-shake Head-thrusting	Head-shake Head-thrusting
(B) *Bond establishment*		
(1) Pre-display warning	Head-shake Shake	
(2) Attracting attention, specifying interest, holding attention, leading away	Repeated calls Grunt-whistle Head-up-tail-up Turn-toward-female Turn-back-of-head Wing-flap Jump-flight Bridling (on land)	Nod-swim Quacks coinciding with male displays
(3) Interest in female, threat toward rival combined	Down-up	
(4) Threat toward rival	Bill-up	
(5) Indicating preference for one male, threat toward other birds		Inciting
(6) Retreat from rival, jockey for position	Nod-swim	
(7) Appease female's hostility	Bill-down	
(C) *Bond maintenance*		
(1) Re-affirm bond		
(a) Mates alone	Preen-behind-wing Belly-preen (on land) Bill-dip; drink	Preen-behind-wing
(b) Other birds present	Grunt-whistle, etc. Down-up	Inciting
(2) Warning and contact in flight	Repeated calls	Repeated quacks
(3) Contact when separated	Repeated calls	Decrescendo call
(4) Post-copulation	Bridling	
(D) *Intention to copulate*	Head-pumping	Head-pumping Prone posture
(E) *Combating rape attempt*	(Fights would-be rapist)	Inciting Repulsion

has an interest in trying to form a bond or a liaison with the female, or that he is exploring these possibilities.†

† Social courtship activity in groups of males only, as reported by Lorenz (1941), Weidmann and Darley (1971b), and others, seems to be abnormal, if not pathological. It can be triggered by homosexual, captive-reared males, but I have seen it occasionally also in wild-caught Green-winged Teal deprived of females in captivity.

This is supported by my recent observations on captive Green-winged Teal. Throughout May and early June 1973, paired males frequently left their mates to perform displays (grunt-whistle, head-up-tail-up, turn-toward-female, turn-back-of-head, bridling) to other females. Once several females began to show pre-nesting activities, males also began raping assaults, but displays continued to be given to females which had not reached this phase. If a male's mate joined such a group, he switched the orientation of his displays to her; thus, at times, four pairs might be swimming together, each male repeatedly giving displays to his own mate.

I interpret these activities as attempts by paired males to establish personal liaisons with females other than their mates. Such relationships could be of benefit in the event that re-pairing with a new mate becomes desirable or necessary, and perhaps they help to increase a male's chances of success in future raping attempts with these females. Occasionally, a male might have an opportunity to acquire two mates simultaneously, and such trios may prove to be commoner than we now think (see Lebret 1961).

Observations of apparent 'pairs' in Mallards in autumn, before social courtship has really begun (Lebret 1961) might suggest that social courtship is not 'necessary' for pairing to occur. It seems more likely that some bonds remain intact after an unsuccessful breeding attempt and occasionally, even in migratory species, like the Mallard, the same birds breed together in successive years (Lebret 1961; Dwyer, Derrickson, and Gilmer 1973). In these instances, strong personal bonds must have existed, but I suspect that they were retested, through competitive courtship, during winter. This phenomenon may well be more common in non-migratory populations and information is badly needed on species reported to have long-lasting bonds.

Bond-maintenance and defence of the mate

In the Green-winged Teal, paired males evidently re-affirm their bonds by performing the same kinds of displays to the mate as they use in pair formation. At other times, however, when the members of a pair are alone together, there are special displays apparently serving this function. 'Preen-behind-wing', 'shake+belly-preen', 'preen-dorsally', 'drinking', and 'bill-dipping' are especially common in *Anas* species. Brief nibbling or touching of the female may also occur ('*Tendieren*'), as described by von de Wall (1965), as a form of 'directed courtship'.

Males perform bursts of repeated calls when separated from the mate and especially when they are searching for her. This is the response given by a paired male to his mate's decrescendo call, and these calls provide the main mechanism for maintaining contact between mates when they are apart.

Males protect their mates, as best they can, by threatening, chasing, or fighting with males attempting to court or rape the mate. As might be expected, the close-range threat displays used are variable in intensity and duration (for example, bill-up in the Green-winged Teal, rab-rab calling in the Mallard) depending on the situation.

Territorial defence

Highly territorial species, such as the Shoveler, have conspicuous threat signals and they use ritualized fighting methods in boundary disputes (McKinney 1970).

Pre-copulatory signals

In most *Anas* species, intention to copulate is signalled by vertical 'head-pumping' movements; in other ducks a variety of ritualized movements are used (see Johnsgard 1965). These displays are usually inconspicuous and rarely are there vocalizations. In general, male ducks perform pre-copulatory displays much more often than their mates and mounting is likely to follow only when the female has taken the initiative, or when she indicates her readiness by responding to the male's displays with similar ones or by assuming the 'prone' posture. Thus the frequency of copulations between the members of a pair appears to be regulated by the female.

In at least some species (for example, Mallard: Raitasuo 1964) copulations occur months before egg-laying begins, and presumably they are part of the pair-bond testing procedure. Just before and during the egg-laying period, well established pairs of captive Shoveler (McKinney 1967) and Mallards (Barrett 1973) copulated only once or a few times each day. Such a rate, governed by the female, is presumably a minimum required to ensure fertilization of the eggs since copulating birds must be especially vulnerable to predators.

After his mate has been assaulted (and perhaps inseminated) by other males, however, it must be to a male's disadvantage to allow her to decide whether or not she will permit him to copulate. In such situations, Barrett (1973) has recorded rape-like, 'forced pair-copulations', involving mated birds, in captive Mallards. Similar behaviour occurred frequently in captive Green-winged Teal, paired males attempting to rape their own mates a few minutes after the latter had been assaulted by other males. Pre-copulatory 'pumping' is absent in these situations where selection for a prompt response by the mated male must be very strong.

Post-copulatory signals

Immediately after an apparently successful copulation, male ducks usually perform one or a series of displays with obvious orientation of the male's body in relation to the female. In *Anas* species, the performance can include such displays as bridling, turn-toward-female, nod-swim, turn-back-of-head, and at least one call is given (for example, a loud whistle accompanying bridling). Attempts to explain the occurrence of these displays in terms of motivational conflict theory (McKinney 1961) or isolating mechanism function (Johnsgard, 1963) have not been very convincing, and it seems more likely that they are functioning to re-affirm individual identity, thereby helping to maintain the pair-bond.

The characteristics of male post-copulatory displays in ducks suggest that they are designed to draw the female's attention to the performer and they seem to demonstrate structural features and voice characteristics that could be used by females to identify their mates individually. There can be no better time for a male to draw attention to his identity than immediately after he has accomplished a successful copulation if, by so doing, he increases his mate's confidence in his competence as a breeding partner and reduces the risk that she will desert him. If this is the case, post-copulatory displays should be especially well developed in birds which can least afford to lose a mate and in those where the danger of doing so is great.

In general terms, comparative evidence seems to support this argument. In swans, geese, and Whistling Ducks, which have long-term pair-bonds and the investment of individuals in their mates must be great, both male and female usually perform post-copulatory displays in unison. In sheld-ducks, where competition for mates may be severest in females, both sexes also perform, but Johnsgard (1965) reports that the female calls loudly, and in *Tadorna ferruginea* and *T. cana* she starts calling before the male begins his displays. In the remaining duck tribes, where competition for mates appears to be greatest among males, post-copulatory displays are usually given exclusively by males.

16.5. Signalling needs of females

Pairing

At times during periods of social courtship in Green-winged Teal, females perform nod-swimming (repeated forward and back movements of the head) and they give loud, single quacks almost coinciding with, but clearly in response to, male displays. Johnsgard (1965) has noted similar calls in other *Anas* species, and Weidmann and Darley (1971a) have analysed the effects of female nod-swimming in Mallards. These movements and calls appear to stimulate and hold the interest of males in the female, although they are by no means essential triggers for male courtship. Moreover, these attention-getting devices could be important in competition between females during pairing.

Pair formation in ducks is usually regarded as a process in which females make a selection from the males that court them. Certainly this is going on, but males also must be exercising choice in their decisions to court certain females. Although there will usually be an excess of males in the flocks when pairing is taking place, this is not the 'cause' of male competition for females. Rather it must increase the intensity of the competition that is inevitably present when the males and females in a population sort themselves out into pairs. Early in the pairing season especially, we should expect to see females 'making advances' toward particularly desirable males and encouraging them to court. Late in the season, when most females have established bonds and yearling males have become active in courtship, competition between males is certainly the most conspicuous aspect of the phenomenon. Nevertheless, the female must be able to change her mind (for example, if the mate becomes sick or injured), and she will need signals indicating her interest in other males.

Instances of active 'courtship' on the part of females might also be expected in captive flocks where choice of mates is limited. Among Shovelers and Blue-winged Teal, A. *discors*, I have occasionally seen females direct displays typical of the male repertoire (belly-preen, lateral dabbling, turn-back-of-the head) at males (McKinney 1970).

For pairing to occur, females must have a method of indicating preference for one male over another so that both favoured and rejected males are informed of the female's decision. This is accomplished in most *Anas* species by 'inciting'—sideways movements of the head performed beside the favoured male with the bill directed away from him, often pointing toward the rejected male. In the full performance, rattling calls accompany these movements. Although ritualized to a considerable degree (Lorenz 1958), there is much variability in the intensity of the movements and calls, and at times the performance grades into threat and even short chases.

Bond-maintenance and contact between mates

Once a female has indicated a preference for one male she re-affirms the liaison by repeatedly performing 'inciting' beside him when other males approach. Especially when the bond has recently formed, females may respond to the ritualized 'preen-behind-wing' of the male by giving the same display while the two birds stand close together. If the mates become separated, the female gives decrescendo calls, the male repeated calls, and the two come together again.

When the pair takes wing, especially as a result of a disturbance, the female gives a series of loud, evenly-spaced quacks ('going-away call' of Lorenz 1953)) and the male often gives repeated calls. These responses to being flushed occur in pairs before they move to the breeding grounds, and they must be functioning mainly in keeping the birds together.

During the breeding season, however, females give much longer bouts of similar evenly-spaced quacks in two contexts: (1) during flights by the pair over nesting cover (or when swimming before flights) during the days immediately before laying of the first egg, and (2) in late incubation or when females are leading broods, associated with mobbing or distraction displays in the presence of a predator.

While there are striking differences in context between the going-away call, 'persistent quacking' during nest-site selection, and the broody calls of an alarmed, defensive female, I suspect that they have a common motivational basis. Clearly, however, they must be serving different functions. The going-away call apparently functions in maintaining contact with the mate, broody calls presumably warn the brood of danger and distract the predator, and perhaps persistent quacking serves both to ensure attendance by the mate and to lure predators during nest-site selection.

Persistent quacking is a loud, conspicuous signal with characteristics suggesting a 'broadcast' function rather than simply a mate-contact function. Since it occurs in both territorial and non-territorial species (for example, Shoveler and Green-winged Teal) it seems not to be advertisement associated with defence of an area. Its frequent occurrence in twilight and at night suggest that it undoubtedly keeps the mate informed of his female's location and, since decrescendo calls are also given during this phase (in Green-winged Teal) and persistent quacking is given especially in flight, it appears important that the male be informed in this way. Indeed, nesting cover seems always to be inspected by the pair as a unit (for example, in Shoveler, Mallard, and Green-winged Teal), and the male must play an important role at this time in being watchful for predators and warning his mate of danger. Also, since persistent quacking does seem to be stimulated by appearance of a potential predator (though often such a stimulus is certainly not present), I believe that it may have a luring effect on mammals, resulting in betrayal of their presence. The decision on exactly where to locate the first nest of the season must be an extremely important one for females, and it would be surprising to find that the decision was not influenced by her knowledge of the local predators.

The male usually accompanies the female very closely during inspection of cover. In the beginning, they are very alert and they often seem to take turns in moving into patches of grass for brief periods. On subsequent days, more and more time is spent moving around in cover (McKinney 1967). In my captive Green-winged Teal, it was not always the female who made the first move into cover; frequently the male seemed highly motivated to do so and he often preceded the female. Evidently these excursions into unknown terrain, where predators are at home, are dangerous and strong co-operation between the mates is favoured.

Incubation, rape assaults, and re-nesting

Once the clutch has been laid females incubate steadily with one or two brief periods off the nest each day to feed, drink, and bathe. During these periods, so long as a strong bond persists, the male is alert for the arrival of his mate on their favourite feeding waters, and he is active in joining her and accompanying her while she feeds. His presence presumably is important to both members of the pair, in warning the female of predators and in guarding her against rape attempts. In the latter respect, however, females are victims of the social system, and there may be little they can do to prevent harassment by alert males intent on raping.

In species where paired males specialize in promiscuous activities, it is not yet clear whether they discriminate between laying and incubating females although, from the point of view of fathering progeny, the latter are inappropriate targets for assault. On average, however, it may be a good strategy for males to chase all females in view of the high rate of predation on eggs and the steady 'supply' of females in the 're-nest interval' phase throughout the breeding season. From the female's point of view, on the other hand, it would seem to be advantageous to signal her broody condition and this appears to be achieved through 'repulsion behaviour' (Lorenz 1953). This involves characteristic hunched, ruffled postures and distinctive loud 'broody' calls. Females give this response to harassing males once they begin to incubate, and their aggressiveness increases throughout incubation, culminating in vigorous attack on conspecifics in defence of the brood. Presumably some would-be rapists are 'repulsed' by such behaviour and, if a choice is available, they would be expected to direct their chases toward females which are not performing in this way, since these are more likely to be laying birds.

The fertilization of females for re-nest clutches is extremely difficult to study in wild birds, and it is still not clear how this is accomplished. There are surely specific differences, since male Shovelers usually remain faithful to their mates throughout the incubation period, and thus will be available to father the eggs in replacement clutches. In other species, such as the Mallard and Pintail, bonds are thought to break in mid-incubation, and females must either re-pair or be inseminated through raping. Sowls (1955) believed that females 'tease' males during the re-nest interval, encouraging males to pursue them, but behavioural changes in females during this phase need further careful study.

16.6. Characteristics of displays with different functions

If the displays of each species serve different functions, each display must have evolved under a different combination of selective pressures. Thus it is not sufficient to suppose that selection has favoured individuals with efficient sets of 'courtship displays'. Each display must have evolved, to some extent, independently, and the ritualization processes determining its present form should reflect its special functions.

Male courtship displays

These displays appear to serve at least five different functions: attracting the female's attention to the performer, specifying which female is being courted, holding the female's attention, leading her away, and helping males to compete with other males.

Attention-getting displays tend to be sudden, startling performances, involving motions that are distinctly different from predictable, everyday activities. Loud, brief calls (for example whistles) usually coincide with movements involving the head, wings, or the whole body. Plumage features often reinforce the optical effect (for example, the Pintail's tail in the head-up-tail-up posture). Many of these displays are derived from, or are related to alert, alarm, or escape behaviour (for example, repeated calls, nod-swimming). These are predictable sources since birds are naturally responsive to signs of alarm in their companions.

Most male courtship displays in *Anas* species are performed with a deliberate orientation component in relation to the female; often the whole body is either broadside or pointing directly at the female, or the head only is turned toward or away from her. This seems to be the main method whereby males specify which female interests them, and water-splashing, as in the grunt-whistle, is the highest development of this phenomenon.

The main technique used in holding the female's attention on the performer seems to be through the use of display-chains. Body-shakes or head-shakes are often used as a 'warning' that a display is likely to be performed. These are often described as being 'introductory to a bout of courtship', but it seems better to regard them as preliminary moves in the attention-getting strategy of each male. This is well illustrated by the linkages of displays associated with the grunt-whistle in the Green-winged Teal (Fig. 16.2). Males often give only preliminary head-shakes, as they 'jockey' for position in the group, waiting until the female and the other males are appropriately placed before proceeding with the major movements in the sequence (grunt-whistle+head-up-tail-up+turn-toward-female+turn-back-of-head or nod-swim). The male can break off the sequence at various points, however, depending on the changing positions and activities of the other birds.

As pointed out by Johnsgard (1960*a*), turn-back-of-head is a very widespread display in ducks of many tribes, performed by a male as the female swims toward him and apparently serving to encourage her to follow him. In the grunt-whistle sequences described above, it comes toward the end, culminating in the ideal procedure of catching-attention+holding-attention+leading-away. Other displays may serve similar functions in other ducks, for example, the jump-flights of *Anas* species and the ritualized short-flights of the Goldeneye, *Bucephala clangula*. The evolution of such displays entailing moving away in front of the female was

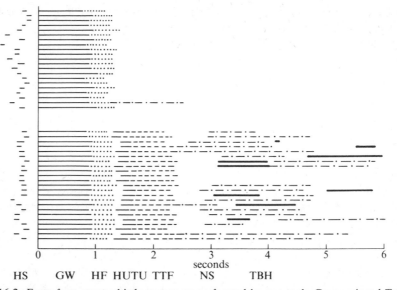

FIG. 16.2. Forty-four grunt-whistle sequences performed by one male Green-winged Teal during one session of social courtship. A preliminary head-shake (HS) is followed by the grunt-whistle (GW) and a head-flick (HF) in all cases. Head-up-tail-up (HUTU) and turn-toward-female (TTF) may (lower sequences) or may not (upper sequences) follow. Occur-rence, order, and duration of nod-swimming (NS) and turn-back-of-head (TBH, shown as a solid line vary greatly depending on the positions and behaviour of the female and other males in the group. (From a 16 mm cine film, exposed at 24 frames per second. See McKinney (1965b) for descriptions and photographs of the displays.)

presumably facilitated by the basic tendency of these social birds to follow others.

Although the idea of competition between males was inherent in the views of Heinroth (1911) and Lorenz (1941), this aspect of duck social courtship has often been neglected. Clues to the presence of hostility between males are subtle in many species and easily missed when attention is focused on the displays. This is especially well illustrated by three types of study on the displays of Goldeneyes, *Bucephala clangula*, with emphasis on the displays as fixed action patterns (Dane, Walcott, and Drury 1959; Dane and van der Kloot 1964), as products of motivational conflict (Lind 1959), and as signals used in pairing (Nilsson 1969). Each approach has its merits, and the results are complementary, but the most recent study by Nilsson is especially valuable in interpreting displays in the light of competition between males. The recent work of Weidmann and his colleagues on the Mallard, and my own studies on Eider Ducks, Blue-winged Ducks, and the Green-winged Teal agree closely on the important role of aggression in duck courtship. In fact, as Heinroth concluded long ago, social courtship in *Anas* makes no sense unless it is viewed as a consequence of competition between males for mates.

Lebret (1961) has argued that social courtship functions not only in pairing but as a method of canalizing aggression during autumn and winter when Mallard Ducks must live in flocks. His careful observations can be interpreted in another way, however, if males are assumed to benefit from extra-pair-bond liaisons. He noted that 'many paired males may take part in social display, while their mates are sleeping on the banks' and 'we may not conclude that a certain drake is unpaired because of the absence of a possible mate'. Such behaviour corresponds to that described earlier for the Green-winged Teal, and is explainable if it is assumed that these males (paired though they may be) are indeed trying to establish liaisons with other individual females. Thus Lebret's idea of a 'canalizing of aggression' function for courtship should be modified: the over-all behaviour of the males in a courting group appears always to be a result of competition between males for the attention of a female but male displays can have different functions. These can be distinguished most easily by comparing the simplest possible situations, namely, when only one male or two males are involved.

Association of the down-up display of the Mallard with hostile situations has been known for a long time (for example, Weidmann 1956), and this is readily confirmed by its linkage with the obviously hostile rab-rab palavers of threatening males. The Green-winged Teal has exact parallels in the bill-up threat display and the down-up display. However, I have never seen these displays performed by a single male to a female; for a male to perform the down-up, he must be close to both a female and another male. In the simplest situations, a male gives this display when his mate is close beside him and another male is nearby, all three birds often having their bodies aligned parallel. I conclude that this display has both threat function in relation to other males, and attention-getting function (including bond-affirmation) for the female.

This phenomenon of a single male display having different signal functions for males and females may well be widespread in ducks, and it may be suspected where there are signs of hostile motivation in close association with a display that is directed toward females. Thus many 'greeting ceremonies' involving the members of a pair include male displays that occur typically during hostile encounters between males (for example, rab-rab in Mallards; hostile pumping in Shovelers).

In addition to performing highly ritualized displays, which appear to have threat function, competing males also perform graded threat signals during social courtship (especially when two males become preoccupied with one another rather than with the female), they swim off or chase one another (especially when provoked by performance of an attention-getting display), and they are constantly jockeying for position, interrupting one another's display-chains, and occasionally fighting vigorously. The tendency to synchronize their displays in 'bursts' is most famous (at least in Mallards and the Green-winged Teal) in the case of the down-up—suggesting again the high level of male–male interaction involved when this display is given. Presumably synchronization is a 'tactic' resulting from the competition between males; a solitary performance of an attention-getting display is likely to be advantageous to a male in capturing the female's undivided attention, a result countered by synchronizing. Repeated calls also tend to occur in bouts during social courtship in these two species—a prime attention-getting display. The grunt-whistle, on the other hand, tends to be solitary performance, each male concentrating on timing and orientating the display precisely in relation to the female, moving further away from her and beginning only when other males seem unlikely to interrupt.

Many of the complexities of social courtship have been omitted here since I have been concerned with arguing a point of view. Each species must be examined carefully to test the validity of the ideas, and many complicating factors will no doubt emerge.

16.7. Factors responsible for specific differences

Sibley (1957) and Johnsgard (1960b, 1963, 1965) have argued that diversity in the display movements, calls, and male plumage patterns of ducks has evolved in response to selection for specific distinctiveness. They point to the ease with which closely related sympatric species can be hybridized in captivity, reasoning that effective isolating mechanisms must be operating to prevent this in wild birds. The male displays used in courtship vary much more between species than do pre-copulatory displays and female vocalizations, as it should be if the former are functioning to prevent hybrid matings. The argument is further supported by the frequent occurrence of reduced sexual dimorphism in plumage on islands (where closely related species are absent) and in the southern hemisphere (where pair bonds are thought to be long, pairing seasons may be more distinct, there may be more ecological separation, and in general there seems to be less danger of hybridization).

There are alternatives to this argument, however, and it does not provide a convincing explanation for many kinds of specific differences. I have tried to show that signalling devices have evolved in response to needs of males and females to communicate varied kinds of information of vital importance in resolving their individual relationships. While some (perhaps all) of these signals undoubtedly do give information on the species and sex of the bird performing them, a great deal more is involved. Signalling needs are different in species with different life-styles and social systems, and display repertories have evidently been moulded by a great variety of selection pressures. Such things as vulnerability to predators (McKinney 1965c), degree of territoriality (McKinney 1970), proportion of courtship performed on land (Kaltenhäuser 1971), characteristics of pair bonding and male promiscuity have evidently been involved. For most duck species, we do not have enough information on signalling needs nor on social systems to judge the full extent of their repercussions on display movements, sounds, or morphology.

There is still much to be discovered about species-recognition and factors influencing mate-choice in ducks. The work of Schutz (1965) on sexual imprinting suggests that preferences for a type of mate are heavily influenced by the social environment during the brood phase, and it now seems likely that male courtship displays may have little to do with prevention of hybrid matings.

Many male displays involve a combination of movements, sounds, and an exhibition of plumage features, and there has been a tendency to think of all three as though they have evolved in concert. Linkages there are to be sure, but it seems more important to recognize that changes in each of these categories have surely been influenced by quite different blends of selective forces, and changes in each have proceeded at different rates.

Judging from their occurrence in many living species, most male courtship display movements seem to have been relatively conservative features, although they have been combined in various sequences and some movements have been dropped from adult repertoires. There have probably been constraints on repertoire size, as suggested by Moynihan (1970), but differences of this type are unlikely to be understood until signalling needs, and the functions of each display have been worked out for each species. It seems particularly dangerous to assume that the presence or absence of a grunt-whistle or a down-up in one species has come about as a result of selection for specifically distinct courtship repertoires, as Johnsgard (1965) proposes for the two sympatric South American Pintails, *A. georgica* and *A. bahamensis*.

The sources of selection that have moulded male vocalizations and plumage patterns now seem likely to be much more difficult to investigate than those involved in the evolution of display movements. We know very little about the biological functions of these characteristics and this must be the first step in working out why they have evolved.

16.8. Conclusions

The displays of ducks are viewed as signalling devices evolved through selection of individuals gaining advantages in their personal social relationships from performance of these behaviour patterns. Signalling is needed for many purposes (for example, pairing, territory defence, flight-intention, copulation-intention, pair-bond maintenance). The signal characteristics required vary with the situation. Males and females often have to communicate different kinds of information. Signalling needs vary from species to species in relation to differences in the social system.

The breeding system involving incubation by the female only has apparently had fundamental repercussions on social behaviour in the Anatinae. Breeding entails different risks for males and females; unbalanced adult sex-ratios are common, and competition for mates and copulations is strong. Seasonal pair bonding is apparently usual in *Anas* species but the extent to which males defend territories, engage in promiscuous matings, and attend the female and brood vary from species to species.

Social courtship in *Anas* species is interpreted as a consequence of competition between males for breeding partners. Highly directional displays, such as the grunt-whistle, may have evolved to enable males to specify which female interests them in situations where they are attempting to establish or maintain bonds or liasons with more than one female. The extent to which males are able to combine promiscuous raping activity with maintenance of a pair bond seems to have been an important factor in the evolution of *Anas* social systems.

In spite of much study, duck displays continue to pose some of the most challenging problems in the fields of animal communication and behaviour evolution. A deeper understanding of the factors that have shaped their evolution may come especially from comparative studies of social systems and investigations of the signalling needs of individuals.

Acknowledgements

Niko Tinbergen's approach has inevitably influenced the way all ethologists think about behaviour evolution and his comparative programme on gull behaviour has provided the model for these duck studies. I am grateful to him for inspiration and encouragement in the fascinating but frustrating task of trying to figure out why each species behaves differently. Julie Barrett, Scott Derrickson, Walter Graul, Cathy Laurie, Roy Siegfried, and Scott Stalheim gave very helpful, detailed criticisms of the manuscript, and Paul Stolen helped in many ways with the Green-winged Teal study. The research was supported by the Graduate School, University of Minnesota, the National Science Foundation (Grant GB-36651X), and the U.S. Atomic Energy Commission (Contract AT (11-1)-1332; Publication COO-1332-98).

References

ABRAHAM, R. L. (1974). Vocalizations of the Mallard (*Anas platyrhynchos*). *Condor* **76,** 401–20.
BARRETT, J. (1973). *Breeding behavior of captive Mallards.* M. S. Thesis, University of Minnesota.

BELLROSE, F. C., SCOTT, T. G. HAWKINS, A. S. and LOW, J. B. (1961). Sex ratios and age ratios in North American ducks. *Bull. Ill. nat. Hist. Survey* **27**, 385–474.

BROWN, J. L. (1964). The evolution of diversity in avian territorial systems. *Wilson Bull.* **76**, 160–9.

—— ORIANS G. H. (1970). Spacing patterns in mobile animals. *A. Rev. Ecol. Syst.* **1**, 239–62.

CROOK, J. H. (1964). The evolution of social organization and visual communication in the weaver birds (Ploceinae). *Behaviour Suppl.* **10**.

DANE, B., WALCOTT, C., and DRURY, W. H. (1959). The form and duration of the display actions of the goldeneye (*Bucephala clangula*). *Behaviour* **14**, 265–81.

—— VAN DER KLOOT, W. G. (1964). An analysis of the display of the Goldeneye duck (*Bucephala clangula* L.) *Behaviour* **22**, 282–328.

DELACOUR, J. and MAYR, E. (1945). The family Anatidae. *Wilson Bull.* **57**, 3–55.

DWYER, T. J., DERRICKSON, S. R., and GILMEŘ, D. S. (1973). Migrational homing in a pair of Mallards. *Auk*, **90**, 687.

HEINROTH, O. (1911). Beitrage zur Biologie, namentlich Ethologie und Psychologie der Anatiden. *Proc. int. orn. Congr.* **5**, 589–702.

HILDÉN, O. (1964). Ecology of duck populations in the island group of Valassaaret, Gulf of Bothnia. *Ann. zool. Fenn.* **1**, 153–279.

HINDE, R. A. (1959). Behaviour and speciation in birds and lower vertebrates. *Biol. Rev.* **34**, 85–128.

HJORTH, I. (1970). Reproductive behaviour in Tetraonidae with special reference to males. *Viltrevy* **7**, 183–596.

JOHNSGARD, P. A. (1960a). Pair-formation mechanisms in Anas (Anatidae) and related genera. *Ibis* **102**, 616–18.

—— (1960b). Hybridization in the Anatidae and its taxonomic implications. *Wildfowl Trust 11th Annual Report*, pp. 31–45.

—— (1960c). A quantitative study of sexual behavior in Mallards and Black Ducks. *Wilson Bull.* **72**, 133–45.

—— (1962). Evolutionary trends in the behaviour and morphology of the Anatidae. *Wildfowl Trust 13th Annual Report*, pp. 130–148.

—— (1963). Behavioral isolating mechanisms in the family Anatidae. *Proc. int. orn. Congr.* **13(1)**, 531–43.

—— (1965). *Handbook of waterfowl behavior*. Cornell University Press, Ithaca, New York.

KALTENHÄUSER, D. (1971). Über Evolutionsvorgänge in der Schwimmentenbalz. *Z. Tierpsychol.* **29**, 481–540.

KEAR, J. (1970). The adaptive radiation of parental care in waterfowl. In *Social behaviour in birds and mammals* (ed. J. H. Crook). Academic Press, New York.

KEITH, L. (1961). A study of waterfowl ecology on small impoundments in south-eastern Alberta. *Wildl. Mongr.* **6**.

LEBRET, T. (1958). The 'Jump-flight' of the Mallard, *Anas platyrhynchos* L., the Teal, *Anas crecca* L. and the Shoveler, *Spatula clypeata* L. *Ardea* **46**, 68–72.

—— (1961). The pair formation in the annual cycle of the Mallard, *Anas platyrhynchos* L. *Ardea* **49**, 97–158.

LIND, H. (1959). Studies on courtship and copulatory behavior in the goldeneye (*Bucephala clangula* (L.)). *Dansk. orn. Foren. Tidssk.* **53**, 177–219.

LOCKNER, R. F. and PHILLIPS, R. E. (1969). A preliminary analysis of the decrescendo call in female mallards (*Anas platyrhynchos* L.) *Behaviour* **35**, 281–7.

LORENZ, K. (1941). Vergleichende Bewegungsstudien an Anatinen. *J. Orn., Lpz.* **89**, 194–294.

Lorenz, K. (1953). Comparative studies on the behaviour of the Anatinae. Reprinted from *Avicultural Magazine*.

—— (1958). The evolution of behavior. *Scient. Am.* **199**, 67–78.

McKinney, F. (1961). An analysis of the displays of the European Eider *Somateria mollissima mollissima* (Linnaeus) and the Pacific Eider *Somateria mollissima v. nigra* Bonaparte. *Behaviour, Suppl.* **7**.

—— (1965a). The comfort movements of Anatidae. *Behaviour* **25**, 120–220.

—— (1965b). The displays of the American Green-winged Teal. *Wilson Bull.* **77**, 112–21.

—— (1965c). The spring behavior of wild Steller Eiders. *Condor*, **67**, 273–90.

—— (1967). Breeding behaviour of captive Shovelers. *Wildfowl Trust 18th Annual Report*, pp. 108–21.

—— (1969). The behaviour of ducks. In *The behaviour of domestic animals* (ed. E. S. E. Hafez) (2nd edn). Ballière, Tindall, and Cox, London.

—— (1970). Displays of four species of blue-winged ducks. *Living Bird* **9**, 29–64.

—— (1973). Ecoethological aspects of reproduction. In *Breeding biology of birds*, pp. 6–21. National Academy of Sciences, Washington, D.C.

Moynihan, M. (1970). Control, suppression, decay, disappearance and replacement of displays. *J. theor. Biol.* **29**, 85–112.

Nelson, J. B. (1970). The relationship between behaviour and ecology in the Sulidae with reference to other sea birds. *Oceanogr. mar. Biol. A. Rev.* **8**, 501–74.

Nilsson, L. (1969). Knipans *Bucephala clangula* beteende under vinterhalvaret. *Vår fågelvärld* **28**, 199–210

Orians, G. H. and Christman, G. M. (1968). A comparative study of the behavior of Red-winged, Tricolored, and Yellow-headed Blackbirds. *Univ. Calif. Publs. Zool.* **84**.

Raitasuo, K. (1964). *Social behaviour of the mallard, Anas platyrhynchos, in the course of the annual cycle*. Papers on Game Research, Helsinki, No. 24.

Sargeant, A. B. (1972). Red fox spatial characteristics in relation to waterfowl predation. *J. Wildl. Mgmt* **36**, 225–36.

Schutz, F. (1965). Sexuelle Prägung bei Anatiden. *Z. Tierpsychol.* **22**, 50–103.

Sibley, C. G. (1957). The evolutionary and taxonomic significance of sexual dimorphism and hybridization in birds. *Condor* **59**, 166–91.

Siegfried, W. R. (1965). The Cape Shoveler *Anas smithii* (Hartert) in southern Africa. *Ostrich* **36**, 155–98.

—— (1968). The Black Duck in the south-western Cape. *Ostrich* **39**, 61–75.

—— (1970). Wildfowl distribution, conservation and research in southern Africa. *Wildfowl* **25**, 33–40.

—— (1974). Brood care, pair bonds and plumage in southern African Anatini. *Wildfowl* **25**, 33–40.

Simmons, K. E. L. and Weidmann, U. (1973). Directional bias as a component of social behaviour with special reference to the Mallard, *Anas platyrhynchos. J. Zool. Lond.* **170**, 49–62.

Smith, R. I. (1968). The social aspects of reproductive behavior in the Pintail. *Auk* **85**, 381–96.

Smith, W. J. (1968). Message-meaning analysis. In *Animal communication* (ed. T. A. Sebeok). Indiana University Press, Bloomington.

Sowls, L. K. (1955). *Prairie ducks*. Wildlife Management Institute, Washington, D.C.

TAMISIER, A. (1972). Rythmes nycthemeraux des sarcelles d'hiver pendant leur hivernage en Camargue. *Alauda* **40,** 109–59.

TINBERGEN, N. (1959). Comparative studies of the behaviour of gulls (Laridae): a progress report. *Behaviour* **15,** 1–70.

—— (1963). On adaptive radiation in Gulls (Tribe Larini). *Zool. Med.* **39,** 209–23.

TRIVERS, R. L. (1972). Parental investment and sexual selection. In *Sexual selection and the descent of man 1871–1971* (ed. B. G. Campbell). Aldine, Chicago.

VON DE WALL, W. (1965). 'Gesellschaftsspiel' und Balz der Anatini. *J. Orn.* **106,** 65–80.

WEIDMANN, U. (1956). Verhaltensstudien an der Stockente (*Anas platyrhynchos* L.). I. Das Aktionssytem. *Z. Tierpsychol.* **13,** 208–71.

—— DARLEY, J. (1971*a*). The role of the female in the social display of mallards. *Anim. Behav.* **19,** 287–98.

—— (1971*b*). The synchronization of signals in the 'social display' of Mallards. *Rev. comp. Anim.* **5,** 131–5.

WELLER, M. W. (1968). Notes on some Argentine Anatids. *Wilson Bull.* **80,** 189–212.

—— (1972). Ecological studies of Falkland Islands' waterfowl. *Wildfowl* **23,** 25–44.

17. Adaptations to colony life on cliff ledges: a comparative study of guillemot and razorbill chicks

B. TSCHANZ AND M. HIRSBRUNNER-SCHARF

17.1. Introduction

ORGANISMS THAT are only distantly related to each other but which live under the same environmental conditions can look alike and act similarly in some ways. It is generally believed that these similar characteristics make organisms more suited to respond to certain demands of their common environment. Such convergent characteristics are called adaptations (MacArthur and Connell 1970).

Depending on how the problem is formulated and studied, adaptation can be understood as an ongoing process or as a result (Schwerdtfeger 1963, 1968; Wickler 1970, 1972). In both cases, an inter-species comparison of characteristics and their effects is necessary to elucidate adaptation. In order to carry out a comparison, information must be obtained with respect to the demands to which an organism should be more suited to respond, and the criteria determining suitability (Curio 1973).

Adaptation manifests itself not only in the above-mentioned convergent characteristics but also in divergent characteristics: if several environmental conditions in the habitat of two closely related species greatly differ from one another and if the species show distinct differences in certain characteristics, it is plausible to interpret these differences as adaptations. On such thoughts are based, for example, Cullen's studies of kittiwakes (1957), Emlen's studies of Herring Gulls (1963), Nelson's studies of gannets (1967, and this volume), and Scherzinger's studies of owls (1971).

Similar thoughts prompted a comparison between data on the behaviour of guillemots (Tschanz 1959, 1968) and razorbills (Ingold 1973); and this comparison, in turn, provoked search for an experimental proof that, during incubation and the chick period, guillemots manifest adaptations to the hazards of colonial breeding on cliffs. The nature of incubation period adaptations has been discussed in detail elsewhere (Tschanz 1972); in this essay, therefore, we shall repeat only the main points.

Guillemots and razorbills belong to the same family, Alcidae, and differ from each other in only a few morphological, reproductive, and ethological characteristics (Kartaschew 1960). Both species breed on the cliffs of coasts and islands; each pair produces a single egg, which the two birds share in incubating without building a nest. The razorbill pairs are spatially separated from one another; guillemots, on the other hand, stay crowded together in breeding groups. Razorbills prefer to nest in caves and crevices; guillemots prefer spots on open ledges that are sometimes narrow and slope toward the sea. For this reason the guillemot's eggs are the more endangered: the eggs can be destroyed by the parents' carelessness, by members of the same species, and by egg predators. This risk of loss is counteracted to a great extent by the guillemots' behaviour: they leave the egg only at times of extreme disturbance, and the partners work closely together in relieving each other during the incubation period, one never leaving before the other is present to take over. Razorbills leave their egg relatively often, and when relieving each other for incubation the incubating bird leaves the egg before the partner replaces him. If the guillemots are forced to behave like razorbills in an experimental situation, their breeding success is greatly diminished (Tschanz 1972). The guillemots' constant presence with the egg, and their relief ceremony (Tschanz 1959), make them more suited to breed in colonies on narrow ledges. Therefore these behavioural patterns would seem to be adaptations.

We can also consider as adaptations the colour and design polymorphism of the guillemot eggs, the different reactions of the birds to the characteristics of their own and to strange eggs, the marked pear-shape of the egg, and the position of the incubation-patch, against which the egg is held during incubation, which is in the median line of the bird's body slightly behind the sternum (Tschanz and Ingold 1969). Adaptations manifest themselves also during the chick-rearing period. Thus they have adaptations which enable them to combat the risk of falling to which the guillemot chicks are especially prone. Also, as we shall show later, they are adapted to survive a higher level of threat from other members of their own species and from predators, than that facing razorbill chicks.

The behavioural adaptations that ensure successful rearing in guillemots will be discussed here. First of all the rearing conditions of the guillemots must be compared with those of the razorbills and the differences specified. Then the question must be asked which of these differences can be considered to be adaptations to the guillemot's rearing conditions. Finally, these interpretations will be examined experimentally in order to test their validity.

From 1967 until 1972 we carried out systematic comparative studies of both of these species on the bird island of Vedøy (67°30′ N, 12° E). On the nearby island of Røst rooms in a fisherman's house were equipped for the purpose of experimentally testing these species under laboratory conditions. Several hundred eggs were incubated to hatching in five 'La Nationale' incubators. The chicks were brought up in indoor rooms and in outdoor enclosures, and then were released on the ocean at between 1 month and 3 months of age. In addition to the apparatus used in the investigations made by Tschanz (1968) and Ingold (1973), three observation tents, two Uher tape recorders, two loud-speakers, and two Nikon cameras with objectives up to f-400 mm were available.

17.2. Comparison between the natural rearing conditions of guillemots and razorbills

The environmental conditions at the time of chick-rearing are identical to the environmental conditions during the incubation period, except that now the adult birds are concerned with young, not with eggs. In view of our formulation of the problem, we shall consider only the differences in behavioural patterns between guillemots and razorbills.

As Table 17.1 shows, the chicks manifest differences in (1) resting position, (2) use of action space, and (3) behaviour when disturbed. These behavioural characteristics might be important for the survival of guillemot chicks.

1. At the beginning of its development the guillemot chick prefers a sitting position. Thus the chick keeps much of its body off the ground, which is often covered with wet dung and peat, and so reduces the chances of getting dirty and the consequent interference with the heat-insulation function of its feathers.

2. By staying between the adult bird and the rock face the chick lessens the risk of reaching the edge of the ledge and so of falling off, as well as the danger of being stepped on, pecked, or pushed out to the edge of the ledge by guillemots walking by. In addition, the chick is better protected against being grabbed by Herring Gulls and ravens, or being attacked by kittiwakes, whose nests are often situated near guillemot sites.

3. If it has been left alone by its parents due to some disturbance, the chick can be equally well protected by being sheltered by a strange adult or by going to a crevice or the rock-face.

Razorbill chicks do not need to be especially protected from getting dirty (1), from strangers of the same species, or from predators and kittiwakes (2, 3). Their breeding sites have less dung on them and are not too wet, there are no strange adults or kittiwakes close by, and the caves and fissures offer protection from predators and the risk of falling. The almost constant presence of one parent at the guillemots' breeding-place offers protection against predators (4) and decreases the possible danger of disadvantageous interaction between chicks and strange adults (6).

TABLE 17.1

Comparison between the natural rearing conditions of guillemots and razorbills

No.	Characteristic	Guillemots	Razorbills
1	Resting position of youngest chicks	Mostly sitting	Mostly lying
2	The chick's movement within the area surrounding the breeding-place in the presence of a parent	Only in the area between the adult birds and the rock-face	Without preference for one particular area
3	Reaction of the young temporally abandoned by their parents	Young chicks leave the breeding-place, search for shelter with strange adults or in cracks or nestle into a rock-face	Young chicks usually remain at the breeding-place; older chicks go to the rear of the cave
4	Presence of one of the parents at the breeding-place	Normally, constant until shortly before the young jumps off the ledge	Normally, the young are occasionally left alone at the breeding-place
5	Reaction of a chick standing at some distance from its parent to its acceptance-call	Mostly, turning towards and approaching the calling parent	Often no reaction
6	Behaviour of adults of the same species towards the young of strangers	Brooding, pecking, stepping on them, pushing, feeding (rarely), stealing food (rarely)	None
7	Feeding		
(a)	Number of fish per feeding	One	Several
(b)	Manner in which the fish is carried by the adult bird	Lengthwise in the beak (Fig. 17.1)	Diagonally in the beak (Fig. 17.2)
(c)	Visibility of the fish in the beak of the adult bird *before* feeding	Poor	Good
(d)	Visibility *during* feeding	Poor; the wings of the adult bird are pushed forward, the head is bent low, the beak is close to the body (Fig. 17.2)	Good; the wings are not pushed forward, the head is only slightly bent (Fig. 17.2)
(e)	Activity of the parent and the chick during feeding	Both are active; their activity is co-ordinated; it is 'ceremonial'	Parent is rather passive; lets the chick act
(f)	Manner in which chick takes the fish	It seizes the fish's tail perpendicularly, lets the fish glide through its beak, grabs at the head, turns the fish around lengthwise, and devours it	Pecking at the fish, pulls it out of the adult's beak, lets it glide through its beak and devours it head- or tail-first

Young guillemots are pecked by some strange adult birds (6) and are more or less regularly sheltered by others (6) (Wehrlin, Jenny, and Tschanz, in preparation), but they are rarely fed by them (6). Nevertheless, the young are not in danger of starving when being sheltered by strange adult birds: they immediately respond to the acceptance-call made by the parent during the fish-presentation at the empty breeding-site. Besides, the young chick is usually hindered from going to a strange guillemot at all by its reaction to the parent's acceptance-call (Tschanz 1959). Guillemots give only one fish per meal (7(a)). It seems appropriate that they bring it to the ledge in such a manner that it is as invisible as possible to the other birds of the same species (7(c), (d)). If the fish were visible, all the neighbouring guillemots, except the breeding-partner, would immediately begin pecking at the fish and trying to seize it for themselves (Figs 17.1 and 17.2).

It is important that the chick adjusts its movements to those of the adult bird (7(e), (f)). Only in this way is it possible for the adult bird to control the fish-transfer at times of disturbance (that is, when neighbours attempt to steal the food) in such a way that the chick nevertheless gets its food.

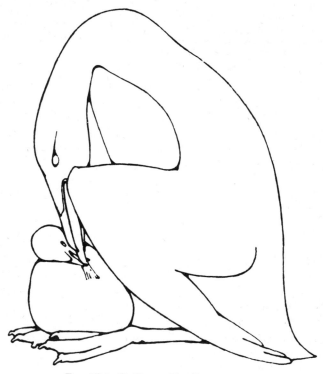

FIG. 17.1. Guillemots' feeding ceremony.

FIG. 17.2. Razorbill parent brings fish.

Due to the behaviour of the parents, the guillemot chick is generally well-protected against interference from strange adults, against predators, and against loss of food. Such protective measures are not necessary for razorbills, for, due to the local isolation of the breeding pairs, there is no danger of food-stealing or of threat from strange adults (Table 17.1, 7(a)–(f), razorbills).

In summary, we can say that razorbills and guillemots differ from each other in 12 behavioural characteristics (Table 17.1, 1–7(f)) which can all be interpreted as adaptations to the particular demands of chick-rearing within a dense breeding group and on ledges.

In order to test the validity of these interpretations, it must be proven that the guillemot's specific behavioural characteristics enable it to have a high breeding success under natural conditions. The genesis of the behavioural patterns can be determined only if chicks are brought-up without adults of their own species and away from their own environmental conditions.

The way in which environmental conditions influence the development of certain behavioural characteristics will be investigated in two series of experiments. In the first series, developmental conditions are identical for the young of both species. In the second series, the outdoor conditions for development are interchanged: razorbill chicks grow up with guillemots and guillemot chicks with razorbills.

Development of the characteristics under laboratory conditions

Rearing and experimental conditions.   Guillemot and razorbill eggs, 22 of each, were artificially incubated at 37·8 °C (100 °F). The chicks were left in the incubator until they were dry (about 12 hours) and then left in a heated box at a temperature of 28 °C for 2 days, each bird in a compartment of its own. Afterwards, the chicks were placed onto a grid 1 metre above the floor. Each chick had its own compartment, the front and the top of which were open (approximately 15 × 15 × 15 cm). The chicks could move around freely, warm themselves at a plastic pipe (40 °C), or leave the compartment and walk around on the ledge (Fig. 17.3). They were fed 5 times a day with whole or cut-up sprats *Clupea sprattus*, depending upon the size of the fish.

Under the ledge newspapers sprinkled with water were spread out on the floor. They ensured that the air remained sufficiently moist, that there was little dust and that cleaning was easy.

As under natural conditions, the 3-week old chicks jumped down from the ledge and were then kept in groups of 4–6 animals in outdoor enclosures. Prior to being released into the sea, the young birds were fed in the water and had to dive for fish.

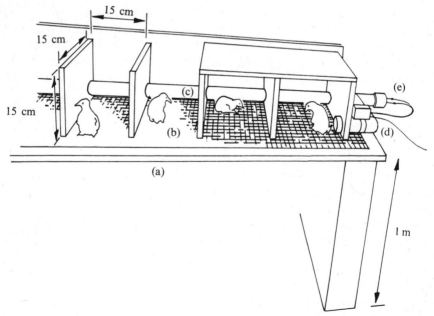

FIG. 17.3. Rearing set-up in the laboratory for guillemot and razorbill chicks: (a) wooden frame; (b) grid; (c) plastic pipe; (d) thermostat; (e) immersion heater.

Various comparisons of the behaviour of guillemot and razorbill chicks were carried out. The first of these utilized the experimental ledges with 10 birds in niches (the cubicle was open in front and on top) and 9 in caves (the cubicle was open in front and closed at the top—see Fig. 17.3). The behaviour of each young bird was recorded in 3 trials each of 30 minutes. The first trial was made 12–24 hours after hatching, the second 2 days after hatching, and the third 4 days after the first one. Thirty-six different activities and 10 different body positions were recorded.

17.2. Results

Resting position

When living in the laboratory situation illustrated in Fig. 17.3 the chicks rest on the experimental ledge, either lying or sitting. Razorbills and guillemots, however, do not lie and sit to the same extent. As indicated in Fig. 17.4, the young guillemots were found in a sitting position, in all 3 trials, more often than the young razorbills; the latter, however, were found to be in a lying position more often than the guillemots. In addition,

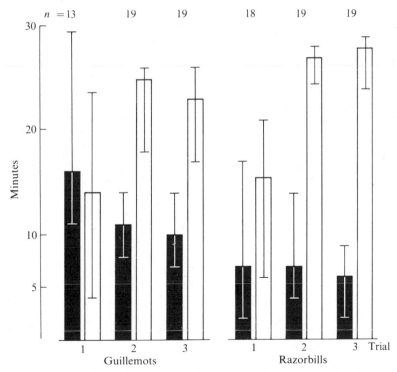

FIG. 17.4. Chicks' resting position. Median time spent resting in two positions for groups of chicks during three 30-minute trials at increasing ages. Lines indicate the first and third quartile of each distribution. Black columns; sitting, breast lifted off the ground. White columns: lying, breast on the ground.

TABLE 17.2
Head-position during resting

	Beak in wing	Beak pointing perpendicularly up
Observations from 19 guillemots	7	5
Observations from 19 razorbills	13	17

Fig. 17.4 shows that the older guillemot chicks lie more often than the 1-day old chicks, which also seems to be the case under natural outdoor conditions according to Wehrlin (personal communication).

The head as well as the body position adopted during resting differs between razorbills and guillemots. As can be seen in Table 17.2, the razorbill chicks more often either turn their heads back sideways so that their beaks rest on their bodies above the wing, or they place the backs of their heads so far down their backs that their beaks point up almost perpendicularly into the air. Significant differences between these species are, however, only found in the frequency of the latter head position $(\chi^2 = 13 \cdot 1, p < 0 \cdot 001)$.

Use of action-space

In addition to the 30-minute trials, records were kept over a period of 2 weeks (5–6 times daily) on the place where the young were found on the experimental ledge. Whereas 17 of 19 razorbills stood at least once outside of the niche, only 6 of 21 guillemots did this $(\chi^2 = 15 \cdot 1, p < 0 \cdot 001)$. Thus they make different use of their action-space.

This difference might be caused by the fact that guillemots tend to avoid light (Tschanz 1959; Wehrlin and Tschanz 1969), and they also have a great preference for body contact and for the proximity of space-dividing structures (Tschanz 1968; Wehrlin and Tschanz 1969). Razorbills, however, show little or none of this need for contact. It is also possible however that razorbill chicks show more locomotor activity and emerge more onto the ledge for this reason.

In order to test the possibilities, we recorded all turns and steps made by newly hatched chicks in a 'box-test'. The experimental situation is illustrated in Fig. 17.5. The box is lit from above but partly shaded by an opaque screen. It thus has a graded intensity of light across its base and the distribution of light intensities is also plotted in Fig. 17.5. Seven razorbills and 7 guillemots, newly hatched, were tested by placing them individually in the box for a 60-minute test (see also Wehrlin and Tschanz 1969). The following records were made:

(1) the orientation of the chick at the beginning of the observation;

(2) the duration of stay of the chick in the light and dark areas;

(3) the area—either light or dark—where the chick was located at the end of the experiment;

(4) the locomotor activity, measured by the number of steps and turns made during the test;

(5) the duration of stay near a wall, near being defined as a distance not greater than three centimetres between the chick's head and the wall.

Results

1. All the guillemot chicks initially turn towards the dark area of the box but the razorbills show no particular preference ($\chi^2 = 5\cdot1$, $p < 0\cdot05$).

2. As indicated in Fig. 17.6 guillemots spend more time in the dark than razorbills (Mann–Whitney U-test, $p = 0\cdot04$).

3. At the end of the 60-minute period, 5 guillemots were in the dark area, but none of the razorbills ($\chi^2 = 5\cdot0$, $p < 5$ per cent).

4. The locomotor activity of guillemots, expressed as the total steps made with or without a change in direction, does not greatly differ from the activity of razorbills. What does differ between the two

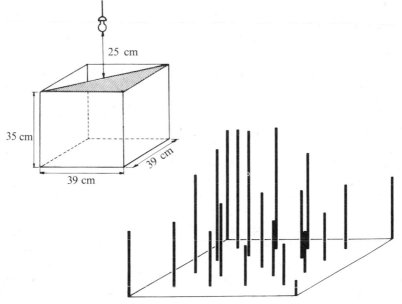

Distribution of light intensity over floor of box

Fig. 17.5. (a) Experimental arrangement to test the phototaxis. (b) Brightness distribution on the floor of the experimental box. Light intensities are expressed on an arbitrary scale using a Pentax–Spotmetre at a distance of 20 centimetres.

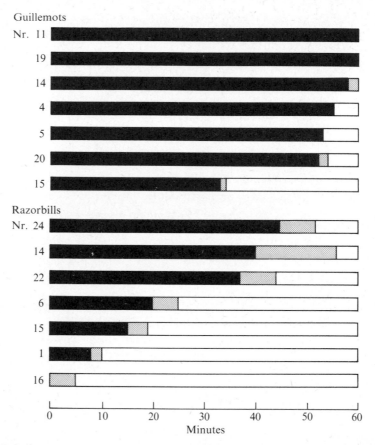

FIG. 17.6. Guillemots' and razorbills' orientation to light during the first hour after hatching. Proportion of 60-minute test spent in dark area (black), intermediate area (grey), and light area (white).

species, however, is their movement pattern (Fig. 17.7) for razorbills turn much more frequently than guillemots.

5. Guillemots first move towards the darkest part of the experimental area, and their turns occur when they are in proximity to a wall. Razorbills, however, tend to keep on turning in the same direction, so that they end up following a spiral path.

6. From Fig. 17.8 it can clearly be seen that razorbills spend far less time in the proximity of a wall than guillemots do (Mann–Whitney U-test, $p = 0.038$).

Therefore, the different use of action-space by guillemot and razorbill chicks is related to their different ways of reacting to light and also, most probably, to their differential responsiveness to walls as structures which divide that space.

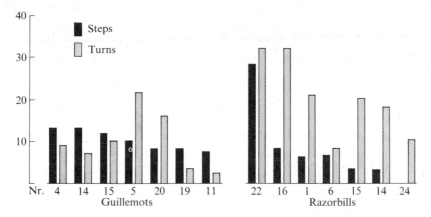

FIG. 17.7. The locomotor activity of individual guillemots and razorbills during the first hour after hatching.

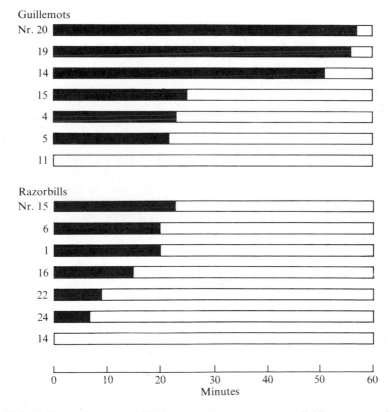

FIG. 17.8. Guillemots' and razorbills' duration of stay near a space-dividing structure during the first hour after hatching expressed as a proportion of a 60-minute test. Dark area: chick's head is at the most 3 centimetres, away from a wall. White area: chick's head is in open space.

Feeding comparisons

The behaviour of chicks of both species was compared when they were fed from an artificial model of the parent's beak. If a fish is placed in a beak-model in the guillemot manner and if one lets it slip out lengthwise at the appropriate speed, the guillemot chick will seize the fish by its tail and make grabbing movements cross-wise at the fish until the fish's head leaves the beak-model and is in the chick's beak. It then turns the fish around lengthwise and swallows it head first. Young razorbills, on the other hand, pick at the fish, but do not let it slide out lengthwise from the beak-model. Instead, they pull it diagonally away from the model towards themselves, transport it through their beak, and usually swallow it head first but sometimes tail first.

When the chicks were fed by hand, the fish was loosed once a chick had seized it. It was then up to the chick to place the fish in a suitable position for swallowing.

In the course of their development, the eating pattern of guillemot chicks increasingly approached that of razorbill chicks. The chicks are, therefore, able to modify those behavioural patterns which aid the food-acceptance process (Oberholzer, dissertation, in preparation).

Reactions to acoustic stimuli

A series of tests were made releasing a chick into an empty room equidistant between two tape recorders. Previous work (Tschanz 1968) had shown that if guillemot chicks had been exposed to a particular call whilst still in the egg, they subsequently approached this call in preference to another. Similarly treated razorbill chicks are less affected and, when exposed to the training call versus a control sound, may remain sitting or may move towards or away from the sound sources (Jenny, in preparation). Nevertheless, if they do approach, razorbills also prefer the training call to control ($\chi^2 = 8 \cdot 2$, $p < 0 \cdot 01$), but far more of them show no reaction ($\chi^2 = 29 \cdot 0$, $p < 0 \cdot 001$). These results are set out in Table 17.3.

TABLE 17.3

Reaction of chicks to training and to control calls

Number and species	Number of reactions to		
	training call	control call	no reaction
27 Guillemots	48	6	0
35 Razorbills	32	12	26

17.3. Summary of laboratory tests

In all but one of the comparisons we have been discussing the results of the laboratory tests have revealed the same species-specific behavioural characteristics as are found on the cliffs in the natural habitat. As described, only the manner of taking food from models of the parent's head is slightly changed in the guillemots.

The over-all correspondence of the behaviour of guillemot chicks or of razorbill chicks that grow up in the field with those that grow up in the laboratory seems to indicate that the development of the behavioural characteristics is more dependent upon internal than external factors.

If this is true, then guillemot chicks which grow up with razorbill parents, and razorbill chicks which grow up with guillemot parents should still develop the behavioural features characteristic of their species. This would make it possible to judge the adaptive effect of guillemot-typical behaviour since if it is missing in razorbill chicks, they should be at a disadvantage growing up on guillemot ledges. In order to test this, we carried out cross-fostering experiments.

17.4. Development of the characteristics under natural conditions with parents of the other species (cross-fostering experiments)

Experimental conditions

To exchange chicks on guillemot ledges requires great experience and much time, and it is only possible at certain places and with certain birds if the chick-rearing is to succeed. For this reason, we did not exchange young between razorbills and guillemots, but eggs.

Guillemots as well as razorbills habituate to colour—and pattern—changes in their own eggs if these changes are made gradually (Tschanz 1959; Ingold 1973). This makes it possible to paint a number of eggs so that they resemble each other and then interchange them so that guillemots brood razorbill eggs, and vice versa. The eggs are painted over a period of 3–7 days, beginning at the top, each time one-third more of the egg-surface is painted. As shown on Table 17.4, both species are about equally successful when incubating eggs of the other.

TABLE 17.4

Hatching success of guillemots and razorbills incubating eggs of the other species

	Eggs	Young hatched	Percentage
Guillemots	51	41	80
Razorbills	53	39	73

Observations on the behaviour of adoptive parents and chicks

Razorbill chicks with adult guillemots. Guillemots treat chicks of the foreign species as if they were their own. With its beak, the guillemot pushes the young razorbill against its body. It shelters the chick under its wing which is spread out and bent low, and it allows the chick to emerge if it tries. If the chick tries to move away, the adult bird produces the acceptance-call and follows it if it does not return after the call has been made. It may try to seize the chick with its beak and to make it keep close. When feeding, the adult bird behaves in the normal species-specific manner. The young razorbill chick also does this. When it is among guillemots, it tries to rest in the species-specific fashion and walks on the ledge in a razorbill manner, thereby trespassing into areas belonging to other adult birds or areas lying between the adults and the edge of the ledge. It only irregularly obeys the foster parent's call, pecks at the fish while being fed, seizes hold of the fish and tries to eat without responding to the behaviour of the adult bird.

Guillemot chicks with adult razorbills. Like guillemots, adult razorbills also treat the young of the other species as if they were their own. While sheltering the chicks, they lie flat on the ground, and only leave them when disturbed or when the partner appears at the entrance to the cave or crevice. At feeding they bring several fish which are placed diagonally in their beaks and wait with bent heads until the young ones have served themselves. Then they eat what is left themselves. The guillemot chicks also retain their typical behaviour. They want to be sheltered often (Wehrlin, in preparation) wait for the fish to be offered to them at feeding time, and at first only eat one of them. If called by the adults, they go to them, provided they do not have to leave the cave or crevice.

Functional analysis

When a bird behaves in a typical razorbill way on guillemot ledges, complications arise. The resting position of young razorbills does not coincide well with the sheltering position of guillemots and thus makes it difficult for the chick to hide itself within the guillemot's feathers. This is particularly disadvantageous because the young razorbills especially need to warm themselves and to rub themselves against the adults ('*frottieren*', Goethe 1953). Their feathers get dirty and sticky on the wet and dungy guillemot ledges, not only because they lie on the ground but also because the adult guillemots are continually forced to fetch the young ones by means of their beaks since the latter often do not obey the acceptance-call. Their warmth requirements which are thus greater than under normal conditions could perhaps be met by providing more food. But the razorbill chicks do not even get that. On the contrary, guillemots which have fish in

TABLE 17.5
Mortality of chicks

			Losses				
Chicks	Adult birds	Hatched	Up to fifth day	Days 6–8	Days 9–10	Days 11–12	Days 13–22
Razorbills	Guillemots	41	14	4	2	1	1
Guillemots	Razorbills	39	3	0	0	0	0

their beaks quickly turn their head away if they are pecked at. They do this also when the young razorbill tries to grasp the fish in its own way, thereby interrupting the feeding process. The foster-parents feeding trials are repeated again and again, until eventually the chick succeeds in grabbing the fish. This frightens the guillemot, which then sits upright and closes its wings, thereby making the fish visible to the neighbouring guillemots. Since the young bird does not often succeed in eating quickly, the fish can be taken away from it.

It is not only the loss of warmth and food which have a disadvantageous influence on the development of the young razorbills. In addition, they expose themselves to the attacks of strange guillemots by their way of walking around on the ledge (Table 17.1), and are in great danger of falling off.

All these factors lead to major losses on the guillemot ledges, particularly among the younger chicks. Out of a total of 22 losses, 18 occurred during the first 8 days of the young birds' lives (Table 17.5).

As they get older, the razorbill chicks become less endangered, not only because they are able to regulate their body temperature themselves, but also because they have learned that they are more likely to be harassed in the area between the adults and the edge of the ledge, and therefore stay close to the rocks. But the guillemots also contribute their share in helping the young birds to thrive better. Their reaction towards the young that peck becomes less frightened, and they also develop a new feeding technique. Instead of letting the fish slowly glide lengthwise out of their beak, they let it go so quickly that the chicks do not even have the chance to peck at it beforehand. If they hurry they catch the fish in the air. Otherwise, they have to pick it up from the ground. The guillemot is not disturbed during this, which makes it possible to protect the chick against neighbouring guillemots. The complications arising at first during feeding, and how they are overcome, are reflected in the change in duration of feeding, as seen in Fig. 17.9. Until the fifth day after hatching, the feeding of razorbill chicks by guillemot parents lasts far longer than feeding between guillemot parents and young, between razorbill parents and young

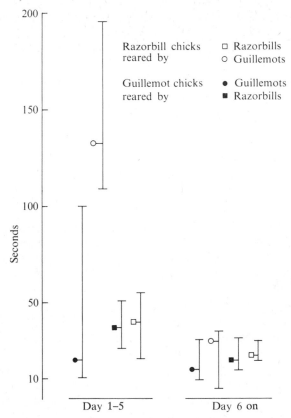

FIG. 17.9. Medium duration of feeding when within and between species over 1–5 days of age, and from the sixth day on. Lines indicate the first and third quartile of each group.

or between guillemot chicks and razorbill parents (White test, $p < 0.01$).

Under natural conditions, approximately 90 per cent of hatched guillemot chicks reach the 'jump-age' (when they leave the ledges), but only 46 per cent of the young razorbills survived so long in the fostering experiments (Table 17.5). The greatest losses occur when the chicks behave in their species-typical way thus making themselves vulnerable to harassment by neighbours. We found that those chicks whose parents were in a rock niche at the edge of a colony faced the least danger. The number of losses decreased as the behaviour of the young razorbills adapted to that of the guillemots. This proves that the behavioural characteristics typical for guillemots fit the individuals for success under their own natural conditions.

Growing up with razorbill foster-parents causes guillemot chicks almost no problems. Only the attempt to reach the adult when it calls and the way the razorbill offers fish caused some difficulties at the beginning. If the calling adult was standing on a step below the nest-site, then the chick was able to get down to the adult but not back up to the nest-site. Chicks that are not yet able to control their own body temperature then cool down and perish. The chicks' behaviour towards cliffs usually saves them from attempting such dangerous excursions (Wehrlin and Tschanz 1969). Razorbill-typical food presentation and guillemot-typical waiting for the parent to initiate feeding creates co-ordination problems at the beginning. These are, however, quickly solved. Razorbills wait patiently until the chick has taken a fish, and the chick quickly gets used to taking the initiative itself and making use of all the abundant offering. Therefore, losses among chicks are small. The co-operation between razorbill parents and guillemot chicks leads to one single deviation from normal behaviour for adults: the parents stand up less often when brooding the adopted chicks than with their own (Fig. 17.10, Wilcoxon–White test, $p < 0.05$).

Wehrlin (1973) was able to show that guillemot chicks strive to remain beneath the adult for much of the time (70–90 per cent of the observation time). This might cause the razorbills to stand up less often.

Owing to the resting position typical for guillemots, the young bird does not have enough space under the adult and is therefore forced to stick its head out between the adult's body and its wings. This, however, does not

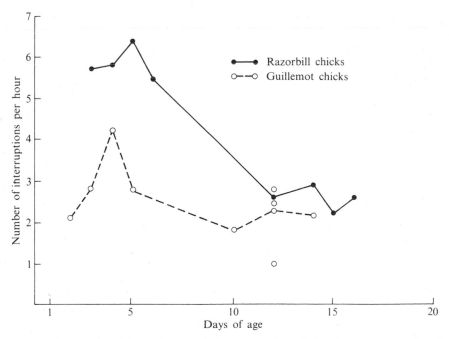

FIG. 17.10. Number of times that razorbill parents get up during brooding per hour of observation.

seem to disturb either the chick or the adult, and in no way limits the breeding success which is as high as 92 per cent. It is thus remarkably higher than the breeding success for razorbill chicks reared by guillemots and the reasons for this difference have already been discussed.

In summary we may review the external conditions under which both species rear their chicks and the behaviour patterns of adults and young: *The open breeding ledges of the guillemots are in strong contrast to the more sheltered breeding sites of razorbills.* For chicks, the danger of falling exists only on the guillemot ledges; the same holds true for the danger of getting dirty, of being threatened by enemies and members of the same species, of food-plundering, and of mixing up the young. These dangers set special demands on the adults as well as on the young birds during rearing time. We conclude from the observations and experiments that guillemots are able to meet these demands thanks to special behavioural patterns; these are summarized in Table 17.6.

The situation is somewhat different for razorbill chicks. They do not need any special behavioural patterns to ensure receiving fish and to avoid the interaction with strange adults. When transferred to guillemot ledges, they must change the behaviour that they usually manifest after hatching (Table 17.6). Thus, for example, they must learn that there are certain areas of the ledge across which they cannot pass without being hurt and that certain adults will peck when they are approached. The razorbill chicks must learn, therefore, to avoid the repetition of bad experiences. Seemingly, they succeed in this only to a certain extent since the species-specific resting position as well as their manner of pecking at the fish persists. If the adults did not modify their behaviour, the young razorbills would lose their food even more often.

The inability to adopt all guillemot-typical behaviour as well as the many dangers encountered during the learning process both put the razorbill chicks at a disadvantage compared with the guillemot chicks. The latter do not have to learn how to cope with dangers. Furthermore, they are often not even forced to modify their species-specific behaviour to any great extent.

17.5. Conclusion

All of the observed behavioural differences were observed from the very earliest age under natural conditions, in the laboratory situation and in the fostering situation. Young birds retain species-specific behavioural characteristics if they grow up under their natural conditions. These behavioural characteristics enable them to satisfy all the demands set on them. That these particular behavioural features are essential if enough young birds are to survive to breed is proved by the fact that chicks that do not possess these characteristics are unable to thrive. Thus, the behavioural patterns observed in guillemots prove to be phylogenetic adaptations (Lorenz 1961).

However, this does not mean that these behavioural features cannot be adaptively modified. The laboratory and exchange experiments have shown that such modifications are possible during development, and the change in the feeding behaviour of the adult guillemots makes it obvious that modifications can also occur in the adult animal and can positively influence the growing-up process of the young.

Thus, adaptations occur as the result of a phylogenetic development as well as an ontogenetic modification. The former are more efficient in ensuring the survival of the species in situations where wrong behaviour immediately causes losses. The latter enable the individuals, within the range of their hereditary possibilities, to survive in atypical habitats, thereby opening up for their species new living-space.

17.6. Summary

Characteristics are called adaptations if they enable the organism to meet better the special demands of its environment. In order to determine if a characteristic is an adaptation, we must compare the form and function of the characteristics of several animal species. Such a comparison has been made with guillemot and razorbill chicks. The chicks' environmental conditions differ from one another insofar as the razorbill pairs breed in spatially isolated crevices or caves, whereas the guillemot pairs breed within densely populated colonies on cliff-ledges. Comparative observations on natural chick-rearing suggest that the chicks' different resting behaviour, their different use of action-space, their different behaviour during disturbance and when fed, as well as their different behaviour towards their parents' acceptance-calls, can all be traced back to special adaptations of the guillemot chicks to their particular rearing conditions. With hand-raised young it was examined whether the characteristics in question are innate. Experiments in the light–dark field indicate that different phototaxis is an important reason for the chicks' different use of action-space.

The effects of the characteristics discussed on survival are shown under natural conditions by letting razorbill chicks grow up with guillemot parents, and guillemot chicks with razorbill parents. The greater losses among the young razorbill chicks living in guillemot colonies were mainly caused by chills, lack of food, and falling from the ledges. With guillemot chicks all these dangers were prevented by the specialized behavioural characteristics examined. The loss of guillemot chicks brought up by razorbill parents is minimal. The guillemot chicks do not need any additional adaptations.

TABLE 17.6

RAZORBILLS   **GUILLEMOTS**

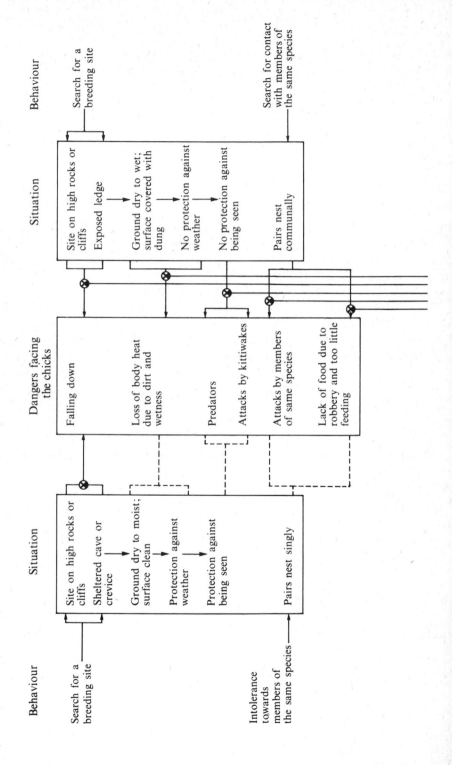

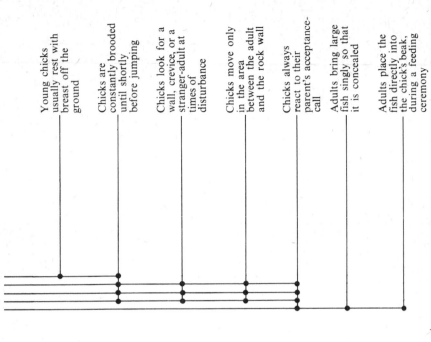

Young chicks usually rest with breast off the ground

Chicks are constantly brooded until shortly before jumping

Chicks look for a wall, crevice, or a stranger-adult at times of disturbance

Chicks move only in the area between the adult and the rock wall

Chicks always react to their parent's acceptance-call

Adults bring large fish singly so that it is concealed

Adults place the fish directly into the chick's beak, during a feeding ceremony

Young chicks usually rest with breast on the ground

Chicks are not constantly brooded

Young chicks remain at the breeding-place at times of disturbance

Chicks move around whole area of nest-site

Chicks do not always react to their parent's acceptance-call

Adults bring several small fish, clearly visible

Chicks pick fish up from the ground

KEY

a ———→ b = situation a leads to result b

a – – – – b = a does not lead to b

a ———→ b = situation or behaviour pattern c tends to reduce
      ⊗   the effect of a upon b
      c

Acknowledgements

We should like sincerely to thank the Schweizerischer Nationalsfond for the grant which financed our work. The owners of Vedøy generously gave permission for us to enter their property to work on the birds, and the inhabitants of Røst gave us much co-operation. We heartily thank all of these people as well as our co-workers who, under unusually severe conditions, devotedly and diligently carried through these field and laboratory investigations that have made a broad ethological comparison between guillemots and razorbills possible.

References

CULLEN, E. (1957). Adaptations in the kittiwake to cliff-nesting. *Ibis* **99**, 275–302.

CURIO, E. (1973). Towards a methodology of teleonomy. *Experientia* **29**, 1045–58.

EMLEN, J. T. (1963). Determinants of cliff edge and escape responses in herring gull chicks in nature. *Behaviour* **22**, 1–15.

GOETHE, F. (1953). Experimentelle Brutbeendigung und andere brutethologische Beobachtungen bei Silbermöwen. *J. Ornithol.* **94**, 160–174.

INGOLD, P. (1973). Zur lautlichen Beziehung des Elters zu seinem Küken bei Tordalken (*Alca torda*). *Behaviour* **45**, 154–190.

KARTASCHEW, N. N. (1960). Die Alkenvögel des Nordatlantiks. *Neue Brehm Büch.* **257.**

LORENZ, K. (1961). Phylogenetische Anpassung und adaptive Modifikation des Verhaltens. *Z. Tierpsychol* **18**, 139–87.

MACARTHUR, H. and CONNELL, J. H. (1970). Biologie der Populationen. *Moderne biologie.* BLV, München.

NELSON, J. B. (1967). Colonial and cliff nesting in the gannet. *Ardea* **55**, 60–90.

SCHERZINGER, W. (1971). Beobachtungen zur Jugendentwicklung einiger Eulen. *Z. Tierpsychol.* **28**, 494–504.

SCHWERDTFEGER, F. (1963). *Autökologie.* Parey, Hamburg.

—— (1968). *Demökologie.* Parey, Hamburg.

TSCHANZ, B. (1959). Zur Brutbiologie der Trottellumme. *Behaviour* **14**, 1–100.

—— (1968). Trottellummen. *Z. Tierpsychol.* **4.** Beiheft.

—— (1972). Verhalten und Arterhaltung. *Verh. schweiz. naturf. Ges.* 28–44.

—— INGOLD, P., and LENGACHER, H. (1969). Eiform und Bruterfolg bei Trottellummen *Uria aalge aalge* Pont. *Ornithol. Beob.* **66**, 25–42.

WEHRLIN, J. and TSCHANZ, B. (1969). Cliff-Response bei Trottellummen. *Rev. suisse Zool.* **76**, 1132–44.

WICKLER, W. (1970). Soziales Verhalten als ökologische Anpassung. *Verh. dt. zool. Ges.* **64**, Tagung, 291–304.

—— (1972). Verhalten und Umwelt. *Kritische Wissenschaft.* Hoffmann und Campe, Hamburg.

Species Index

Author Index

Subject Index